Inositol & its Phosphates:

Basic Science to Practical Applications

Authored by

A.K.M. Shamsuddin

The University of Maryland School of Medicine
10 S. Pine Street, MSTF-700
Baltimore, MD 21201-1116
USA

Guang-Yu Yang

Northwestern University Feinberg School of Medicine
303 E Chicago Ave, Ward 4-115
Chicago, IL 60611
USA

Dedication

To Our Parents

CONTENTS

CHAPTERS

Part I. Nature, Biosynthesis, Bioavailability and Metabolism

Part II. Health Impact of Inositol and its Phosphates

contd…..

Part III. Mechanisms of Action of Inositol and its Phosphates

contd…..

Part IV. Industrial and Other Uses of Inositol and Inositol Phosphates

Foreword

Interest in the role of nutritional factors and cancer began in earnest following the 1975 Key Biscayne, FL International Symposium entitled, *Nutrition in the Causation of Cancer*. Since then, literally thousands of publications have appeared ranging from epidemiological and animal model to genomic studies. These studies have ranged from the cancer preventing effects of food groups to specific foods, and to specific biologically active agents in foods, such as carotenoids, curcumin and isoflavones.

Clearly, however, over the past 3 decades, the preponderance of attention paid by the US National cancer Institute (NCI) has been, not to dietary factors, but to genetic factors, despite the fact that genetic factors have been shown to affect roughly only 5%-10% of all cancers. With this discrepancy in mind, Christopher Wild, Director of the International Agency for Research on Cancer, introduced the term the exposome, which includes the total spectrum of the environment i.e., UV radiation, industrial toxins, pollutants and the foods we consume and warned that in order to significantly reduce cancer incidence and mortality more attention must be paid to environmental agents.

This brings us to the contents of this eBook. Until coming upon Dr. Shamsuddin's work, my interest in inositol and its various metabolites was limited to its role as an essential nutrient for culturing cells *in vitro*. The complex metabolism of inositol compounds has been the subject of numerous reviews. In this fascinating and revealing eBook, Dr. Shamsuddin has diligently culled the literature dealing with InsP_6 popularly known as IP6, a phosphorylated derivative of inositol, in relation not only to cancer but to other diseases including osteoporosis, diabetes, kidney stones, heart disease *etc*. This review covers evidence gleaned from animal model, cell culture and gene expression studies and a limited number of clinical trials, suggesting that IP6 has numerous health benefits besides industrial uses.

Not surprisingly, none of the major granting agencies or research organizations has been supportive of this research. The anachronistic standards of the US National Cancer Institute, which require injecting agents into the circulation and measuring blood levels, are standards that do not apply to orally available agents such as IP6. In sum, despite the fact that there is a vast literature on the metabolism of inositol compounds, there is no authoritative account extant of the role of inositol phosphates in health and disease. Dr. Shamsuddin, who has devoted much of his career to IP6 research, has filled this void with a very readable, well-documented and balanced review. This eBook will, no doubt, prove of interest to basic scientists in natural and biological sciences as well as to practicing physicians.

Leonard A Cohen, *Ph.D*
Editor, Nutrition and Cancer: An International Journal
American Health Foundation, Valhalla, NY, USA

Preface

Since discovered in the mid 1800's in plant seeds, inositol and its hexaphosphate (IP_6, or more accurately $InsP_6$), particularly $InsP_6$ have enjoyed varying degree of reputation or infamy. Initially $InsP_6$ was considered an important component of seeds; then the acid form's metal chelating property was appreciated by chemist, the same attribute was then incriminated for mineral deficiency by nutritionist, for a very long time. Then came the discovery of inositol triphosphate ($InsP_3$) and tetraphosphate ($InsP_4$) in yeasts; and $InsP_6$'s anticancer and immune boosting action in the 1980's. All through this, the knowledge has been compartmentalized or boxed in so to speak. In other words, there has not been much, if any interdisciplinary communication and; perhaps a lack of in-depth research in the prevailing scientific literature. For instance, that large doses of $InsP_6$ was successfully used for prevention of kidney stone back in the 1950's (from institutions no less than Harvard University and Massachusetts General Hospital in Boston, and published in the *New England Journal of Medicine*) remained largely unknown, even to the researchers in the field, and worse, those from the same institution, as gathered from review of literatures and omission of citations. Likewise, the earlier reports on the biochemistry and metabolism of inositol phosphates ($InsP_s$) in eukaryotes make no mention of their presence in the plant kingdom; and till now, many simply fail to comprehend and therefore chose to ignore the anticancer property. This scenario is akin to the proverbial description of an elephant by the visually impaired, each of whom has an idea of the parts they have felt, albeit in a limited manner; but the whole is not understood.

Addressing this lack of interdisciplinary knowledge, thus the purpose of this eBook is to give as comprehensive an account of inositol and its phosphates, especially $InsP_6$ - the 'elephant in the room' as possible.

The contents of the eBook are for a wide range of readers consistent with the wide and divergent functions and applications of inositol and its phosphates. To cater to a broader audience from physical sciences as well as biological sciences, we have often used terms and information that some may find too basic while others not; this has also been deliberate. Owing to the vast amount of literature on the chemistry, metabolism and biosynthesis of $InsP_s$ we have chosen to limit the discussion on that, but focus more on areas that we think are poorly represented in the literature.

There are certain overlaps in presentation of the materials that are not only unavoidable, but to some extent intentional: unavoidable because the mechanisms of actions are related to practical applications, especially in biological fields; and it was intentionally done to readily recall the connection between the cause and mechanistic effect, as the two are often inseparable, to get the whole picture to satisfy our intellectual curiosity without having to go back and forth to the relevant chapters.

Finally, we hope that this eBook will help more interdisciplinary communication and research to further enhance our understanding and usage of inositol and its phosphates for betterment of life in this planet.

ACKNOWLEDGEMENTS

Declared none.

CONFLICT OF INTEREST

Professor Shamsuddin is the inventor of several patents related to InsP_6 and hexacitrated InsP_6.

A.K.M. Shamsuddin
The University of Maryland School of Medicine
10 S. Pine Street, MSTF-700
Baltimore, MD 21201-1116
USA

Guang-Yu Yang
Northwestern University Feinberg School of Medicine
303 E Chicago Ave, Ward 4-115
Chicago, IL 60611
USA

2

CHAPTER 1

Inositol Compounds: Natural Sources and Chemistry

Abstract: Considerable evidence has accumulated indicating a significant role of dietary inositol and its phosphates, and phospholipid derivatives in human health. In a typical American diet, the amount of *myo*-inositol per 2500 kcal diet is approximately 900 mg, and most of the inositol compounds are phospholipid derivatives. Virtually all of the ingested *myo*-inositol (99.8%) is absorbed from the human gastrointestinal tract. In general a normal circulating fasting plasma *myo*-inositol concentration is approximately 0.03 mM and this material turns over with a half-life of 22 minutes. Contrary to the misconception held by some, inositol hexaphosphate (InsP_6) likewise is also rapidly absorbed (79 ± 10%) and distributed widely throughout the body. While there are interspecies variations, in general they all point to rather rapid absorption. Following an InsP_6-poor diet, the basal plasma value in humans is 0.07 ± 0.01 mg.L^{-1} that rises to 0.26 ± 0.03 mg.L^{-1} after ingestion of InsP_6-normal diet.

Keywords: Inositol hexaphosphate, inositol pyrophosphates, InsP_6, isomers, phytate, phytic acid, polyphosphate.

INTRODUCTION

From a chemical perspective, structurally inositol is similar to glucose and is thus considered a "sugar"; it is a carbohydrate, and essentially tastes as sugar though it is not a classical sugar and is assayed at half the sweetness of table sugar (sucrose). Many foods are rich in inositol as either inositol phosphates or inositol phospholipids. Several fruits, especially cantaloupe and oranges, are rich in inositol and its phosphates [1]. Inositol as phospholipid form exists in certain animal and plants as lecithin; in that form it is well-absorbed and relatively bioavailable [2-4]. In general, lecithin is designated as any group of yellow-brownish fatty substances rich in phospholipids, including phosphatidylcholine, phosphatidylethanolamine, and phosphatidylinositol. Another common natural inositol compound is inositol hexakisphosphate (InsP_6) also known as phytate (as salt form; and phytic acid for inositol hexaphosphoric acid) amongst nutritional scientists. InsP_6 was discovered in the mid 1800's. It is a saturated cyclic acid. Beans and grains, as seeds, contain large amounts of InsP_6 [1]. Most notably, it serves as the principal store of phosphorus and energy and source of cations and *myo*-inositol (a cell wall precursor). Relatively modest changes to its chemical structure - for instance the addition of phosphate groups (which lead to the creation of a group called inositol phosphates or inositol polyphosphates - IPP) - can produce an array of biochemical effects. In this chapter we will highlight the chemistry and natural occurrence or sources of inositol compounds.

CHEMISTRY

Inositol

Chemically, inositol is a class of cyclitols with molecular formula $C_6H_{12}O_6$ (Fig. **1.1**). Inositol was originally named 'inos' (muscle in *Greek*) by Josef Scherer a German chemist, who more than 150 years ago isolated the new molecule from muscle tissue. To date nine inositol stereoisomers have been identified based on the position of OH groups on the ring and the chemical structures of these isomers are shown in Fig. (**1.2**). Among these isomers, five are identified as naturally occurring inositol namely *myo*-inositol, D-*chiro*-inositol, *scyllo*-inositol, *muco*-inositol, and *neo*-inositol; the others are L-*chiro*-inositol, *cis*-inositol, *epi*-inositol, and *allo*-inositol; only *chiro* inositol has the D and L conformations. The D- and L-prefixes of the *chiro*-inositol isomers indicate that the compounds are optically either dextrorotatory (D-*chiro*-) or laevorotatory (L-*chiro*-); however, only the D-*chiro*-inositol phosphates appear to exist in nature.

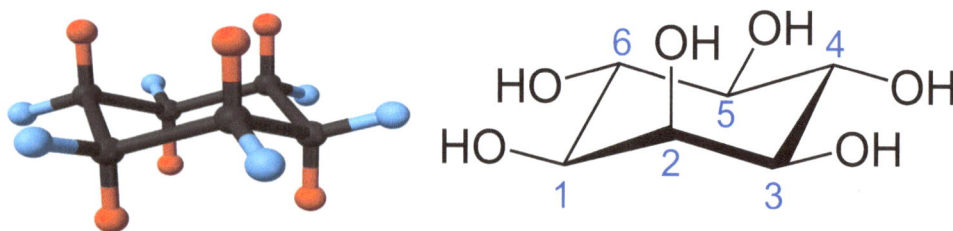

Fig. (**1.1**). Chemical structure of *myo*-inositol.

Among the possible isomers, *myo*-inositol is the most abundant and is identified as a key component of eukaryotic cells, exists as either salt form such as inositol phosphates (Ins*P*s), or lipid form such as phosphatidylinositol (PI) and phosphatidylinositol phosphate (PIP) lipids.

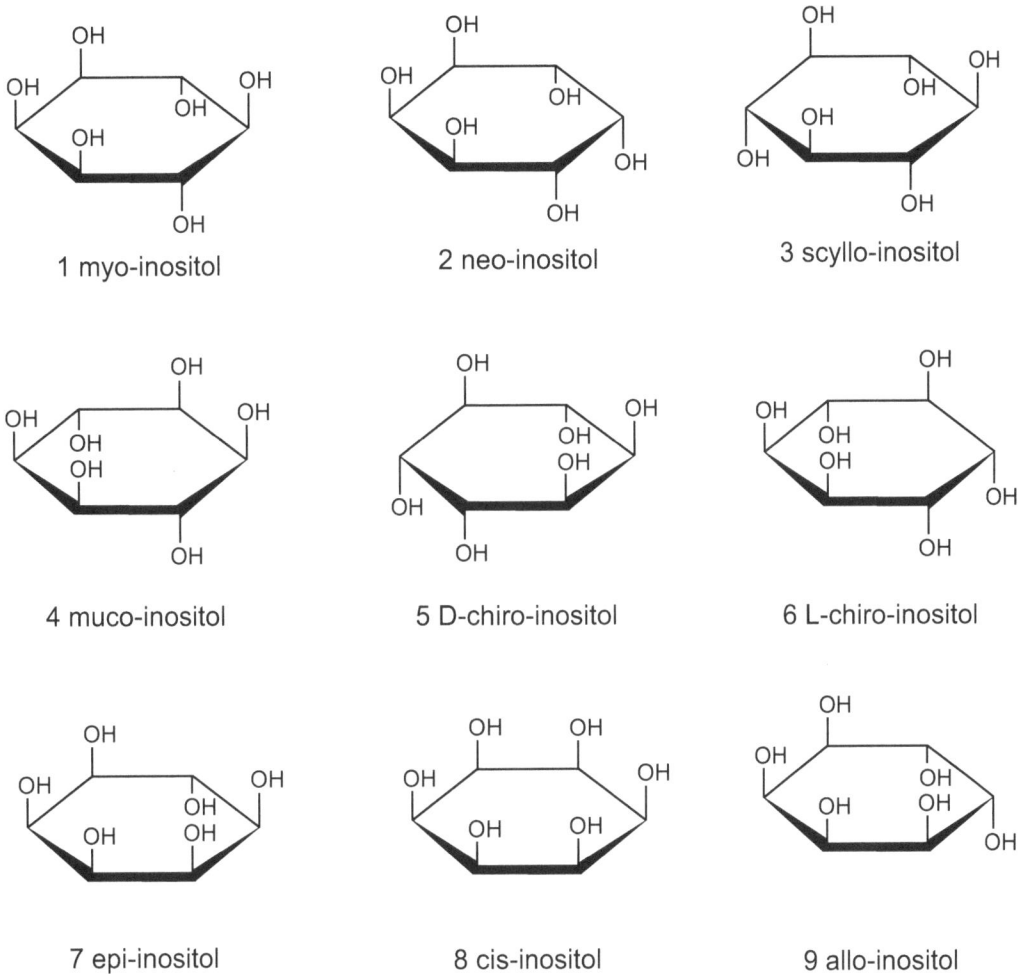

1 myo-inositol   2 neo-inositol   3 scyllo-inositol

4 muco-inositol   5 D-chiro-inositol   6 L-chiro-inositol

7 epi-inositol   8 cis-inositol   9 allo-inositol

Fig. (1.2). Stereoisomers of inositol.

Inositol phosphates are a group of *mono-* to polyphosphorylated inositols. They play crucial roles in diverse cellular functions, particularly as a message involving the regulation of cell growth, apoptosis, cell migration, endocytosis, and cell differentiation [5]. The group of inositol phosphates comprises $InsP_1$, $InsP_2$, $InsP_3$, $InsP_4$, $InsP_5$ and $InsP_6$ as illustrated in Fig. (**1.3**); and inositol pyrophosphates discussed later in the chapter.

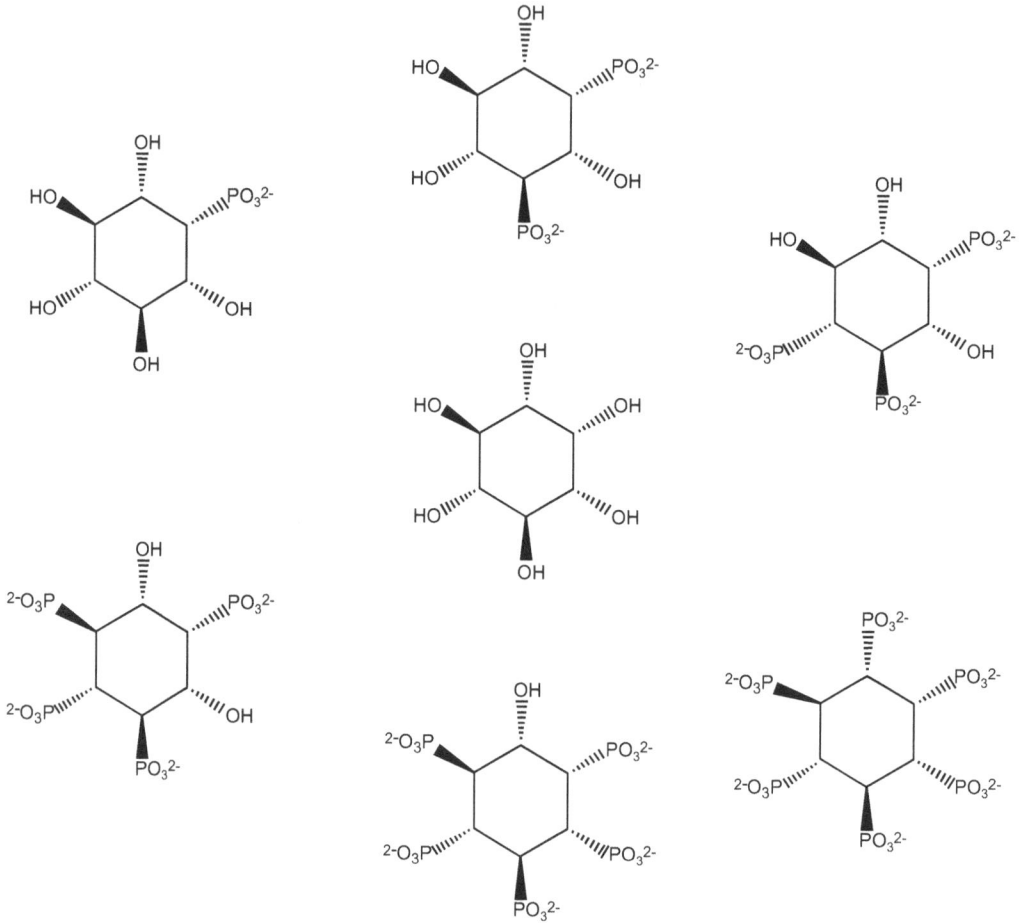

Fig. (1.3). Inositol and its phosphorylation to InsP_{1-6}.

Phosphatidylinositol Phosphate

Phosphatidylinositol phosphate lipid is an important lipid, either as a key membrane constituent or as a participant involved in essential metabolic processes in all plants and animals. Chemically it is an acidic (anionic) phospholipid with a phosphatidic acid backbone in essence that links to inositol (hexahydroxycyclohexane). In most organisms, the stereochemical form of the last is D-*myo*-inositol with one axial hydroxyl group in position 2 with the remainder equatorial. The 1-stearoryl-2-arachidonoyl molecular species, which is of considerable biological importance in animals is illustrated in Fig. (**1.4**).

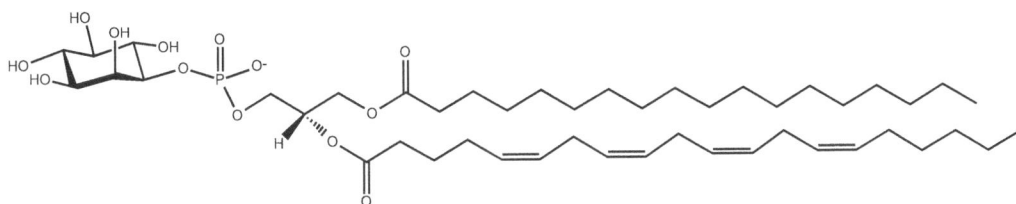

Fig. (1.4). 1-stearoryl-2-arachidonoyl molecular species.

There are polar and non-polar regions in phosphatidylinositol (PI) that makes the lipid an amphiphile. PI is a glycerophospholipid and contains a glycerol backbone, two non-polar fatty acid tails, a phosphate group and an inositol group; the phosphate group links with an inositol polar head group. Phosphorylated forms of PI are called phosphoinositides. The inositol ring can be phosphorylated by a variety of kinases on the three, four and five hydroxyl groups in seven different combinations. However, the two and six hydroxyl groups are typically not phosphorylated due to steric hindrance. Stearic acid and arachidonic acid are the most common fatty acids in phosphoinositides, in which stearic acid is in the SN_1 position and arachidonic acid is in the SN_2 position. Hydrolysis of phosphoinositides yield one mole of glycerol, two moles of fatty acids, one mole of inositol and one, two, or three moles of phosphoric acids, depending on the number of phosphates on the inositol rings. Thus, phosphoinositides are classified as the most acidic phospholipid.

Lecithin: A Major Source of Inositol Phosphatides in the Diet

Lecithin is a yellow-brownish fat that exists in animal and plant tissues and is a major source of inositol phosphatides composed of 20-21% inositol phosphatides and is a major phospholipid component of cell membrane. Several foods including soybeans, eggs, milk, marine sources, rapeseed, cottonseed, sunflower, *etc.*, are usual sources of lecithin [6, 7]. Lecithin can easily be extracted chemically and is an excellent emulsifier with low solubility in water. Phosphatidylinositol is one of the main phospholipids in lecithin from soybean and sunflower, and other lipids include phosphatidylcholine, phosphatidylethanolamine, and phosphatidic acid. These lipids in lecithin are often abbreviated as PI for phosphatidylinositol, PC for phosphatidylcholine, PE for phosphatidyl ethanolamine, and PA for phosphatidic acid.

Since lecithin is well absorbed and relatively bioavailable, and contains abundant inositol, it is an important source of inositol. Generally, lecithin is nontoxic which leads to its being widely used with food as an additive or in food preparation and in foods requiring a natural emulsifier or lubricant. In addition, lecithin derived from egg is not usually a concern for people who are allergic to eggs. Commercially available egg lecithin usually is highly purified and lacks the allergenic egg proteins.

InsP_6

Chemistry

InsP_6 was originally discovered in different plant seeds in 1855-1856 by T. Hartig, and was named "phytin" owing to its plant origin [8, 9]. In 1897 Winterstein named it inosite-phosphoric acid [10]. Various molecular structures were proposed since then resulting in, as one can imagine some controversy till 1914 when Anderson presented it as *myo*-inositol-1,2,3,4,5,6, hexakis dihydrogen phosphate, a structure that has stood the test of time [11]. The prefix "hexa*kis*" instead of "hexa-" indicates that the phosphate groups are not internally connected.

Between the pH ranges of 0.5-10.5, InsP_6 is sterically stable with 5 equatorial and 1 axial (5eq/1ax) conformation; at higher pH it flips on the reverse to 1 equatorial 5 axial conformation. Lower inositol phosphates maintain the stable 5eq/1ax conformation over the entire pH range. At 5eq/1ax conformation, the molecule can be viewed as a turtle [12], the four limbs and tail of the turtle are coplanar and represent the five equatorial hydroxyl groups. With the turtle's head erect, the head represents the axial hydroxyl group. Looking down at the turtle from above, the numbering of the turtle begins at the right paw and continues past the head to the other limbs, in counterclockwise (D) direction [13].

The process of preparation of InsP_6 was patented by Posternak (US Patent # 1,313,014 awarded to Ciba) in 1919 who named it phytic acid - a label that has stuck since then [14, 15]. As obvious from the name, InsP_6 has 6 phosphate groups attached to the inositol ring and can bind up to 12 protons in total. Therefore, inositol hexaphosphoric *acid* can bind with cations such as Ca^{++}, Mg^{++}, Zn^{++}, Fe^{++} *etc.* forming the salt - inositol hexaphosphate (a.k.a. phytate) which is how it exists in nature (calcium magnesium InsP_6, Fig. **1.5**). At low pH, the affinity of the acid for different cations is as follows: $Cu^{2+} \geq Zn^{2+} \sim Cd^{2+} > Mn^{2+} > Mg^{2+} > Co^{2+} > Ni^{2+}$. The preceding refers to chemical reactions in test-tubes whereby inositol hexaphosphoric *acid* binds with cations, which may have been the 'reason' at least in part for the grossly exaggerated and unfounded claim of the salt form's chelation *in vivo* (more on this in Chapter 22 Safety).

In Plants

The salt form of InsP_6 (calcium magnesium InsP_6 or 'phytin') is the principal storage form of phosphorus and inositol in many plant tissues. It is found within the hulls of nuts, seeds, and in grains - especially in the bran part. It is stored in electron dense spherical particles named globoids which are localized predominantly in the aleurone layer of wheat and barley or, in the embryo of maize. They are compartmentalized inside protein storage vacuoles in the seeds.

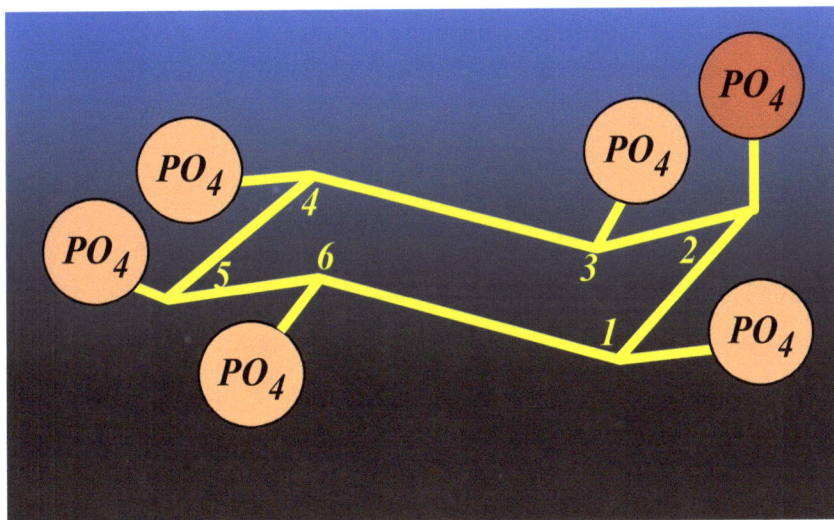

Fig. (1.5). Chemical structure of inositol hexaphosphoric acid a.k.a. phytic acid; five phosphates at C1,3,4,5,6 are axial and one at C2 (red) equatorial conformation (pH 5-12).

InsP_6 accumulates during seed development until the seeds reach maturity and accounts for 60%~90% of total phosphorous content in cereals, legumes, nuts and oil seeds. In the developing mung bean [*Phaseolus aureus*] seeds InsP_6 was identified as a phosphate donor for ADP phosphotransferase and conversion from ADP to ATP by transferring a phosphate group from 2 position of InsP_6 to ADP. Subsequently, inositol 1,3,4,5,6-pentakisphosphate 2-kinase a key enzyme in InsP_6 synthesis that catalyzes the conversion of ATP from ADP in germinating seed was isolated from soybean. Future discoveries of InsP_6 function in ATP regeneration in mammals seems like just a matter of time.

In Soil

It would seem logical that the seeds and plants, and then the grains receive their supply of $InsP_6$ from the soil. The majority of soil organic P exists as orthophosphate monoesters; inositol phosphates comprise ~60% of orthophosphate monoesters, and $InsP_6$ comprises the bulk. However, the amount of $InsP_6$ varies between soils in different parts of the planet. Typical proportions of the inositol phosphate esters in the soil are as follows: $InsP_6$ = 83%, $InsP_5$ = 12%, $InsP_4$ = 4%, $InsP_3$ = 1%, $InsP_2$ = trace; the presence and the amount of $InsP_1$ in soil has been controversial [16]. *Myo-* isomer is the commonest form constituting approximately 90% of the total soil $InsP_6$; however, the presence of D-*chiro-*, *neo*, and *scyllo-* $InsP_6$, forms that rarely exist elsewhere in nature are found in the soil at approximately similar ratios [16]. Since plants contain only the *myo-* isomer, the presence of other isomers in the soil is interesting; it is believed that the soil microbial enzymes are responsible for the conversion.

In Mammalian Cells

Using the same rationality of transfer of $InsP_6$ from the soil to the grains, it would not be illogical to think that other life forms, especially animals and humans that eat plants and grains would be presented with $InsP_6$ to their digestive system. Whether or not they absorb $InsP_6$ into their cells is another matter that had been doubted for a very long time; but logic and science have prevailed (discussed in Chapter 5).

In mammalian cells, *myo*-inositol polyphosphates are ubiquitous, and in here too $InsP_6$ is the most abundant with concentrations ranging from 10 to 100 µM, depending on cell type and developmental stage [17, 18]. $InsP_6$ is found in ten-fold higher concentrations in the brain than the kidney in rats. In humans the basal level in plasma is 0.07 ± 0.01 mg L^{-1} or 0.106 ± 0.015 µmol L^{-1} in volunteers on an $InsP_6$ poor diet. Consuming an $InsP_6$-normal diet results in the plasma level to raise 3-5 fold to 0.26 ± 0.03 mg L^{-1} or 0.393 ± 0.045 µmol L^{-1} [19]. The details of $InsP_6$ metabolism will be discussed in Chapter 5 (Pharmacokinetics).

$InsP_6$ in Other Organisms

As can be seen from the preceding, inositol and its phosphates especially $InsP_6$ is ubiquitous. Their role is rather well defined in most of the places they have been found. In various organisms they act as important signal transduction molecules;

we discuss that in detail in Chapter 3. Interestingly it has also been found extracellularly in the hydatid cyst wall of the parasitic cestode *Echinococcus granulosus* wherein it is present as calcium salt. Ca-InsP_6 in the form of deposits was observed as 20-80 nm diameter electron dense granules. The deposits of Ca-InsP_6 were also observed inside membrane vesicles in cells of the germinal layer - the inner cellular layer of the hydatid cyst wall, indicating that InsP_6 precipitates with calcium within a cellular vesicular compartment and is then secreted to the laminated layer [20]. Ca-InsP_6 constitutes about one-third of the total dry weight of the hydatid cyst wall. Histological studies demonstrate inflammatory cells around the *E. granulosus* cysts can phagocytose the Ca-InsP_6 deposits. However its role is unclear at this time. The presence of InsP_6 in other organisms is further described in Chapter 3.

Other Sources of Inositols in the Diet

Although inositol is not considered an essential nutrient at this time, it is an important component of our diet. Inositol content in our common foods is summarized in Table **1.1**. The amount of *myo*-inositol present in the 2500 kcal North American diet is approximately 900 mg, of which 56% is lipid-form and others are predominantly InsP_6 [21]. Importantly, all of the ingested *myo*-inositol (99.8%) is absorbed from the human gastrointestinal tract [21]. The circulating fasting plasma *myo*-inositol concentration has been found to be approximately 0.03 mM and turns over with a half-life of 22 min [21]. This information on dietary content of inositol and its concentration in the blood is not only important to develop diets with high *myo*-inositol contents that could be significant for evaluating and monitoring its therapeutic and/or prophylactic efficacy in various diseases, but also to determine the relationship between dietary *myo*-inositol intake and disease development, as well as the role of inositol in the pathogenesis of cancer, diabetes mellitus and neuropsychiatric diseases, *etc.* discussed in Part II (Chapters 6-15).

Inositol Pyrophosphates (IPP)

Myo-inositol is present in nature either unmodified or in more complex phosphorylated derivatives. More recently, inositol pyrophosphates (IPP: InsP_7, InsP_8) have been discovered first in the yeast *Saccharomyces cerevisiae* and the amoeba *Dictyostelium discoideum*, and later in mammalian cells. The diphosphoinositol polyphosphates (PP-InsPs), known as "inositol pyrophos-

Table 1.1. Food sources of InsP_6 [22-24].

Food	% Minimum Dry	% Maximum Dry
Linseed	2.15	2.78
Sesame seeds flour	5.36	5.36
Almonds	1.35	3.22
Brazil nut	1.97	6.34
Coconut	0.36	0.36
Hazelnut	0.65	0.65
Peanut	0.95	1.76
Walnut	0.98	0.98
Corn	0.75	2.22
Oat	0.42	1.16
Oat Meal	0.89	2.40
Brown rice	0.84	0.99
Polished rice	0.14	0.60
Wheat	0.39	1.35
Wheat flour	0.25	1.37
Wheat germ	0.08	1.14
Whole wheat bread	0.43	1.05
Beans, pinto	2.38	2.38
Chickpeas	0.56	0.56
Lentils	0.44	0.50
Soybeans	1.00	2.22
Tofu	1.46	2.90
Soy beverage	1.24	1.24
Soy protein concentrate	1.24	2.17
New potato	0.18	0.34
Spinach	0.22	NR

phates" possess the most crowded three-dimensional array of phosphate groups. InsP_5 and InsP_6 are the precursors of inositol pyrophosphate molecules that contain one or more pyrophosphate bonds [25]. These 'high energy' pyrophosphates have been found ubiquitously in eukaryotic cells. IPPs are generated by phosphorylation of InsP_6 and have been linked to a diverse range of cellular functions. It appears that energy metabolism, and thus ATP production is closely regulated by these molecules [26].

Phosphorylation of InsP_6 generates diphoshoinositolpentakisphosphate (InsP_7 or PP- InsP_5) and bisdiphoshoinositoltetrakisphosphate (InsP_8 or (PP)2- InsP_4), as shown in Fig. (**1.6**) [27]. InsP_6-kinase and PP-InsP_5-kinases, the two distinct classes of enzymes, are responsible for inositol pyrophosphate synthesis, which are highly conserved throughout evolution [28]. Inositol pyrophosphates have been isolated from all eukaryotic organisms and are found throughout nature.

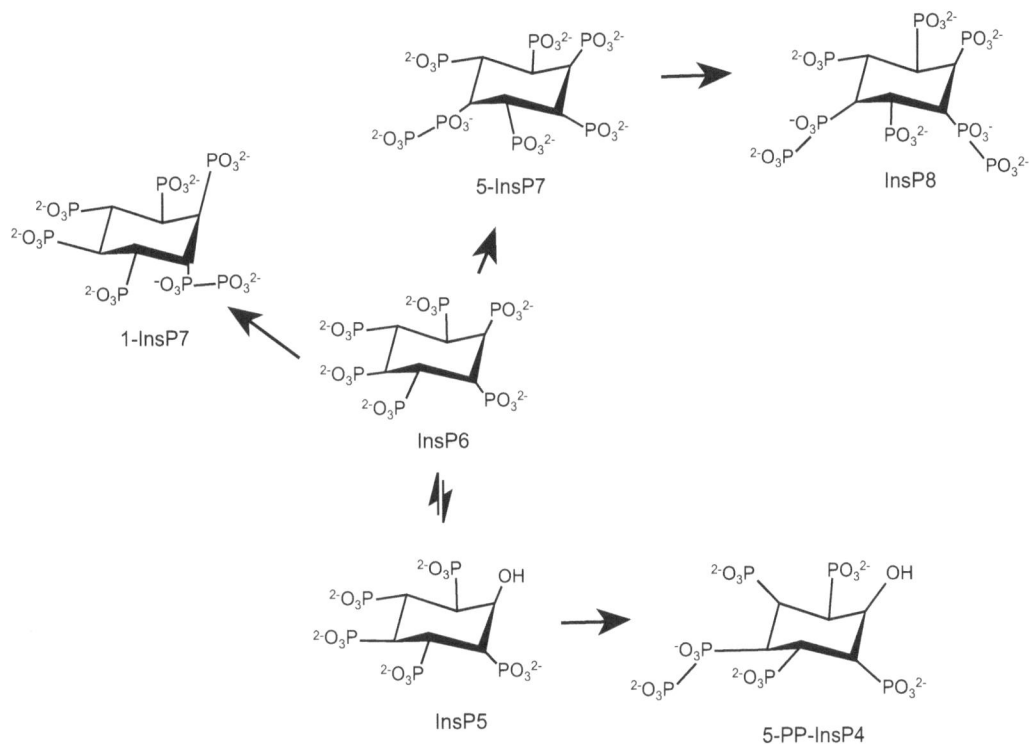

Fig. (1.6). Chemical structures and biosynthesis pathways of InsP6/InsP5 derived inositol pyrophosphates.

IPPs have been implicated in a variety of diverse activities, such as apoptosis, autophagy, chemotaxis, embryonic development, telomere maintenance *etc*. Levels of InsP_7 as well as InsP_6 were greatly increased in hepatocytes from 10-month-old wild-type mice compared with 2-month-old mice; on the other hand, hydrogen peroxide treatment of wild-type yeast resulted in a reduction in InsP_7 levels indicating a role in oxidative stress. These molecules play a signaling role in immunological functions, insulin homeostasis, obesity *etc.* IPPs have also been considered to be master regulators of cellular energy metabolism. As fascinating as InsP_6 is, these equally intriguing roles and functions of IPPs have become a very rapidly growing field of investigation. It would be impossible to do a fair discussion of the subject lest it becomes quickly outdated. Interested readers are therefore referred to the latest articles as they become published; a relatively recent review is by Wilson *et al.,* [26].

CONCLUDING REMARKS

Inositol and its phosphates InsP_{1-8} are widely distributed in nature. While the purpose and function of some of these have been known, those of others are yet to be determined.

REFERENCES

[1]   Clements RS Jr., Darnell B. Myo-inositol content of common foods: development of a high-myo-inositol diet. Am J Clin Nutr 1980; 33: 1954-67.
[2]   Blank ML, Nutter LJ, Privett OS. Determination of the structure of lecithins. Lipids 1966. 1: 132-5.
[3]   O'Brien BC, Corrigan SM. Influence of dietary soybean and egg lecithins on lipid responses in cholesterol-fed guinea pigs. Lipids, 1988; 23; 647-50.
[4]   Koven WM., Kolkovski S, Tandler A *et al.* The effect of dietary lecithin and lipase, as a function of age, on n-9 fatty acid incorporation in the tissue lipids of Sparus aurata larvae. Fish Physiol Biochem, 1993; 10: 357-64.
[5]   Croze ML, Soulage CO. Potential role and therapeutic interests of myo-inositol in metabolic diseases. Biochimie, 2013; 95: 1811-27.
[6]   Blesso CN, Andersen CJ, Barona J *et al.* Whole egg consumption improves lipoprotein profiles and insulin sensitivity to a greater extent than yolk-free egg substitute in individuals with metabolic syndrome. Metabolism, 2013; 62: 400-10.
[7]   Yang F, Ma M, Xu J *et al.*, An egg-enriched diet attenuates plasma lipids and mediates cholesterol metabolism of high-cholesterol fed rats. Lipids, 2012; 47: 269-77.
[8]   Hartig, T., Weitere Mitteilungen, das Kleibermehl (Aleuron) betreffend. Bot Ziet, 1856. 14: 257-69.
[9]   Hartig, T., Über das Klebermehl. Bot Ziet, 1855. 13: 881-2.
[10]   Winterstein, E., Über einen phosphorhaltigen Pflanzenbestandteil, welcher bei der Spaltung Inosit liefert, Berichte Deutsch. Chem. Ges., 1897. 30: 2299-302.
[11]   Anderson, R.J., A contribution to the chemistry of phytin,. J.Biol. Chem., 1914. 17: 171-90.
[12]   Agranoff BW. Textbook errors - Cyclitol confusion. Trends Biochem Sci 1978; 3: N283-5
[13]   Bohn L, Meyer AS, Rasmusen SK. Phytate: impact on environment and human nutrition. A challenge for molecular breeding. J Zhejiang Univ Sci B 2008; 9: 165- 191.
[14]   Posternak S. Organic phosphorus compound from plants. US Patent #1,313,014, 1919. August 12.
[15]   Posternak S. Sur la synthèse de l'acide inosito-hexaphosphorique,. Helv. Chim. Acta, 1921. 4: p. 150-165.
[16]   Turner BL. Papházy MJ, Haygarth PM *et al.* Inositol phosphates in the environment. Phil Trans R Soc London B 2002; 357: 449-69.
[17]   Szwergold BS, Graham RA, Brown TR. Observation of inositol pentakis- and hexakis-phosphates in mammalian tissues by 31P NMR. Biochem Biophys Res Commun, 1987; 149: 874-81.

[18] Sasakawa N, Sharif M, Hanley MR. Metabolism and biological activities of inositol pentakisphosphate and inositol hexakisphosphate. Biochem Pharmacol 1995; 50: 137-46.

[19] Grases F, Simonet BM, Vucenik I *et al.* Absorption and excretion of orally administered inositol hexaphosphate (IP_6 or phytate) in humans. Biofactors 2001; 15: 53-61.

[20] Irigoin F, Casaravilla C, Iborra F *et al.* Unique precipitation and exocytosis of a calcium salt of myo-inositol hexakisphosphate in larval Echinococcus granulosus. J Cell Biochem, 2004; 93: 1272-81.

[21] Clements RS Jr, Reynertson R. Myoinositol metabolism in diabetes mellitus. Effect of insulin treatment. Diabetes 1977; 26: 215-21.

[22] Phillippy BQ, Bland JM, Evens TJ, Ion Chromatography of Phytate in Roots and Tubers. J Agric Food Chem 2003; 51: 350-3.

[23] Macfarlane BJ, Bezwoda WR, Bothwell TH *et al.* Inhibitory effect of nuts on iron absorption. Am J Clin Nutr 1988; 47: 270-4.

[24] Gordon DT, Chao LS, Relationship of components in wheat bran and spinach to iron bioavailability in the anemic rat. J Nutr 1984; 114: 526-35.

[25] Bennett M, Onebo SM, Azevedo C *et al.* Inositol pyrophosphates: metabolism and signaling. Cell Mol Life Sci 2006; 63: 552-64.

[26] Wilson MS, Livermore TM, Saiardi A. Inositol pyrophosphates: between signalling and metabolism. Biochem J 2013; 452: 369-79.

[27] Barker CJ, Illies C, Gaboardi GC *et al.* Inositol pyrophosphates: structure, enzymology and function. Cell Mol Life Sci 2009; 66: 3851-71.

[28] Shears SB. Diphosphoinositol polyphosphates: metabolic messengers? Mol Pharmacol, 2009. 76: 236-52.

CHAPTER 2

Biosynthesis and Metabolism of Inositol and Inositol Phosphates

Abstract: Water-soluble and non-soluble inositol phosphates and phosphatidylinositol are key constituents of all plants and animals, and participate in their essential metabolic processes. In animal tissues, although the dietary route is one of the three pathways to maintain inositol homeostasis, the receptor mediated salvage pathway involving IMPase 1 and a *de novo* biosynthetic pathway involving inositol synthase play crucial role in cell signaling and maintaining inositol homeostasis, particularly the polyphosphorylated inositol. Water-soluble inositol polyphosphates including inositol pyrophosphates (up to 60 different possible compounds and at least 37 of these have been found in nature) are extremely important biologically in cell signaling, DNA repair *etc*. Phosphatidylinositol is the primary source of the arachidonic acid required for biosynthesis of eicosanoids, including prostaglandins *via* the action of the enzyme phospholipase A_2, as well as the main source of diacylglycerols that serve as signaling molecules in animal and plant cells. Phosphatidylinositol is converted to polyphosphoinositides with important signaling and other functional activities in animal cells, and involve a number of different kinases, particularly 3-phosphorylated forms.

Keywords: Biosynthesis, diacylglycerol, inositol pyrophosphates, IP7, IP8, metabolism. phosphatidylinositol, phospholipase.

INTRODUCTION

Inositol and its phosphorylated derivatives - inositol polyphosphates are present in cells as either phosphatidylinositol (an important lipid) or water-soluble inositol phosphates. Inositol polyphosphates function as the basis for a number of signaling and secondary messenger molecules (discussed in Chapter 3, Signal Transduction); and are involved in several biological processes. In this chapter, we mainly highlight their synthesis and metabolism in the cells.

BIOSYNTHESIS AND METABOLISM OF *MYO*-INOSITOL

Biosynthesis of *myo*-inositol is an important intracellular process although the major source of *myo*-inositol for cells is from diet through a sodium-dependent uptake process. Essentially, biosynthesis of *myo*-inositol is influenced by glucose level or, is part of the metabolism of glucose. Mammalian inositol synthase was first identified by Eisenberg from crude rat tissue homogenates [1]. Biosynthesis of *myo*-inositol in mammalian cells takes place in two stages utilizing glucose-6-

A.K.M. Shamsuddin and Guang-Yu Yang

phosphate (G-6-P) as substrate or starts with G-6-P, which is converted to *myo*-inositol 1-phosphate by inositol 1-phosphate synthase; followed by dephosphorylation of *myo*-inositol 1-phosphate to free *myo*-inositol by inositol monophosphatase, as shown in Fig. (**2.1**) [2]. The process of this synthesis requires an oxidized nicotinamide adenine dinucleotide (NAD) and is stimulated by NH$_4$CL and MgCL$_2$. Biosynthetic formation of one molecule of free inositol is accompanied by the formation of one molecule of D-inositol 3-phosphate by inositol synthase [2]. The removal of phosphate from inositol phosphates is a required step to recycle inositol in the cell by inositol monophosphatase, particularly for inositol (1,4) P_2 and inositol (1,4,5)P_3 breakdown, as shown in Fig. (**2.1**).

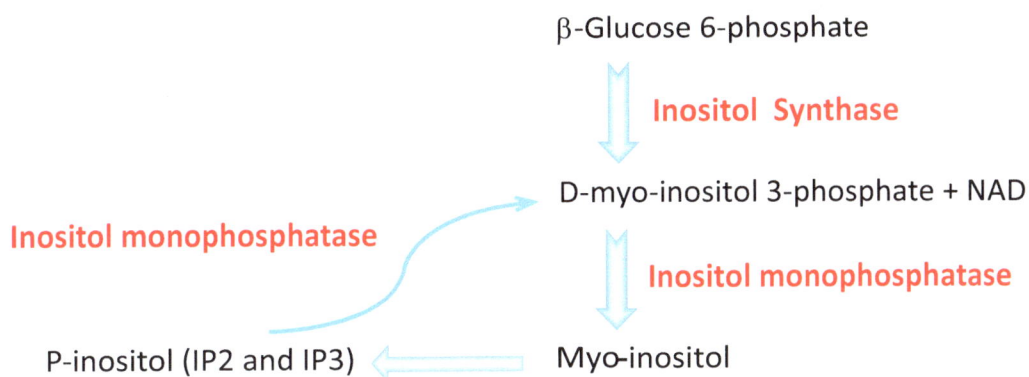

β-Glucose 6-phosphate

Inositol Synthase

D-myo-inositol 3-phosphate + NAD

Inositol monophosphatase

Inositol monophosphatase

P-inositol (IP2 and IP3)   Myo-inositol

Fig. (2.1). The pathway of biosynthesis of free *myo*-inositol in mammalian tissues.

In humans the synthetic process of *myo*-inositol is solely in the kidneys, and typically there are a few grams of *myo*-inositol produced per day [3].

Myo-inositol to *chiro*-inositol conversion is another important step in mammalian cells, particularly in the kidney [4]. *Myo*-inositol oxygenase (MIOX) catalyzes the first committed step in the only pathway of *myo*-inositol catabolism, which occurs predominantly in the kidney. The enzyme is a non-haem-iron enzyme that catalyzes the ring cleavage of *myo*-inositol with the incorporation of a single atom of oxygen.

BIOSYTHESIS AND METABOLISM OF WATER-SOLUBLE INOSITOL POLYPHOSPHATES AND PYROPHOSPHATES

The inositol polyphosphates (simply IPs), or "inositol phosphates", are composed of a *myo*-inositol core that is phosphorylated at different position. This phosphorylation process is a crucial step to generate the significant signaling molecules such as myo-inositol (1,4,5)-trisphosphate (InsP_3 or simply IP3). It has been well known that the released soluble InsP_3 is through the hydrolysis of the phosphodiester of PI(4,5)P_2 by phospholipase C (PLC). InsP_3 is the quintessential biologically active inositol polyphosphates due to its prominent role as a second messenger in intracellular calcium release, which was first reported in 1983 [5, 6].

There are 63 possible isomers of IPs structure, at least half of which have been identified as metabolites in various biological systems, but only few of them with the definitive biological functions have been elucidated [7]. The most abundant of Ins$P$ isomers are myo-inositol (1,2,3,4,5,6)-hexakisphosphate (InsP_6 that has been known as "phytate") and, myo-inositol(1,3,4,5,6)-pentakisphosphate (InsP_5), both of which are achiral meso structures for which concentrations are generally maintained between 10-100 mM in the cell [8]. InsP_6 has been implicated in widespread activities including neurotransmission, immune responses, regulation of protein kinases and phosphatases, and activation of calcium channels [9]. InsP_6 is essentially synthesized from Ins(1,4,5)P_3 *via* a sequential-step route of several kinases and a phosphatase in eukaryotic organisms, including 1) the initiation step through an Ins(1,4,5)P_3 3-kinase (InsP_3K), and 2) the requirement of inositol phosphate "multi-kinase" (IPMK) which is responsible for phosphorylating Ins(1,3,4,6)P_4 to Ins(1,3,4,5,6)P_5, and Ins(1,3,4,5,6)P_5 to Ins (1,2,3,4,5,6)P_6, as shown in Fig. (**2.2**).

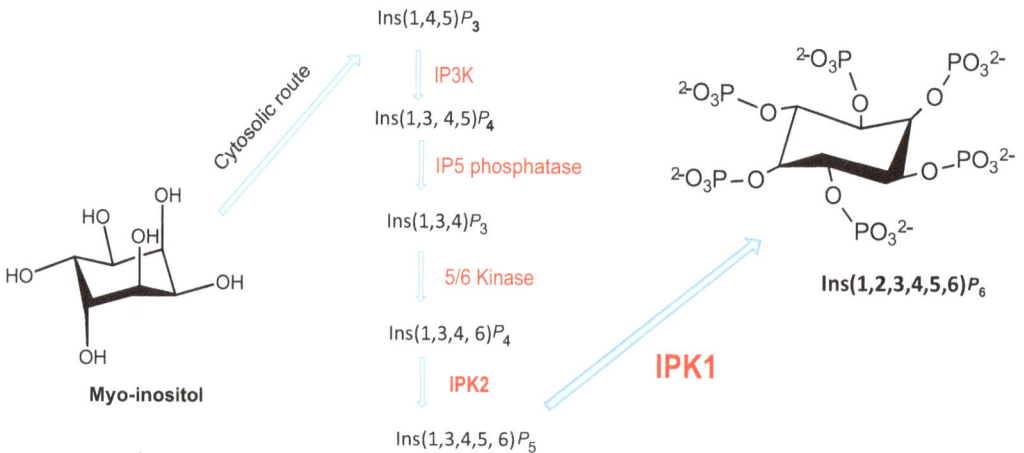

Fig. (2.2). The pathway of biosynthesis of InsP_6 *via* InsP_3. IPMK is referred as inositol phosphate multi-kinase including inositol phosphate kinase 1 and 2 (called IPK1 and IPK2).

Inositol Pyrophosphates

Another important family of compounds are the diphosphoinositol polyphosphates (PP-Ins*P*s) and the bis-diphosphoinositol polyphosphates ([PP]2-Ins*P*s), which possess one or two diphosphate groups introduced onto an Ins*P* backbone, commonly referred to as "inositol pyrophosphate polyphosphates" or simply "inositol pyrophosphates." InsP_6-kinases and PP-InsP_5-kinases, the two distinct classes of enzymes, are responsible for the synthesis of inositol pyrophosphate using InsP_5 and InsP_6 [10, 11]. Both enzymes are highly conserved throughout evolution. Essentially, inositol pyrophosphates contain high-energy diphosphate moieties on the inositol ring. As seen in Fig. (**2.3**), the InsP_6-kinases possess an enormous catalytic flexibility, converting InsP_5 and InsP_6 to PP-InsP_4 and InsP_7 respectively, and subsequently, by using these products as substrates, promote the generation of more complex molecules [11]. PP-InsP_5-kinases convert InsP_7 to InsP_8 and PP-InsP_4 to PP$_2$-InsP_3 [11]. Inositol pyrophosphates have been isolated from all eukaryotic organisms and are found throughout nature.

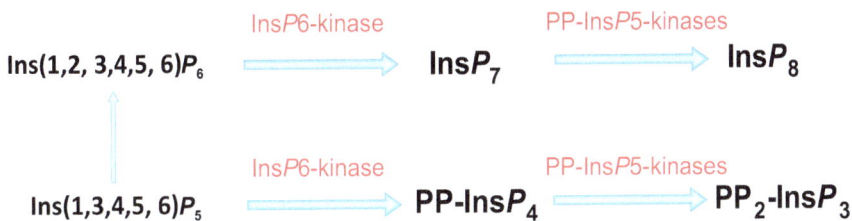

Fig. (2.3). The pathway of biosynthesis of Inositol pyrophosphates using InsP_5 and InsP_6.

SYNTHESIS AND METABOLISM OF PHOSPHATIDYLINOSITOL

Phosphatidylinositols are a minor component on the cytosolic side of cell membranes; and its synthesis and metabolism are essential metabolic processes in both plant and animal cells, either directly or *via* their metabolites. Phosphatidylinositol is biosynthesized through reaction of cytidine diphosphate diacylglycerol with inositol catalyzed by the enzyme CDP-diacylglycerol inositol phosphatidyltransferase ('phosphatidylinositol synthase') [12, 13]. In this synthetic process, the other product of the reaction is cytidine monophosphate (CMP). The CDP-diacylglycerol inositol phosphatidyltransferase is located in the endoplasmic reticulum and almost entirely on the cytosolic side of the bilayer [13, 14]. The final product of phosphatidylinositol is then delivered to other membranes either by vesicular transport or *via* specific transfer proteins [13, 14].

Since phosphatidylinositol is the primary source of the arachidonic acid required for biosynthesis of eicosanoids, including prostaglandins, *via* the action of the enzyme phospholipase A for releasing arachidonic acid from position *sn*-2, understanding this process is important [15-18]. Lysophosphatidylinositol is a by-product of eicosanoid formation, or as an intermediate is part of the normal cycle of deacylation-acylation of phosphatidylinositol in tissues catalyzed by phospholipase A [15, 16, 18, 19]. Furthermore, A membrane-bound *O*-acyltransferase is specific for lysophosphatidylinositol with a marked preference for arachidonoyl-CoA in inflammatory cells such as neutrophils, implying that free arachidonic acid and eicosanoid levels are regulated [18, 19]. When cells receive inflammatory stimuli, phosphatidylinositol containing two molecules of arachidonate is produced in this reaction.

Phosphatidylinositol and the phosphatidylinositol phosphates also serve as the main source of diacylglycerols and water-soluble inositol phosphates [13]. These products are crucial signaling molecules in animal and plant cells [20, 21]. The important signaling molecules are released by the action of the enzyme phospholipase C on phosphatidylinositol and the polyphosphoinositides, especially phosphatidylinositol-4,5-bisphosphate [20].

In addition, the inositol phospholipids (including the phosphatidylinositol phosphates or polyphosphoinositides) have crucial roles in interfacial binding of proteins and in the regulation of proteins at the cell interface [20, 21]. Phosphoinositides are polyanionic, and are involved very effectively in non-specific electrostatic interactions with proteins. They are also especially efficient in specific binding to so-called 'PH' domains of cellular proteins.

METABOLIC PROCESS OF PHOSPHATIDYLINOSITOL PHOSPHATES

Phosphatidylinositol is converted to polyphosphoinositides with important signaling and other functional activities in animal cells [20, 21]. Phosphorylation of this lipid involves a number of different kinases that place the phosphate moiety on positions 3, 4 and 5 of the inositol ring [20]. There are seven different isomers, all of which have distinct biological activities [20]. Phosphatidylinositol 4-phosphate and phosphatidylinositol 4,5-bisphosphate as well as 3-phosphorylated forms are the most significant and extremely important in quantitative and possibly biological terms [20]. Each phosphoinositide may have its own role - the 'lipid code' hypothesis. This distinct lipid act as labels for each cellular membrane to maintain the orderly flow required for the complexities of membrane trafficking and spatiotemporal signaling reactions [22]. Phosphatidylinositol 4-phosphate, phosphatidylinositol 4,5-bisphosphate, phosphatidylinositol 3-phosphate and phosphatidylinositol 3,5-bisphosphate are mainly identified on the Golgi, plasma membrane, early endosomes and late endocytic organelles, respectively - regarded as landmarks for these compartments [22]. The specific distributions of phosphoinositides in cellular compartments are important to selectively recruit effector proteins and to bind specifically to each type of phosphoinositide [22]. The interaction between phosphoinositide-binding proteins and various enzymes leads to either phosphorylation *via* kinases or dephosphorylation to remove phosphates *via* hydrolysis in the various organelles [20].

Sequential series of phosphorylation and dephosphorylation reactions of inositol are crucial intracellular process, and are controlled by specific kinases and phosphatases as well as phospholipase [20]. This process consumes a significant proportion of cellular ATP production, and occurs in different intracellular compartments for distinct and independently regulated functions with differing target enzymes. In mammals, the complexity is such that 18 phosphoinositide inter-conversion reactions have been identified to date, and these are mediated by at least 19 phosphoinositide kinases and 28 phosphoinositide phosphatases. The polyphosphinositide metabolism and the related enzymes are illustrated in Fig. (**2.4**). The detailed signaling role of these metabolites is discussed in Chapter 3.

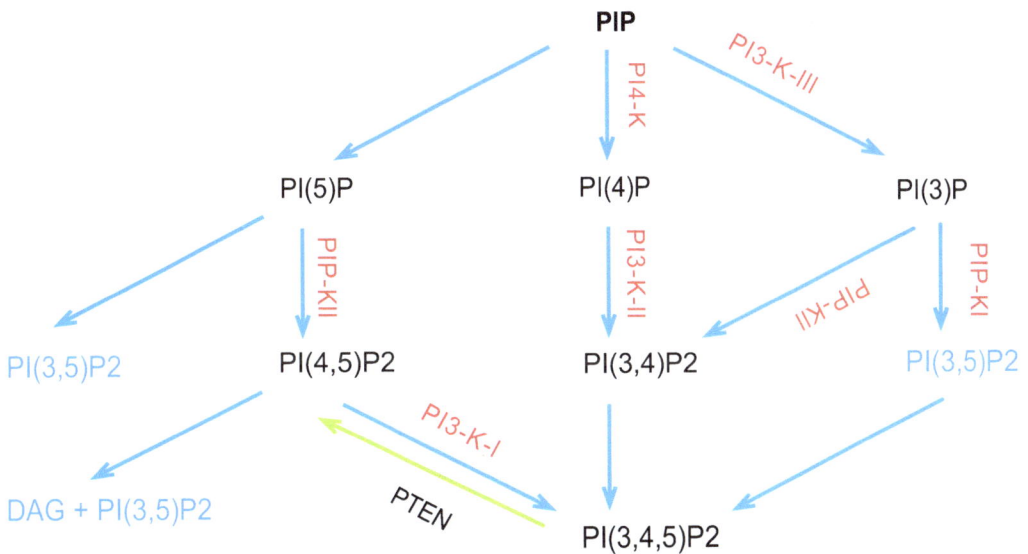

Fig. (2.4). Polyphosphoinositide metabolism.

CONCLUDING REMARKS

Water-soluble and lipid form inositol and its phosphates are crucial cellular components, and their biosynthesis and metabolism are widely involved in numerous cell functions, particularly for signal transduction. While the purpose and function of some of these have been known, those of others are yet to be determined.

REFERENCES

[1]   Eisenberg F Jr. D-myoinositol 1-phosphate as product of cyclization of glucose 6-phosphate and substrate for a specific phosphatase in rat testis. J Biol Chem 1967; 242: 1375-82.

[2]   Eisenberg F Jr, Parthasarathy R. Measurement of biosynthesis of myo-inositol from glucose 6-phosphate. Methods Enzymol 1987; 141: 127-43.

[3]   Parthasarathy LK, Seelan RS, Tobias C *et al*. Mammalian inositol 3-phosphate synthase: its role in the biosynthesis of brain inositol and its clinical use as a psychoactive agent. Subcell Biochem 2006; 39: 293-314.

[4]   Pak Y, Huang LC, Lilley KJ. *et al*. *In vivo* conversion of [3H]myoinositol to [3H]chiroinositol in rat tissues. J Biol Chem 1992; 267: 16904-10.

[5]   Streb H, Schulz I. Regulation of cytosolic free Ca2+ concentration in acinar cells of rat pancreas. Am J Physiol 1983; 245: p. G347-57.

[6]   Streb H, Irvine RF, Berridge MJ *et al*. Release of Ca2+ from a nonmitochondrial intracellular store in pancreatic acinar cells by inositol-1,4,5-trisphosphate. Nature 1983; 306: 67-9.

[7]   Irvine RF, Schell MJ. Back in the water: the return of the inositol phosphates. Nat Rev Mol Cell Biol 2001; 2: 327-38.

[8]   Sasakawa N, Sharif M, Hanley MR. Metabolism and biological activities of inositol pentakisphosphate and inositol hexakisphosphate. Biochem Pharmacol 1995; 50: 137-46.

[9]   Shears SB., Assessing the omnipotence of inositol hexakisphosphate. Cell Signal 2001; 13: 151-8.

[10]   Shears SB. Diphosphoinositol polyphosphates: metabolic messengers? Mol Pharmacol 2009; 76: 236-52.

[11]   Wilson MS, Livermore TM, Saiardi A. Inositol pyrophosphates: between signalling and metabolism. Biochem J 2013; 452: 369-79.

[12]   Mueller-Roeber B, Pical C. Inositol phospholipid metabolism in Arabidopsis. Characterized and putative isoforms of inositol phospholipid kinase and phosphoinositide-specific phospholipase C. Plant Physiol 2002; 130: 22-46.

[13]   Gardocki ME, Jani N, Lopes JM. Phosphatidylinositol biosynthesis: biochemistry and regulation. Biochim Biophys Acta 2005; 1735: 89-100.

[14]   Mayinger P. Phosphoinositides and vesicular membrane traffic. Biochim Biophys Acta 2012; 1821: 1104-13.

[15]   Corda D, Zizza P, Varone A *et al*. The glycerophosphoinositols and their cellular functions. Biochem Soc Trans 2012; 40: 101-7.

[16]   Jungalwala FB, Evans JE, McCluer RH. Compositional and molecular species analysis of phospholipids by high performance liquid chromatography coupled with chemical ionization mass spectrometry. J Lipid Res 1984; 25: 738-49.

[17]   Patton GM, Fasulo JM, Robins SJ. Separation of phospholipids and individual molecular species of phospholipids by high-performance liquid chromatography. J Lipid Res 1982; 23: 190-6.

[18]   Tanaka T, Iwawaki D, Sakamoto M. *et al*. Mechanisms of accumulation of arachidonate in phosphatidylinositol in yellowtail. A comparative study of acylation systems of phospholipids in rat and the fish species Seriola quinqueradiata. Eur J Biochem 2003; 270: 1466-73.

[19]   Yashiro K, Kameyama Y, Mizuno-Kamiya M *et al.* Substrate specificity of microsomal 1-acyl-sn-glycero-3-phosphoinositol acyltransferase in rat submandibular gland for polyunsaturated long-chain acyl-CoAs. Biochim Biophys Acta1995; 1258: 288-96.

[20]   Sasaki T, Takasuga S, Sasaki J. *et al.* Mammalian phosphoinositide kinases and phosphatases. Prog Lipid Res 2009; 48: 307-43.

[21]   Skwarek LC, Boulianne GL. Great expectations for PIP: phosphoinositides as regulators of signaling during development and disease. Dev Cell 2009; 16: 12-20.

[22]   Vicinanza M, D'Angelo G, Di Campli A *et al.* Function and dysfunction of the PI system in membrane trafficking. EMBO J 2008; 27: 2457-70.

CHAPTER 3

Signaling Role of Inositol and its Phosphates

Abstract: Inositol and its phosphates are integral parts of cellular signaling. InsP_6 is plentiful in organisms as diverse as yeast, actinobacteria, mammals and plants. In the plant kingdom, InsP_6 accumulates during seed development; at the time of germination it is broken down into lower inositol phosphates and micronutrients to maintain seedling. InsP_6 is the most abundant of intracellular inositol phosphates in eukaryotes. In mammals InsP_6 maintains homeostasis, stores phosphate and acts as strong anti-oxidant and neurotransmitter. InsP_6 and other inositol phosphates including the pyrophosphates, play critical roles in cellular pathways involved in signal transduction, control of cell proliferation and differentiation, RNA export, DNA repair, energy transduction, ATP regeneration *etc.* By enabling communication between nucleus, cytoplasm, and the outside environment, InsP_6 and the other inositol phosphates play a crucial role in many aspects of cell biology.

Keywords: Apoptosis, ATP, auxin, cell proliferation, clathrin, coated pits, differentiation, DNA repair, eukaryote, exocytosis, FGF, MAPK, nuclear pore, PKC, prokaryote, RNA export, signal transduction, syndecan, zinc finger.

INTRODUCTION

In order to survive and fulfill their specialized functions all cells must efficiently communicate with their environments. Intracellular signaling is a complex mechanism used by cells to transmit and coordinate changes sensed from outside. The inositol polyphosphate family represents phylogenetically ubiquitous messengers with particularly dynamic turnover implicated in rapid cellular signaling. InsP_6 is the most abundant polyphosphate in nature [1]. Originally it had been found as a rich source of phosphorous in plants [2, 3]. Later, it became obvious that it is also the most abundant inositol phosphate in eukaryotic cells. The discovery of calcium mobilizing properties of inositol triphosphate (InsP_3) began the career of inositol phosphates in the field of cell biology [4]; and a few years later in plants [5]. In plants, Ca^{2+} mobilization is the effect of binding of Ins(1,4,5)P_3 or Ins(2,4,5)P_3-phytase (phytase - a phosphatase specific for InsP_6) complex to its putative receptor [6]. As shown in Chapter 1, there are several isoforms of inositol phosphates; the *myo*-isoform and enzymes associated with their rapid turnover are well conserved evolutionarily. More recently, growing research interest on these enigmatic inositol phosphates as signal transduction molecules has led to fascinating new discoveries reviewed recently [7]; we will primarily discuss the various roles of $InsP_6$ and to a lesser extent the lower phosphorylated inositols in cellular signal transduction in different organisms.

$InsP_6$ IN PROKARYOTES

In prokaryotes, inositols were rarely found. They are parts of the lipid component, phosphatidylinositol (PI) of the cell wall of organism such as *Mycobacterium* [8], *Treponema* [9], *Myxobacteria* [10]; and in *Escherichia coli*, *Pseudomonas* and *Erwinia* [11]. Still not much is known about the synthesis of PI in those organisms. Since *scyllo*-inositol found in soil is a highly selective inhibitor of PI synthase, it was believed that it could serve as a promising candidate for anti-tuberculosis therapy [12]. In contrast, antioxidant actions of inositols seem to be universal and highly conserved throughout species. Anti-oxidative functions for *myo*-inositol were found in highly aerobic actinobacteria, which makeup the most common soil life [13]. In contrast to plants where only *myo*-inositol hexaphosphate has been identified, soil life contains *neo*-, *chiro*-, and *scyllo*-inositol hexaphosphates [14]. Signal transduction properties for $InsP_3$ were found for both prokaryotic and eukaryotic cells [15].

Studies of *Clostridium difficile* toxins show that cellular $InsP_6$ induces an autocatalytic cleavage of the toxins, releasing an N-terminal glucosyltransferase domain into the host cell cytosol. X-ray crystal structure of the domain shows that $InsP_6$ is bound in a highly basic pocket [16]. The role of $InsP_6$ in mediating bacterial toxin-related activation/inactivation has also been demonstrated in *Vibrio cholerae*. This pathogen damages the actin cytoskeleton *via* proteins released from its toxin. These proteins are products of autoproteolysis, catalyzed by an embedded cysteine protease domain (CPD). Binding of $InsP_6$ stabilizes the CPD structure, allowing the formation of enzyme-substrate complex [17].

$InsP_6$ IN EUKARYOTES

Eukaryotic cells are larger with more elaborate intracellular systems than their prokaryotic counterparts. Accompanying this highly complicated structure are also the multiple tasks the cells/organisms carry out, requiring intricate communication between the various components.

$InsP_6$ in Yeasts

The budding yeast *Saccharomyces cerevisiae* has been a favorite not only to the beer-brewers and bread-bakers, but also as a model of eukaryotes for cell biologists. Many of the early discoveries of eukaryotic function and mechanisms, including signal transduction, have been studied in *S. cerevisiae*; indeed, early work on phosphoinositides and inositol polyphosphates were carried out in this model. Steiner *et al.,* found that inositol comprises a big portion of *S. cerevisiae* cell lipids, forming the basic structure of the phytoglycolipids [18]. Monoclonal antibody specific for phosphatidylinositol 4,5-bisphosphate, when introduced into cells results in inhibition of growth of *S. cerevisiae* [19]. There is little accumulation of $InsP_3$ in yeast due to its rapid conversion to polyphosphates [20]. So far, no $InsP_3$ receptors have been found in the yeast genome; however a study done on *Neurospora crassa* suggests that the fungi must express novel types of receptors with little or no homology to the $InsP_3$ receptors found in mammals [21]. In *S. cerevisiae* $InsP_3$ has later been found to act as a second messenger for glucose-induced calcium signaling [22].

Most, if not all organisms respond to environmental alterations. Nutrient availability is one of those environmental factors critical for the organisms' survival; *S. cerevisiae* is no exception, for it has to decide whether or not to grow, divide or activate other mechanisms for its survival. Inorganic phosphate (Pi) is an essential nutritional component for building nucleic acids, membrane phospholipids, *etc.* Under Pi-sufficient nutritional condition, the Pho85 kinase phosphorylates Pho4 excluding it from the nucleus and resulting in repression (*i.e.*, lack of transcription) of PHO genes. When yeasts are deprived of Pi from their food source, genes required for foraging are activated. Pho4 transcription factor through its migration in and out of the nucleus brings about this activation. This Pi-starvation signal is transmitted by inositol pyrophosphate $InsP_7$ [23]; more on pyrophosphates to follow.

Signaling Role of $InsP_6$ in Plants

In plants, minerals such as Mg, K and Ca are stored in forms of salt with $InsP_6$ (it had been known as 'phytic acid' or 'phytate' - the salt form) in storage vacuoles called globoids, which are localized predominantly in the aleurone layer in wheat and barley, or in the embryo in maize, and create the main storage of the total plant phosphorous. During the germination process the enzyme phytase breaks down the salts and minerals. $InsP_6$ and its lower phosphate derivatives are released as well, creating the nutrition pool necessary for the developing plant. An additional role for $InsP_6$ in plants is conserving water, and ensuring plant survival by involvement in the processes by which the drought stress hormone abscisic acid induces stomatal closure. In response to abscisic acid in *Solanum tuberosum* stomatal guard cells, InsP_6 inactivates the plasma membrane. This process is dependent on calcium released from endomembrane stores [24]. Recent studies on auxin (indole-3-acetic acid or IAA), an important plant hormone involved in growth and plant development, revealed InsP_6 as a crucial co-factor in communication between auxin and its receptor TIR1, the F-box protein subunit of ubiquitin ligase [25, 26].

There has been a recent trend in certain agricultural sectors to reduce InsP_6 content through lpa mutants *via* knock-out of genes involved in InsP_6 biosynthesis [27]. However Doria *et al.,* [28] have shown that reduction of InsP_6 content through lpa mutation could actually be harmful to the seeds as evidenced by lower germination capacity, increased carbonylation and DNA damage and decreased (~50%) γ-tocopherol content. Further on this may be found in the review by Murgia *et al.,* [29].

The plant hormone auxin regulates virtually every aspect of plant growth, from embryo patterning, all the way to fruit development [30]. As in animals, auxin achieves these feats through activation of hundreds of genes *via* transcriptional factors called auxin response factors (ARF). The receptor for auxin is an F-box protein named TIR1 (Transport Inhibitor Response 1), the crystal structure of which shows InsP_6 in the core of TIR1 LRR domain, right beneath the auxin binding site [26] (Fig. **3.1**). That two of the critical residues which form the floor of the auxin-binding site are buttressed by InsP_6 suggests that InsP_6 is an important structural component of auxin receptor TIR1, and therefore aids in its function. However, at this time we are not sure what if any specific signaling function InsP_6 has *vis-à-vis* auxin and TIR1 [31].

InsP_6 IN MAMMALS

The concentration of InsP_6 in mammalian cells range between 10 μM and 1 mM [32]. InsP_6 has been shown to be rapidly internalized and metabolized to lower inositol phosphates [33-35]. InsP_6 in mammalian tissues is most abundant in the brain. Together with its high-affinity membrane binding receptor, it acts as a potent neural stimulator. InsP_6's actions in the central nervous system and the anterior pituitary gland seem to imitate excitatory neurotransmitter glutamate receptor [36, 37]. Studies demonstrate that inositol 1,3,4-trisphosphate 5/6-kinase (ITPK1), which is a crucial enzyme in the synthesis of InsP_6, is also essential for neural tube and axial mesoderm development [38]. High affinity binding sites for

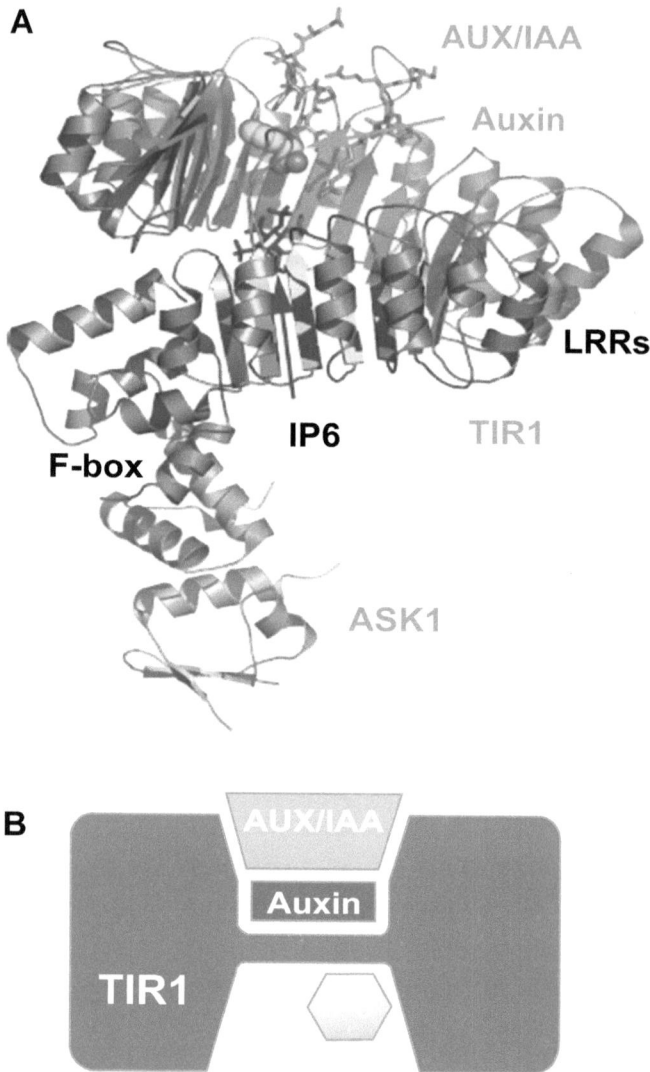

Fig. (3.1). Structural and mechanistic relationship of auxin, its receptor and InsP_6 (IP$_6$). A. Crystal structure of Arabidopsis SKP1 (ASK1)-TIR1-IP6 (InsP_6) in complex with auxin and an AUX/IAA domain II peptide. Both ASK1 and TIR1 are shown in ribbon. Auxin is shown in space-filled model. The AUX/IAA peptide and InsP_6 are shown in sticks. B: Schematic diagram showing how auxin is sensed by TIR1 with the assistance of InsP_6 and how the hormone promotes TIR1-AUX/IAA interactions. LRRs, leucine-rich repeats. Reproduced with permission from Dr. Ning Zheng [26].

[^{3}H]-InsP_6 have been found in other tissues. In heart InsP_6 and InsP_4 are produced in response to stimulation of cardiac α 1-adrenoreceptors. Studying InsP_6 receptors in mitochondrial and sarcoplasmic reticulum fractions of heart Rowley *et al.,* found KD ranging from 22 ± 1.9 nM in the right atrium to 35 ± 2.6 nM in the interventricular septum. The maximal number of binding sites (Bmax) ranged from 5.1 ± 0.48 to 12 ± 1.8 pmol mg-1 protein in the left atrium and left ventricle, respectively [39]. Further studies on InsP_6 receptors in other cells revealed InsP_6 binding sites in a low-density membrane fraction from human platelets [40] and in human neutrophil membrane preparations [41]. Eggleton *et al.,* showed an important modulatory role for InsP_6 on neutrophil functions. Initial stimulation of neutrophils with InsP_6 had no effect on the production of reactive oxygen intermediates, however subsequent exposure substantially enhanced neutrophils' phagocytic properties [42] accompanied by increase in IL-8 secretion by stimulated cells [43]. An entirely different role for InsP_6 has been found in *Xenopus laevis* oocyte maturation where the levels of InsP_5 and subsequently InsP_6 rose by 6-fold suggesting regulation of oogenesis and oocyte maturation [44]. Changes in the cellular levels of InsP_5 and InsP_6 were shown during the cell cycle of rat thymocytes [45]. At the beginning of the cell cycle, levels of InsP_5 and InsP_6 increased followed by decreased intracellular concentration of both the compounds as the cell cycle progressed, suggesting a role for these compounds in cell cycle progression.

INOSITOL POLYPHOSPHATES AND THEIR RECEPTORS

The success of inositol polyphosphates in signal transduction conglomerate is related to their rapid turnover mediated by different classes of enzymes. Protein kinases and phosphatases by adding or deleting phosphate groups can finely tune activation or deactivation signals. The GTP (guanosine 5'-triphosphate)-binding proteins create a class of molecular switches; the large trimeric GTP-binding proteins are commonly called G proteins. As a response to a variety of environmental signals, the inositol-phospholipid signaling pathway is activated *via* G proteins, specifically G_q and activation of phospholipase C-β (PLC-β). The activated PLC-β cleaves the inositol-phospholipid phosphatidylinositol 4,5-biphosphate (PIP$_2$) in the plasma membrane lipid bilayer to generate diacylglycerol (DAG) and inositol 1,4,5-triphosphate (InsP_3). InsP_3 which is still in the plasma membrane, diffuses rapidly through the cytosol to bind to its receptor proteins (InsP_3Rs) in the endoplasmic/sarcoplasmic reticulum causing the Ca^{2+} stored at high concentrations in the endoplasmic/sarcoplasmic reticulum to enter the cytoplasm. As a consequence, the local Ca^{2+} concentration can increase 10-20 folds and trigger Ca^{2+} responsive proteins in the cell [35, 46]. InsP_3Rs are large structures composed of four subunits and appear to have different affinity for InsP_3 [47]. InsP_3Rs are almost ubiquitously expressed in mammalian tissues. $InsP_3$ is required for proper function of the $InsP_3Rs$, and their activation is tightly regulated by the Ca^{2+} concentration at their cytosolic surface. $InsP_3Rs$ can simultaneously bind to a multitude of diverse proteins such as transcription factor Bcl-2 and enzymes like protein kinase A that determine their cellular location and functionality [48]. Therefore, in addition to modulation of cellular activities through Ca^{2+} channels, $InsP_3Rs$ can act as connection points to bring multiple signal transduction pathways together.

Inositol 1,3,4,5-tetrakisphosphate $(InsP_4)$ is also considered to be a signaling molecule and seems to have a regulatory role in maintaining levels of $InsP_3$-sensitive Ca^{2+} pools [46]. Interestingly, $InsP_3$ serves not only as a substrate for kinases to produce other inositol derivatives but also as an activator for $InsP_5$ 1-phosphatase an enzyme that synthesizes $InsP_4$ [49]. Phosphatidylinositol 3,4,5 triphosphate can also be generated as a result of PI-3 kinase action, and it also plays an important role in the regulation of protein trafficking, cell growth and cell survival, cytoskeletal organization *etc.*

The role of $InsP_6$ as a signaling molecule became apparent in the late 1980's and early 1990's after it was found to stimulate Ca^{2+} uptake in cultured cerebellar neurons [50] and anterior pituitary cells [51]. Moreover, the low concentration of $InsP_6$ increases both intracellular free Ca^{2+} and prolactin secretion in perifused pituitary cells. Subsequently, $InsP_6$-binding sites - the $InsP_6$ receptors, were isolated [52-55]. $[^3H]$-$InsP_6$ bound to specific and saturable recognition sites in membranes and in all subcellular fractions including the mitochondria prepared from cerebral hemispheres, anterior pituitaries and cultured cerebellar neurons. It became obvious that $InsP_6$ may also act as an intracellular regulator of Ca^{2+} homeostasis just like $InsP_3$. Shamsuddin *et al.*, have shown that $InsP_6$ treatment of K562 human erythroleukemia cells results in a 57% increase in intracellular Ca^{2+} ($p<0.02$) and surprisingly a 41% increase in $InsP_3$ ($p<0.05$) with concomitant decrease in cell proliferation and increased cell differentiation. This was unexpected as it was contrary to the dogma that cell division is associated with increased intracellular Ca^{2+} and increased $InsP_3$ [56].

It appears that the binding affinity for $InsP_6$ receptor is different. For instance, Theibert *et al.*, [54] had reported that $InsP_6$ receptor comprises a protein complex of 115-kDa, 105-kDa and 50-kDa subunits. On the other hand, the concept that the $InsP_6$ binding site in the membrane may not necessarily be protein but non-protein entities such as phospholipid has been advanced by Poyner *et al.*, [57]. It has been shown that $InsP_6$ regulates both receptor endocytosis and receptor signaling, and binds with higher affinity than its lower phosphate derivatives to AP-2 and, blocks arrestin binding to receptors and ion channels preventing signal inactivation [58] (Fig. **3.2**).

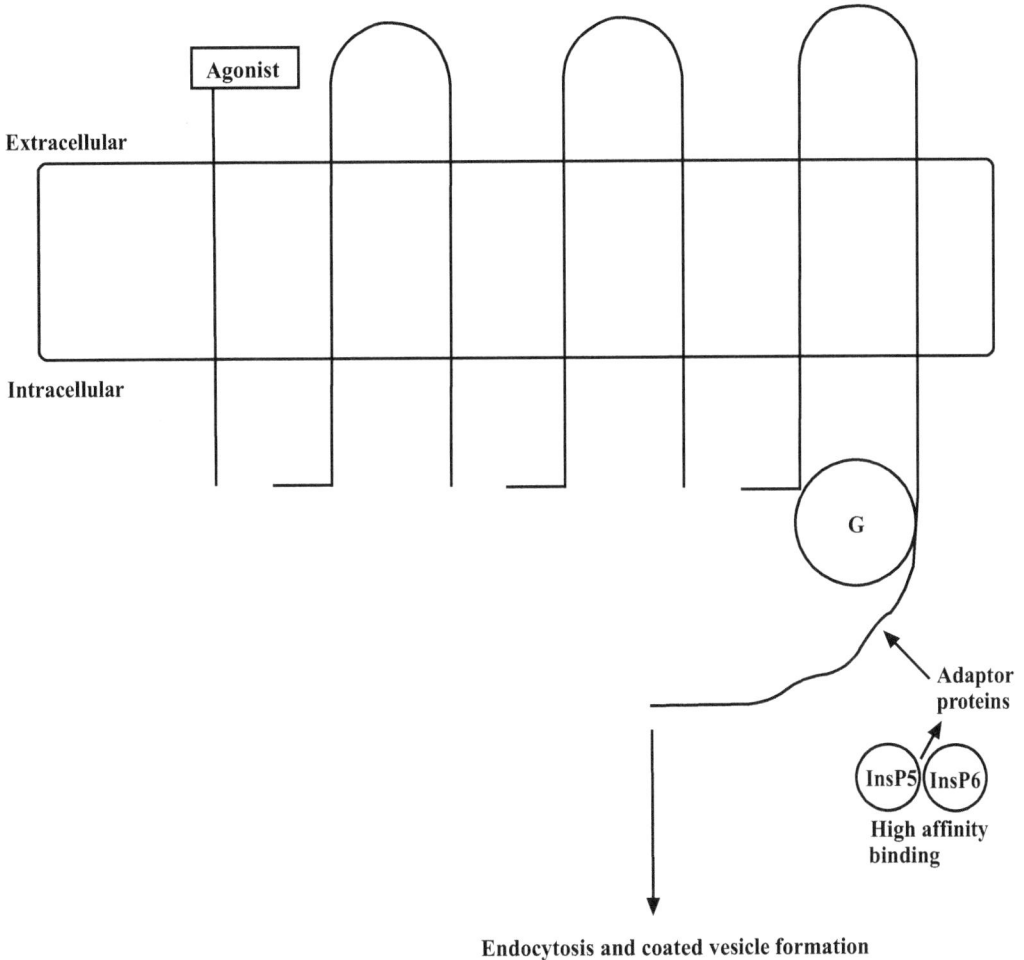

Fig. (3.2). $InsP_6$'s high-affinity binding to AP-2.

ATP REGENERATION

ATP regeneration properties of $InsP_6$ are well established in plant kingdom. The original idea of Morton and Raison more than three decades ago suggesting a link between $InsP_6$ and ATP, and its possible role in seed development and germination [59] was further empirically confirmed by Biswas *et al.,* [60]. In developing mung bean [*Phaseolus aureus*] seeds, InsP_6 was identified as a phosphate donor for ADP phosphotransferase; and conversion from ADP to ATP by transferring a phosphate group from 2 position of InsP_6 to ADP. A later study on inositol 1,3,4,5,6-pentakisphosphate 2-kinase isolated from the soybean, a key enzyme in InsP_6 synthesis, catalyzes the conversion of ATP from ADP in germinating seeds [61, 62]. Future discoveries of InsP_6 function in ATP regeneration in mammals seems like just a matter of time.

VESICLE TRAFFICKING

Exocytosis

Eukaryotic cells release secretory proteins in the process called exocytosis. InsP_6 has been shown to regulate glucose level by stimulation of insulin exocytosis in pancreatic β cells [63]. In pancreatic insulin secreting β cells, glucose stimulation increased InsP_6 concentration which correlated well with the initial rise in intracellular Ca^{2+} [64]. These transient changes in InsP_6 concentrations were most pronounced at the site of exocytosis and at the proximity of Ca^{2+} channels. The physiological level of InsP_6 in the un-stimulated cells is quite stable and reaches concentrations of 40-54 μM. A 10 μM rise in concentration is sufficient to inhibit the serine/threonine protein phosphatase activity and increase the current through voltage-gated L-type Ca^{2+} channel activity. L-type Ca^{2+} channels have a crucial role in tissues such as myocardium, smooth muscles and the brain; thus slight changes in InsP_6 concentrations can have a big impact on the function of these cells, both in physiological as well as in pathological conditions. It has been shown that prolonged opening of L-type Ca^{2+} channels mediates apoptosis of the β cells; therefore, higher InsP_6 concentration could stimulate apoptosis, an effect much desired in cancer cells. As expected, a high level of InsP_6 or prolonged exposure to InsP_6 can induce apoptosis in various cancer cell lines [65, 66]. Studies in Shamsuddin laboratory on HT-29 human colon cancer cell line demonstrated increased intracellular Ca^{2+} level by 3-4 folds within 10 seconds of InsP_6 exposure. Along with this increased intracellular Ca^{2+} there was a concomitant down-regulation of cyclin E - a protein that helps control progression from one stage of the cell cycle to the next [67].

Thus, disruption in InsP_6 metabolism could result in dramatic functional changes in cells possessing L-type Ca^{2+} channels like the pyramidal cells of the hippocampus associated with memory. This would suggest a potential role for InsP_6 in the prevention and treatment of Alzheimer's disease. As shown by Lu *et al.,* InsP_6 may act as a modulator of neurotransmitter by altering the state of synaptotagmin-phospholipid interaction [68]. Synaptotagmin I, a Ca^{2+} sensor for rapid exocytosis has two large cytoplasmic domains: C2A and C2AB. $InsP_6$ induces conformational changes in C2AB in the presence of lysosome and weakens the Ca^{2+}-dependent C2AB-membrane interactions. Grases *et al.,* also believe that $InsP_6$ may have a beneficial function in Alzheimer's disease, although the mechanism they propose is a different one [69].

Endocytosis

Macromolecules are engulfed inside the cell by the process called endocytosis which depends on PIP_2, clathrin and clathrin adapters, the guanosine triphosphatase dynamin I, synaptojanin 1 and the amphiphysin dimer. Dynamin I, a substrate for calcineurin, is responsible for membrane fission during endocytosis in the pancreatic beta cell. Activation of dynamin is dependent on $InsP_6$ dephosphorylation, leading to the activation of protein kinase C and inhibition of the phosphoinositide phosphatase synaptojanin [70].

GROWTH FACTORS' SIGNALING

Receptor-dependent communication of the cell with the outer environment begins with receptor dimerization, activation of intrinsic receptor tyrosine kinase, autophosphorylation of the receptor at the carboxyl terminus, and tyrosine phosphorylation of intracellular signaling molecules such as Shc and PI-3 K (phosphatidylinositol 3-kinase). As the effect of ligand-induced endocytosis, the erbB receptor is down-regulated *via* clustering into clathrin-coated pits on the plasma membrane [71]. An important structural component of coated pits is the clathrin lattice anchored to the cytoplasmic surface of the membrane by plasma membrane clathrin-associated protein complex 2 (AP2) and is necessary for receptor-mediated endocytosis. Zi *et al.,* [72] report that $InsP_6$ interferes with both receptor-mediated and fluid-phase endocytosis resulting in inhibition of mitogenic signals associated with growth and proliferation of human prostate cancer cell line DU145. $InsP_6$ interacts with AP2 and inhibits PI-3 K and PI3-K-AKT signaling pathway as an upstream response and has effects on the inhibition of fluid-phase endocytosis. $InsP_6$ has a potent effect on transforming growth factor α (TGFα)-induced binding of activated erbB1 receptor to AP2. One of the mechanisms by which $InsP_6$ inhibits growth of DU145 human prostate cancer cells is by increasing the levels of activated erbB1 receptor which in consequence impairs ligand-induced Shc phosphorylation [71]. As reported by Voglmaier *et al.,* [73] there is a striking homology between the InsP_6 receptor protein, which contains a series of subunits, and the α-subunit of the clathrin assembly protein AP-2.

Fibroblast Growth Factors (FGF)

FGF, one of the factors important in tumor cell growth and angiogenesis is activated by both heparin-like extracellular matrix and transmembrane tyrosine kinase receptor sites. InsP_6 inhibits basic FGF (bFGF) binding to heparin and protects bFGF from degradation by trypsin. Moreover, InsP_6 in a dose-dependent manner has strong inhibitory effects on the cellular binding of bFGF and other FGF family members such as acidic FGF (aFGF) and K-FGF. This was very well demonstrated in bFGF- and K-FGF-transfected NIH/3T3 cells where low concentrations (100 μM) of InsP_6 could inhibit cell growth as well as bFGF-induced DNA synthesis in AKR-2B fibroblasts [74].

Insulin-Like Growth Factor II/Mannose-6-Phosphate (IGF II/Man-6-P)

The trafficking of lysosomal enzymes partially depends on transmembrane-anchored IGF II/Man-6-P receptors. Similar to signal transduction *via* erbB, IGF II/Man-6-P receptor also works *via* AP-2, creating a complex, which is further transported in coated vesicles to the lysosomes. In rat brain InsP_6 binding sites are co-localized with IGF II binding sites; thus InsP_6 competes for IGF II [75]. This finding was confirmed by studies on HT-29 human colon cancer cells where the InsP_6-mobilization of intracellular Ca^{2+} was blocked by incubation with IGF-II [67].

Syndecan-4

Syndecan-4 is a heparan sulfate proteoglycan localized in cellular membranes and regulates cell-matrix interactions. Its cationic motif in cytoplasm binds inositol phospholipids like PIP$_2$ and phosphatidylinositol 4-phosphate which with high affinity promote syndecan-4 oligomerization. InsP_6 has a similar function and is considered a potential down-regulator of syndecan-4 signaling. In fibroblasts, InsP_6 could block syndecan-4 - dependent focal adhesion and microfilament bundle formation [76]. In human mammary cancer cell lines, cell adhesion to extracellular matrix was decreased after InsP_6 treatment suggesting a role of InsP_6 in inhibition of cancer cell metastasis [77, 78].

CELL CYCLE REGULATION AND NORMALIZATION OF CELL PROLIFERATION

Cells division rate in normal cells is tightly controlled by several checkpoints. During malignant processes checkpoints are ignored, and uncontrolled replication leads to tumor formation. InsP_6 has been proven to normalize uncontrolled multiplication in all the different cancer cell lines tested [79-81]. In breast cancer cells InsP_6 has been shown to inhibit two major pro-proliferative and pro-survival pathways: ERK's and PI-3 K/Akt, and up-regulate cyclin-dependent kinase inhibitor p27^{Kip1} [82]. Studies in Shamsuddin laboratory showed that InsP_6 down-regulated proliferation marker PCNA (proliferating cell nuclear antigen) [83]. Interestingly, it did not affect the rate of proliferation of normal cells either *in vivo* or *in vitro*. InsP_6 normalizes cell proliferation by blocking uncontrolled cell division and forcing malignant cells either to differentiate or to go into apoptosis. These have been described further in Chapter 18; in here we will discuss the pertinent aspects as related to signal transduction.

One of the cell cycle checkpoints disrupted in cancer cells is the progression of cells through G_1 and entry to the S-phase, a process which requires hyperphosphorylation of retinoblastoma protein (pRb). In human mammary cancer cell line a marked reduction in pRb phosphorylation by InsP_6 has been reported [84]. InsP_6 controls the progression of the cells through the cell cycle [85]. The intracellular concentrations of InsP_6 are high in G_1 and G_2/M phases of cell cycle, and the level becomes one-half in S phase [45, 86]. InsP_6 treatment significantly decreased S-phase and arrested the human colon and breast cancer cells in G_0/G_1 phase. pRb/p107 and pRb2/p130 belong to a subfamily of pRb family of proteins and regulate cell cycle progression through G_1 phase. E2F family of transcription factors induce the transcription of genes needed for progression to S-phase. In quiescent cells (in G_0 phase), the pRb2/p130-E2F complex is the most prevalent. InsP_6 increased the level of hypophosphorylated Rb-related proteins pRb/107 and pRb2/p130 in DU145 human prostate cancer cell line. InsP_6 moderately decreased E2F4, but increased its binding to both pRb/p107 and pRb2/p130 [87]. In human leukemia cell lines InsP_6 showed a dose-dependent cytotoxic effect by arresting in G_2M phase of cell cycle (as opposed to G_0/G_1 phase in breast cancer cells [85]) [88]. cDNA micro array analysis showed a down-modulation of genes involved in transcription and cell cycle regulation (c-myc, HPTPCAAX1, FUSE, cyclin H) and an up-regulation of cell cycle inhibitors such as CKS2, p57 and Id-2. Additionally genes involved in important signal transduction pathways STAT-6 and MAPKAP, were also down-regulated [88]. Using DU145 human prostate cancer cell line, Singh *et al.,* [87] studied the cell cycle progression and the involvement of G_1 cell cycle regulators as well as apoptosis. A significant dose- and time-dependent growth inhibition of $InsP_6$-treated cells was associated with an increase in cells in G_1. $InsP_6$ markedly increased the expression of cyclin-dependent kinase inhibitors (CDKIs) - Cip1/p21 and Kip1/p27 without any noticeable changes in G_1 CDKs and cyclins, except a slight increase in cyclin D2. $InsP_6$ inhibited kinase activities associated with CDK2, 4 and 6, and cyclin E and D1. Further studies showed an increased binding of Kip1/p27 and Cip1/p21 with cyclin D1 and E. At higher doses and longer treatment times, $InsP_6$ caused a marked increase in apoptosis, which was accompanied by increased levels of cleaved PARP and active caspase 3. $InsP_6$ modulated the CDKI-CDK-cyclin complex, and decreased CDK-cyclin kinase activity [87].

INDUCTION OF DIFFERENTIATION

In order to fulfill their biological function cells need to mature. In other words, they need to become differentiated. Differentiated features depend on the cell type. For example hemoglobin is a marker of differentiation for mature red blood cells, prostate specific acid phosphatase for prostatic epithelial cells, lactalbumin for mammary cells, and myoglobin for muscle cells and so on, further described in Chapter 18.

Induction of specific differentiation was first demonstrated in K-562 human erythroleukemia cells, which showed increased hemoglobin production following $InsP_6$ treatment [56], and later in human colon carcinoma HT-29 cells [83, 89], prostate cancer cells [90], breast cancer cells [91], and rhabdomyosarcoma cells [92]. In all these studies $InsP_6$ has been proven to initiate differentiation of malignant cells to the normal phenotype. The molecular mechanisms for this action of $InsP_6$ are poorly understood. PI-3 K [phosphoinositide 3-kinase] plays an important role in granulocytic differentiation of HL-60 leukemia cells and might be one of the target molecules [93]. It has been shown that during chemotactic stimulation of HL-60 cells the intracellular concentration of $InsP_6$ (and $InsP_5$) is elevated by about two orders of magnitude [94] suggesting requirement of elevated dosage of $InsP_6$ in differentiated functions. However, Deliliers *et al.,* did not observe any differentiating ability of $InsP_6$ on HL-60 cells [88].

APOPTOSIS STIMULATION

As discussed above, $InsP_6$ can regulate cell cycle and cell proliferation followed by differentiation, resulting in restoration of the normal phenotype in all different cancer models. All of those processes without a substantial increase in cell death are induced by $InsP_6$ at concentrations up to 2-5 mM. Except for HeLa cells which have low dosage requirements, a higher dosage of $InsP_6$ or prolonged treatment induces apoptosis in cancer cell lines [65, 87]. Conditioning of HeLa cells with tumor necrosis factor or insulin stimulates the Akt-nuclear factor κB (NFκB), cell survival signal pathway involving the phosphorylation of Akt and IκB, nuclear translocation of NFκB, and NFκB-luciferase transcription activity. $InsP_6$ can effectively interfere with all of these cellular events. It starts with mitochondrial permeabilization, followed by cytochrome c release and in consequence activation of the apoptotic machinery: caspase 9, caspase 3 and poly (ADP-ribose) polymerase (PARP). Pro-apoptotic function of $InsP_6$ is further proven by findings that inositol hexaphosphate kinase-2 is a physiologic mediator of cell death [95]. In promoting apoptosis in malignant glioblastoma cells $InsP_6$ has been demonstrated to down-regulate cell survival factors BIRC-2 (baculovirus inhibitor-of-apoptosis repeat containing-2) and telomerase, and up-regulate calpain and caspase-3 activities [96]. Additionally $InsP_6$ has been shown to regulate the pro-apoptotic BCL-2 family of genes [97]. Singh and Gupta demonstrated $InsP_6$'s induction of apoptosis in DMBA exposed mouse epidermis *via* regulation of p53, Bcl-2 expression and caspase activity [98]. Inositol hexaphosphate kinase-2, by generating diphosphoinositol pentaphosphate ($InsP_7$) from $InsP_6$, provides physiologic regulation of the apoptotic process [95].

Enigmatic is the fact that $InsP_6$ influences apoptosis in cancer cells, but prevents cell death in conditions where apoptosis is harmful. It has been demonstrated that $InsP_6$ protected against iron-induced apoptosis in immortalized rat mesencephalic dopaminergic cell model (N27 cells) and subsequent neuronal degeneration [99] to protect the cells and prevent disease. In contrast, in a TRAMP mouse model and in DU-145 human prostate cancers cells $InsP_6$ inhibited telomerase activity, crucial for cells to gain immortality and cell survival [100].

MODULATION OF KINASE ACTIVATION

In contrast to the lower inositol phosphates, in particular $InsP_3$ or $InsP_4$ which are both active signal transduction molecules, $InsP_6$ operates *via* a direct control of protein phosphorylation [101]. $InsP_6$ can modulate activation of several intracellular signaling molecules such as Ras, mitogen-activated protein kinases (MAPK) [72, 102-104], protein kinase C (PKC) [63, 103-105], phosphatidyl-inositol-3 kinase (PI-3 K) and activating protein-1 (AP-1) [102].

The Ras family of monomeric GTPases consists of two subfamilies: one transduces the signal from the cell-surface receptor to the actin cytoskeleton (Rho family), and the other regulates the traffic of intracellular transport vesicles (Rab family). Ras is activated by GTP and deactivated by GDP. A signal initially passed through Ras is further transduced along different pathways downstream, one of the most important being the serine/threonine phosphorylation cascade. Mitogen-activated protein kinase (MAPK) belongs to the serine/threonine kinase family and is a key enzyme, which passes the signal to the nucleus. It can initiate cell division through activation of genes encoding G_1 cyclins. In mammals, phosphorylation of MAPK is completed by MAP-kinase-kinase (also called MEK) and requires phosphorylation of both a tyrosine and a threonine residue. Activation of MEK involves MAP-kinase-kinase-kinase, called Raf. InsP_6 affects the activity of MAPK including Erks (extracellular signal-regulated kinases), JNKs (c-Jun N-terminal kinases) and p38 kinases. InsP_6 inhibited the activities of Erks and JNKs, but not of the p38 kinases in human skin, prostate and breast cancer cells [72, 102, 103, 106]. The blocking of this cellular to nuclear signaling pathway appears as an important mechanism of the anticancer action of InsP_6.

Protein Kinase C (PKC)

Protein kinase C is another enzyme belonging to serine/threonine kinase family. It has an important role in diverse cellular processes such as cell proliferation, differentiation, apoptosis, gene expression, tumor promotion, *etc*. InsP_6-activated PKC stimulates insulin secretion and primes Ca^{2+}-induced exocytosis in pancreatic βcells [63]. This dose-dependent stimulation of exocytosis depends highly on PKC activity. PKC has several isoforms including PKCα, β, ξ, ε isomers *etc*. Studies using antisense oligonucleotides directed against specific PKC isoforms reveal that InsP_6-induced exocytosis involves PKC-ε isoform since negative PKC-ε abolishes InsP_6-evoked exocytosis [107]. In mouse skin tumor, 12-*O*-tetradecanoylphorbol-13-acetate (TPA)-induced ornithine decarboxylase (ODC) activity is important in tumor promotion. In HEL-30 cells, a murine keratinocyte cell line and SENCAR mouse skin, Nickel and Belury investigated the effect of InsP_6 on TPA-induced ODC activity [105]. ODC activity was significantly down-regulated by InsP_6 both *in vitro* and *in vivo*. Likewise, the expression of TPA-induced c-myc mRNA was significantly inhibited by the same

InsP_6 treatments in HEL-30 cells and CD-1 mouse skin. No changes in PKC isoforms α and ζ expression and phorbol dibutyrate binding due to InsP_6 treatment were found in HEL-30 cells. These results indicate that InsP_6 reduces TPA-induced ODC activity independent of PKCα and ζ expression [105]. Vucenik *et al.,* [84] similarly reported that treatment of human breast cancer cells with InsP_6 resulted in no changes in the expression of PKCα, β or ξ isomers. However, there was a 3.1-fold increase in the expression of PKCδ. Along with the increased activity, there was translocation of the enzyme from the cytosol to the membrane [82]. Thus, not only several isomers of PKC may be involved in different pathways, but InsP_6 may also modulate them differentially.

Phosphotidylinostiol-4,5-Bisphosphate 3-Kinase (PI-3 K)

Another crucial molecule in cellular signal transduction is the enzyme phosphotidylinostiol-4,5-bisphosphate 3-kinase (PI-3 Kinase or PI-3 K), involved in a variety of cellular processes, including those affected by InsP_6. Indeed it is PI-3 K that causes phosphorylation of the D-3 position of the inositol ring of phosphoinositides to produce phosphatidylinositol-3-phosphate. In mouse epidermal cell line JB6, InsP_6 markedly blocked epidermal growth factor-induced PI-3 K activity in a dose-dependent manner. This blocking of PI-3 K activity by InsP_6 profoundly impaired epidermal growth factor phorbol ester-induced JB6 cell transformation and extracellular signal-regulated protein kinases activation, as well as activation of the transcription factor activator protein 1 (AP-1) [102].

RNA EDITING AND TRANSPORT

On the RNA level it has been shown that InsP_6 contributes to RNA processing control, RNA transport and localization control, translational control, and mRNA degradation control. Cocco *et al.,* first demonstrated that the envelop-deprived nuclei could synthesize both phosphatidyl 4-phosphate and phosphatidyl 2-phosphate [PIP$_2$], and inositol derivatives present in cytoplasm have also been found in the nucleus [108-111]. In the nucleus, DNA information is transcribed to messenger RNA (mRNA) which is then transported by a complex series of events through pores into the cytoplasm. Proper and efficient export of mRNA from the nucleus to the cytoplasm is dependent on the enzyme phospholipase C and two proteins that influence the generation of InsP_6 [112]. InsP_6 specifically binds Gle1 - a key mRNA export factor [113]. This Gle1/InsP_6 complex, but not Gle1 alone, is essential for activation of the DEAD-box protein Dbp5 [114]. Dbp5 might act as helicase to unwind RNA duplexes or to displace proteins from mRNA [115, 116].

In yeast, three genes are identified that are involved in the inositol signaling pathway: *PLC1* encoding phospholipase C (which converts PIP$_2$ to InsP_3 and DAG), *IPK1* encoding the inositol polyphosphate kinase Ipk1p (which converts I InsP_5 to InsP_6) and *IPK2* encoding Ipk2p (that converts InsP_3 to InsP_4 and InsP_5). A mutation in any of these genes blocks export of mRNA from the nucleus to the cytoplasm. InsP_6, being the end product of this metabolic pathway, is therefore the most likely effector molecule controlling mRNA export. The *IPK2* gene is identical to the yeast gene *ARG8* which encodes Arg82p, a pleiotropic kinase that regulates processes as diverse as response to stress, sporulation, and mating. Incidentally, in yeast, stress increases InsP_6 level [117] causing increased export of certain mRNAs that when translated into proteins would counteract the stressful stimuli. Arg82 has a predicted molecular mass similar to the yeast InsP_3 - InsP_4 kinase activity. Along with Arg80, Arg81, and Mcm1, Arg82p is also an essential component of ArgR-Mcm1 transcription complex essential for proper transcriptional control that activates or represses genes involved in arginine metabolism. Odom *et al.,* showed that this arginine production and breakdown was dependent on the kinase activity of Ipk2p and the generated InsP_4 and InsP_5 [118]. Deamination of adenosine to produce inosine in RNA in the nucleus is carried *via* RNA (ADARs) or tRNA (ADATs) adenosine deaminases. InsP_6 is required for proper enzyme activity and RNA editing [119]. As shown by Zhang and Carmichael, hyper-edited RNA is held in the nucleus [120]. InsP_6-controlled mRNA and tRNA editing would result in more stable splice variants with better export ability [120].

NF-κB (NUCLEAR TRANSCRIPTION FACTOR-κB)

NF-κB proteins are transcriptional factors regulating gene expression. The transcriptional responses to many different stimuli are essential to the proper functioning of the mammalian cells. Five NF-κB proteins have been identified: RelA, RelB, c-Rel, NF-κB1 and NF-κB2 creating a constellation of different homodimers and heterodimers each of which activates different sets of genes. The most profuse dimer binding DNA is a p50/p65 heterodimer. Dimers are held in check by inhibitory proteins called IκB within large protein complexes in the cytoplasm. They prevent activation of NF-κB and its trans-localization from the nucleus and binding to the DNA. Several different stimuli including inflammatory cytokines such as tumor necrosis factor α (TNF-α), interleukin-1 (IL-1) can trigger phosphorylation, ubiquitylation and finally degradation of IκB by 26S proteasome complex. Following degradation of IκB, NF-κB moves into the nucleus and stimulates the transcription of specific genes such as those involved in cell cycle regulation, cell adhesion, apoptosis *etc*.

In some cancers NF-κB is constitutively active, especially in advanced stage. Studies by Agarwal *et al.,* [121] on advanced and androgen-independent human prostate cancer cell line DU145 cells demonstrated the biological relevance of InsP_6 in causing inhibition of proliferation and induction of apoptosis. Treatment of cells with 1 and 2 mM InsP_6 resulted in a strong inhibition of NF-κB activation and decrease in nuclear levels of NF-κB sub-unit proteins p65 and p50. Additionally InsP_6-treated cells also showed a strong inhibition in phospho-IκBα protein levels, increase in total IκBα levels and a significant inhibition in IκBα kinase (IKKα) activity [121]. Recent studies by Kapral *et al.,* [122] showed that InsP_6 primarily influenced p65 subunit of NFκB and its inhibitor IκBα in human colorectal cancer cell line Caco-2. Certain chemotherapeutic agents such as paclitaxel can activate NFκB. InsP_6 has been shown not only to act synergistically in inhibiting squamous cell cancers of the oral cavity, but it also inhibited paclitaxel-mediated increase in NFκB [123].

In a similar manner, in mouse epidermal JB6 cells, InsP_6 strongly blocked UVB-induced AP-1 and NF-κB transcriptional activities in a dose-dependent fashion. InsP_6 also suppressed UVB-induced AP-1 and NFκB-DNA binding activities and inhibited UVB-induced phosphorylation of extracellular signal-regulated protein kinases (Erks) and c-Jun NH2-terminal kinases (JNKs). InsP_6 also blocked UVB-induced phosphorylation of IκB-alpha, which is known to result in the inhibition of NF-κB transcriptional activity. Contrary to the epidermal growth factor-induced PI-3 K activity [57], InsP_6 did not block UVB-induced PI-3 K activity, suggesting that the inhibition of UVB-induced AP-1 and NκF-B activities by InsP_6 is not mediated through PI-3 kinase [106].

Zinc Finger Motif

Zinc Finger Motifs are small protein domains that bind DNA and inactivate transcription factors unless zinc is bound. DNA binding motifs consist of a α helix and a β sheet held together by one or more zinc atoms. These are often found as a cluster with additional zinc fingers arranged one after the other. This arrangement allows the α helix of each zinc finger to come in contact with the major groove of DNA for a considerable length along the groove. InsP_6's ability to bind the zinc atoms from the zinc finger motifs has been discussed. It has been speculated that by removing zinc, $InsP_6$ could inhibit thymidine kinase, an enzyme essential for DNA synthesis [124].

IMMUNE SYSTEM

The effect of $InsP_6$ on the immune system has not been studied adequately. Natural killer (NK) cells are important anti-cancer fighters. In many cancer models, tumor progression is correlated with a decrease in NK activity. Both *in vitro* and *in vivo*, inositol has been shown to boost natural killer cell activity and suppress carcinogenesis [125, 126]. $InsP_6$ and its metabolite inositol were equally efficient *in vivo*, however the lowest incidence of cancer and greatest enhancement in NK cell activity were observed in animals treated with a combination of both [125], a finding which needs further investigation. "Neutrophils, which as a part of the body's innate immune system form a first line of defense," are also affected by $InsP_6$ (further discussed in Chapter 17). $InsP_6$ functions as a neutrophil priming agent and appears to up-regulate a number of diverse neutrophil functions [42]. Not much is known about $InsP_6$'s actions on cytokines except that Eggleton demonstrated an increase of IL-8 production by neutrophils stimulated by $InsP_6$ [43]. In certain tumors, TNFα and its receptor's (TNFRI and TNFRII) activity are dysregulated. Cholewa *et al.,* [127] demonstrated that $InsP_6$ influenced transcription of genes and coding for TNFα and its receptors in human colon cancer Caco-2 cells. $InsP_6$ up-regulated TNF receptor I (TNFRI), and decreased TNFα and TNFRII.

INOSITOL PYROPHOSPHATES (IPP)

The concentration of diphosphoinositol pentakisphosphate (PP-IP$_5$ or $InsP_7$) in most mammalian tissue is 1-5 μM. Illies *et al.,* [128] demonstrated that pancreatic β cells maintain high basal concentrations of $InsP_7$; overexpression of inositol hexakisphosphate kinases (IP6Ks) that can generate $InsP_7$, stimulated exocytosis of insulin-containing granules from the readily releasable pool. Exogenously applied $InsP_7$ dose-dependently enhanced exocytosis at physiological concentrations. Thus, maintenance of high concentrations of $InsP_7$ in the pancreatic β cell may enhance the immediate exocytotic capacity [128].

Compared to $InsP_7$, the role(s) of $InsP_8$ (bisdiphosphoinositol tetrakisphosphate or (PP)2-InsP4) in cellular signal transduction is yet to be determined. Most of the studies of IPPs have been derived from targeted deletions of the $InsP_6$ kinases that generate them. One of the kinases, InsP_6K2 is a mediator of p53-determined cell death, and *IP$_6$K2*-deleted mice are predisposed to upper respiratory and upper digestive tract carcinoma [129, 130].

Mechanism of Signal Transduction by IPPs

Binding with Proteins

It seems that IPPs mediate cellular signaling by binding to or pyrophosphorylation of other cellular components [131]. IPPs bind to the membrane-targeted PH domains of several proteins; PH domains bind phospholipids such as phosphatidylinositol 3,4,5-trisphosphate (PIP$_3$) and phosphatidylinositol 4,5-bisphosphate (PIP$_2$) and thus recruit signaling proteins to membranes. In *Dictyostelium*, PIP$_3$ binds the PH domain of CRAC (cytosolic regulator of adenylate cyclase) protein, which mediates cyclic AMP (cAMP)-related chemotaxis; InsP_7 competes with PIP$_3$ for binding to CRAC and prevents the chemotactic response [132].

Pyrophosphorylation of Proteins

Due to the high free energy of hydrolysis of their pyrophosphate bond(s), which in PP-InsP_5 is comparable with that of adenosine diphosphate (ADP), IPPs may act as phosphorylating agents resulting in non-enzymatic pyrophosphorylation [133]. These exciting emerging data point to protein pyrophosphorylation inhibiting protein-protein interactions [134].

CONCLUDING REMARKS

The preceding has been an attempt to show numerous pathways through which inositol phosphates especially InsP_6 plays critical roles in signal transduction throughout various life forms. This should form the basis for understanding the various biological actions in subsequent chapters. Some of the topics such as cell proliferation, differentiation, apoptosis, DNA damage and repair *etc.* have been considered here in light of signaling within the cell; their other mechanistic pathways are discussed in Chapters 18 and 19. The rapidly evolving field of inositol pyrophosphates is just as fascinating and needs to be followed closely.

REFERENCES

[1]   Kersting MC, Boyette M, Massey JH, *et al*. Identification of the inositol isomers present in Tetrahymena. Journal of Eukaryotic Microbiology 2003; 50: 164-8.

[2]   Cosgrove DJ. The Isolation of Myoinositol Pentaphosphates from Hydrolysates of Phytic Acid. Biochem J 1963; 89: 172-5.
[3]   Raboy V, Gerbasi P. Genetics of *myo*-inositol phosphate synthesis and accumulation. Sub-Cellular Biochem1996; 26: 257-85.
[4]   Putney JW, Jr. Formation and actions of calcium-mobilizing messenger, inositol 1,4,5-trisphosphate. Am J Physiol 1987; 252: G149-157.
[5]   Samanta S, Dalal B, Biswas S, *et al*. *Myo*-inositol tris-phosphate-phytase complex as an elicitor in calcium mobilization in plants. Biochem Biophys Res Comm 1993; 191: 427-434.
[6]   Dasgupta S, Dasgupta D, Sen M, *et al*. Interaction of myoinositoltrisphosphate-phytase complex with the receptor for intercellular Ca2+ mobilization in plants. Biochemistry 1996; 35: 4994-5001.
[7]   Shamsuddin AK, Bose S. IP$_6$ (inositol hexaphosphate) as a signaling molecule, Current Signal Trans Ther 2012; 7: 291-306.
[8]   Goren MB, Dhariwal KR, Jenkins ID. Concerning hydrolysis of mycolate esters, of phthiocerol dimycocerosates and of related mycobacterial lipids: an anecdotal account. J Chromat A 1988; 440: 487-498.
[9]   Belisle JT, Brandt ME, Radolf JD, *et al*. Fatty acids of Treponema pallidum and Borrelia burgdorferi lipoproteins. J Bact 1994; 176: 2151-7.
[10]   Elsbach P, Weiss J. Phagocytosis of bacteria and phospholipid degradation. Biochim Biophys Acta 1988; 947: 29-52.
[11]   Kozloff LM, Turner MA, Arellano F, *et al*. Phosphatidylinositol, a phospholipid of ice-nucleating bacteria. J Bacter 1991; 173: 2053-60.
[12]   Salman M, Lonsdale JT, Besra GS, *et al*. Phosphatidylinositol synthesis in mycobacteria. Biochim Biophys Acta 1999; 1436: 437-50.
[13]   Fahey RC. Novel thiols of prokaryotes. Ann Rev Microbiol 2001; 55: 333-56.
[14]   Reddy NR, Sathe SK, Salunkhe DK. Phytates in legumes and cereals. Advan Food Res 1982; 28: 1-92.
[15]   Walsh D, Grantham J, Zhu XO, *et al*. The role of heat shock proteins in mammalian differentiation and development. Environmental Medicine: Annual Report of the Research Institute of Environmental Medicine, Nagoya University 1999; 43: 79-87.
[16]   Pruitt RN, Chagot B, *et al*. Structure-function analysis of inositol hexakisphosphate-induced autoprocessing in Clostridium difficile toxin A. J Biol Chem 2009; 284: 21934-40.
[17]   Prochazkova K, Shuvalova LA, Minasov G, *et al*. Structural and molecular mechanism for autoprocessing of MARTX Toxin of Vibrio cholerae at multiple sites. J Biol Chem 2009; 284: 26557-68.
[18]   Steiner S, Smith S, Waechter CJ, *et al*. Isolation and partial characterization of a major inositol-containing lipid in baker's yeast, mannosyl-diinositol, diphosphoryl-ceramide. Proc Natl Acad Sci U S A 1969; 64: 1042-8.
[19]   Uno I, Fukami K, Kato H, *et al*. Essential role for phosphatidylinositol 4,5-bisphosphate in yeast cell proliferation. Nature 1988; 333: 188-90.
[20]   Robinson KS, Wheals AE, Rose AH, *et al*. Unusual inositol triphosphate metabolism in yeast. Microbiology 1996; 142 (Pt 6): 1333-4.
[21]   Silverman-Gavrila LB, Lew RR. An IP3-activated Ca2+ channel regulates fungal tip growth. J Cell Sci 2002; 115: 5013-25.
[22]   Tisi R, Belotti F, Wera S, *et al*. Evidence for inositol triphosphate as a second messenger for glucose-induced calcium signalling in budding yeast. Curr Genet 2004; 45: 83-9.
[23]   Nishizawa M, Komai T, Katou Y, *et al*. Nutrient-regulated antisense and intragenic RNAs modulate a signal transduction pathway in yeast. PLoS Biol 2008; 6: 2817-30.
[24]   Lemtiri-Chlieh F, MacRobbie EA, Webb AA, *et al*. Inositol hexakisphosphate mobilizes an endomembrane store of calcium in guard cells. Proc Natl Acad Sci USA 2003; 100: 10091-5.
[25]   Guilfoyle TJ, Hagen G. Auxin response factors. Current Opinion in Plant Biology 2007; 10: 453-60.
[26]   Tan X, Calderon-Villalobos LI, Sharon M, *et al*. Mechanism of auxin perception by the TIR1 ubiquitin ligase. [see comment]. Nature 2007; 446: 640-5.
[27]   Bohn L, Meyer AS, Rasmussen SK. Phytate: impact on environment and human nutrition. A challenge for molecular breeding. Journal of Zhejiang University Science B 2008; 9: 165-91.

[28]   Doria. Phytic acid prevents oxidative stress in seeds: evidence from a maize (Zea mays L.) low phytic acid mutant. Journal of Experimental Botany 2009; 60: 967-78.

[29]   Murgia I, Arosio P, Tarantino D, *et al*. Biofortification for combating 'hidden hunger' for iron. Trends Plant   Sci   2011   [Epub   ahead   of   print] http://www.sciencedirect.com/science/article/pii/S1360138511002251

[30]   Woodward AW, Bartel B. Auxin: regulation, action, and interaction. Ann Bot (Lond) 2005; 95: 707-735.

[31]   Tan X, Zheng N. Hormone signaling through protein destruction: a lesson from plants. Am J Physiol Endocrinol Metab 2009; 296: E223-7.

[32]   Shamsuddin AM. Metabolism and cellular functions of IP_6: a review. Anticancer Research 1999; 19: 3733-6.

[33]   Sasakawa N, Sharif M, Hanley MR. Metabolism and biological activities of inositol pentakisphosphate and inositol hexakisphosphate. Biochemical Pharmacology 1995; 50: 137-46.

[34]   Berridge MJ, Irvine RF. Inositol phosphates and cell signalling. Nature 1989; 341: 197-205.

[35]   Menniti FS, Oliver KG, Putney JW, Jr., *et al*. Inositol phosphates and cell signaling: new views of InsP5 and InsP6. Trends in Biochemical Sciences 1993; 18: 53-6.

[36]   Vallejo M, Jackson T, Lightman S, *et al*. Occurrence and extracellular actions of inositol pentakis- and hexakisphosphate in mammalian brain. Nature 1987; 330: 656-8.

[37]   Hanley MR, Jackson TR, Vallejo M, *et al*. Neural function: metabolism and actions of inositol metabolites in mammalian brain. Philosophical Transactions of the Royal Society of London - Series B: Biological Sciences 1988; 320: 381-98.

[38]   Wilson HC, Bielinska M, Nicholas P, *et al*. Neural tube defects in mice with reduced levels of inositol 1,3,4, -trisphosphate 5/6-kinase. Proc Amer Assoc Cancer Res 2009.

[39]   Rowley KG, Gundlach AL, Cincotta M, *et al*. Inositol hexakisphosphate binding sites in rat heart and brain. British Journal of Pharmacology 1996; 118: 1615-20.

[40]   O'Rourke F, Matthews E, Feinstein MB. Isolation of InsP4 and InsP6 binding proteins from human platelets: InsP4 promotes Ca2+ efflux from inside-out plasma membrane vesicles containing 104 kDa GAP1IP4BP protein. Biochemical Journal 1996; 315: 1027-34.

[41]   Kitchen E, Condliffe AM, Rossi AG, *et al*. Characterization of inositol hexakisphosphate (InsP6)-mediated priming in human neutrophils: lack of extracellular [3H]-InsP6 receptors. British Journal of Pharmacology 1996; 117: 979-85.

[42]   Eggleton P, Penhallow J, Crawford N. Priming action of inositol hexakisphosphate (InsP6) on the stimulated respiratory burst in human neutrophils. Biochimica et Biophysica Acta 1991; 1094: 309-16.

[43]   Eggleton P. Effect of IP6 on human neutrophil cytokine production and cell morphology. Anticancer Research 1999; 19: 3711-5.

[44]   Ji H, Sandberg K, Baukal AJ, *et al*. Metabolism of inositol pentakisphosphate to inositol hexakisphosphate in Xenopus laevis oocytes. Journal of Biological Chemistry 1989; 264: 20185-8.

[45]   Guse AH, Greiner E, Emmrich F, *et al*. Mass changes of inositol 1,3,4,5,6-pentakisphosphate and inositol hexakisphosphate during cell cycle progression in rat thymocytes. Journal of Biological Chemistry 1993; 268: 7129-33.

[46]   Berridge MJ. Inositol trisphosphate and calcium signaling. Annals of the New York Academy of Sciences 1995; 766: 31-43.

[47]   Taylor CW, Genazzani AA, Morris SA. Expression of inositol trisphosphate receptors. Cell Calcium 1999; 26: 237-51.

[48]   Roderick HL, Bootman MD. Bi-directional signalling from the InsP3 receptor: regulation by calcium and accessory factors. Biochemical Society Transactions 2003; 31: 950-3.

[49]   Ho MW, Yang X, Carew MA, *et al*. Regulation of Ins(3,4,5,6)P(4) signaling by a reversible kinase/phosphatase. [see comment]. Current Biology 2002; 12: 477-82.

[50]   Nicoletti F, Bruno V, Fiore L, *et al*. Inositol hexakisphosphate (phytic acid) enhances Ca2+ influx and D-[3H]aspartate release in cultured cerebellar neurons. Journal of Neurochemistry 1989; 53: 1026-30.

[51]   Sortino MA, Nicoletti F, Canonico PL. Inositol hexakisphosphate stimulates 45Ca2+ uptake in anterior pituitary cells in culture. Eur J Pharmacol 1990; 189: 115-8.

[52]   Copani A, Bruno V, Cavallaro S, *et al.* Receptors for inositolhexakisphosphate in neurons and anterior pituitary cells. Pharmacol Res 1990; 22 Suppl 1: 83-4.

[53]   Nicoletti F, Bruno V, Cavallaro S, *et al.* Specific binding sites for inositolhexakisphosphate in brain and anterior pituitary. Mol Pharmacol1990; 37: 689-93.

[54]   Theibert AB, Estevez VA, Ferris CD, *et al.* Inositol 1,3,4,5-tetrakisphosphate and inositol hexakisphosphate receptor proteins: isolation and characterization from rat brain. Proceedings of the National Academy of Sciences of the United States of America 1991; 88: 3165-9.

[55]   Huisamen B, Lochner A. Inositolpolyphosphates and their binding proteins--a short review. Molecular & Cellular Biochemistry 1996; 157: 229-32.

[56]   Shamsuddin AM, Baten A, Lalwani ND. Effects of inositol hexaphosphate on growth and differentiation in K-562 erythroleukemia cell line. Cancer Letters 1992; 64: 195-202.

[57]   Poyner DR, Cooke F, Hanley MR, *et al.* Characterization of metal ion-induced [3H]inositol hexakisphosphate binding to rat cerebellar membranes. J Biol Chem 1993; 268: 1032-8.

[58]   Palczewski K, Pulvermuller A, Buczylko J, *et al.* Binding of inositol phosphates to arrestin. FEBS Letters 295: 195-9, 1991.

[59]   Morton RK, Raison JK. A Complete Intracellular Unit for Incorporation of Amino-Acid into Storage Protein Utilizing Adenosine Triphosphate Generated from Phytate. Nature 1963; 200: 429-33.

[60]   Biswas S, Maity IB, Chakrabarti S, *et al.* Purification and characterization of myo-inositol hexaphosphate-adenosine diphosphate phosphotransferase from Phaseolus aureus. Arch Biochem Biophys 1978; 185: 557-66.

[61]   Josefsen L, Bohn L, Sorensen MB, *et al.* Characterization of a multifunctional inositol phosphate kinase from rice and barley belonging to the ATP-grasp superfamily. Gene 2007; 397: 114-25.

[62]   Phillippy BQ, Ullah AH, Ehrlich KC. Purification and some properties of inositol 1,3,4,5,6-Pentakisphosphate 2-kinase from immature soybean seeds. [erratum appears in J Biol Chem 1995 Mar 31; 270(13): 7782]. Journal of Biological Chemistry 1994; 269: 28393-9.

[63]   Efanov AM, Zaitsev SV, Berggren PO. Inositol hexakisphosphate stimulates non-Ca2+-mediated and primes Ca2+-mediated exocytosis of insulin by activation of protein kinase C. Proceedings of the National Academy of Sciences of the United States of America 1997; 94: 4435-9.

[64]   Larsson O, Barker CJ, Sjoholm A. *et al.* Inhibition of phosphatases and increased Ca2+ channel activity by inositol hexakisphosphate. Science 1997; 278: 471-4.

[65]   Ferry S, Matsuda M, Yoshida H, *et al.* Inositol hexakisphosphate blocks tumor cell growth by activating apoptotic machinery as well as by inhibiting the Akt/NFkappaB-mediated cell survival pathway. [erratum appears in Carcinogenesis. 2003 Jan; 24(1): 149]. Carcinogenesis 2002; 23: 2031-41.

[66]   Singh RP, Sharma G, Mallikarjuna GU, *et al. In vivo* suppression of hormone-refractory prostate cancer growth by inositol hexaphosphate: induction of insulin-like growth factor binding protein-3 and inhibition of vascular endothelial growth factor. Clin Cancer Res 2004; 10: 244-50.

[67]   Cole KE, Smith M, Xu J-F, *et al.* Modulation of the intracellular calcium signal in human colon cancer cells by the novel antineoplastic agent inositol hexaphosphate. Anticancer Res; 1997; 17: 4070-1.

[68]   Lu YJ, He Y, Sui SF. Inositol hexakisphosphate (InsP$_6$) can weaken the Ca^{2+}-dependent membrane binding of C2AB domain of synaptotagmin I. FEBS Letters 2002; 527: 22-6.

[69]   Grases F, Costa-Bauza A, Prieto RM. A potential role for crystallization inhibitors in treatment of Alzheimer's disease. Med Hypotheses 2009; 74: 118-9.

[70]   Hoy M, Efanov AM, Bertorello AM, *et al.* Inositol hexakisphosphate promotes dynamin I- mediated endocytosis. Proc Natl Acad Sci USA 2002; 99: 6773-7.

[71]   Lamaze C, Baba T, Redelmeier TE, Schmid SL. Recruitment of epidermal growth factor and transferrin receptors into coated pits *in vitro*: differing biochemical requirements. Molecular Biology of the Cell 1993; 4: 715-27.

[72]   Zi X, Singh RP, Agarwal R. Impairment of erbB1 receptor and fluid-phase endocytosis and associated mitogenic signaling by inositol hexaphosphate in human prostate carcinoma DU145 cells. Carcinogenesis 2000; 21: 2225-35.

[73]   Voglmaier SM, Keen JH, Murphy JE, Ferris CD, Prestwich GD, Snyder SH, Theibert AB. Inositol hexakisphosphate receptor identified as the clathrin assembly protein AP-2. Biochem Biophys Res Com 1992; 187: 158-63.

[74]   Morrison RS, Shi E, Kan M, *et al.* Inositolhexakisphosphate (InsP6): an antagonist of fibroblast growth factor receptor binding and activity. *In Vitro* Cell Dev Biol Anim 1994; 30A: 783-89.

[75]   Kar S, Quirion R, Parent A. An interaction between inositol hexakisphosphate (IP6) and insulin-like growth factor II receptor binding sites in the rat brain. Neuroreport 1994; 5: 625-28.

[76]   Couchman JR, Vogt S, Lim ST, *et al.* Regulation of inositol phospholipid binding and signaling through syndecan-4. J Biol Chem 2002; 277: 49296-303.

[77]   Tantivejkul K, Vucenik I, Shamsuddin AM. Inositol hexaphosphate (IP6) inhibits key events of cancer metastasis: I. *In vitro* studies of adhesion, migration and invasion of MDA-MB 231 human breast cancer cells. Anticancer Res 2003; 23: 3671-79.

[78]   Tantivejkul K, Vucenik I, Shamsuddin AM. Inositol hexaphosphate (IP6) inhibits key events of cancer metastasis: II. Effects on integrins and focal adhesions. Anticancer Res 2003; 23: 3681-9.

[79]   Vucenik I, Shamsuddin AM. Cancer inhibition by inositol hexaphosphate (IP6) and inositol: from laboratory to clinic. J Nutr 2003; 133: 3778S-84S.

[80]   Shamsuddin AM, Vucenik I, Cole KE. IP6: a novel anti-cancer agent. Life Sci 1997; 61: 343-54.

[81]   Shamsuddin AM, Ullah A, Chakravarthy AK. Inositol and inositol hexaphosphate suppress cell proliferation and tumor formation in CD-1 mice. Carcinogenesis 1989; 10: 1461-3.

[82]   Vucenik I, Ramakrishna G, Tantivejkul K, *et al.* Inositol hexaphosphate (IP6) blocks proliferation of human breast cancer cells through a PKCdelta-dependent increase in p27Kip1 and decrease in retinoblastoma protein (pRb) phosphorylation. [see comment]. Br Cancer Res Treat 2005; 91: 35-45.

[83]   Yang GY, Shamsuddin AM. IP$_6$-induced growth inhibition and differentiation of HT-29 human colon cancer cells: involvement of intracellular inositol phosphates. Anticancer Res 1995; 15: 2479-87.

[84]   Vucenik I. Antiproliferative effects inositol hexaphosphate (IP6) in breast cancer cells is mediated by increase in p27 and decrease in Rb protein phosphorylation. Proc Amer Assoc Cancer Res 2000.

[85]   El-Sherbiny YM, Cox MC, Ismail ZA, *et al.* G0/G1 arrest and S phase inhibition of human cancer cell lines by inositol hexaphosphate (IP6). Anticancer Res 2001; 21: 2393-403.

[86]   Barker CJ, Wright J, Kirk CJ, *et al.* Inositol 1,2,3-trisphosphate is a product of InsP6 dephosphorylation in WRK-1 rat mammary epithelial cells and exhibits transient concentration changes during the cell cycle. Biochem Soc Trans 1995; 23: 169S.

[87]   Singh RP, Agarwal C, Agarwal R. Inositol hexaphosphate inhibits growth, and induces G1 arrest and apoptotic death of prostate carcinoma DU145 cells: modulation of CDKI-CDK-cyclin and pRb-related protein-E2F complexes. Carcinogenesis 2003; 24: 555-563.

[88]   Deliliers GL, Servida F, Fracchiolla NS, *et al.* Effect of inositol hexaphosphate (IP(6)) on human normal and leukaemic haematopoietic cells. Br J Haematol 2002; 117: 577-87.

[89]   Sakamoto K, Venkatraman G, Shamsuddin AM. Growth inhibition and differentiation of HT-29 cells *in vitro* by inositol hexaphosphate (phytic acid). Carcinogenesis 1993; 14: 1815-19.

[90]   Shamsuddin AM, Yang GY. Inositol hexaphosphate inhibits growth and induces differentiation of PC-3 human prostate cancer cells. Carcinogenesis 1995; 16: 1975-9.

[91]   Shamsuddin AM, Yang GY, Vucenik I. Novel anti-cancer functions of IP6: growth inhibition and differentiation of human mammary cancer cell lines *in vitro*. Anticancer Res 1996; 16: 3287-92.

[92]   Vucenik I, Kalebic T, Tantivejkul K, *et al.* Novel anticancer function of inositol hexaphosphate: inhibition of human rhabdomyosarcoma *in vitro* and *in vivo*. Anticancer Res 1998; 18: 1377-84.

[93]   Bertagnolo V, Neri LM, Marchisio M, *et al.* Phosphoinositide 3-kinase activity is essential for all-trans-retinoic acid-induced granulocytic differentiation of HL-60 cells. Cancer Res 1999; 59: 542-6.

[94]   Pittet D, Schlegel W, Lew *et al.* Mass changes in inositol tetrakis- and pentakisphosphate isomers induced by chemotactic peptide stimulation in HL-60 cells. J Biol Chem 1989; 264: 18489-93.

[95]   Nagata E, Luo HR, Saiardi A, *et al.* Inositol hexakisphosphate kinase-2, a physiologic mediator of cell death. J Biol Chem 2005; 280: 1634-40.

[96]   Karmakar S, Banik NL, Ray SK. Molecular mechanism of inositol hexaphosphate-mediated apoptosis in human malignant glioblastoma T98G cells. Neurochem Res 2007; 32: 2094-102.

[97]   Diallo JS, Betton B, Parent N, *et al.* Enhanced killing of androgen-independent prostate cancer cells using inositol hexakisphosphate in combination with proteasome inhibitors. Br J Can 2008; 99: 1613-22.

[98]   Singh J, Gupta KP. Inositol hexaphosphate induces apoptosis by coordinative modulation of p53, Bcl-2 and sequential activation of caspases in 7,12 dimethylbenz(a)anthracene exposed mouse epidermis. J Environ Pathol Toxicol Oncol 2008; 27: 209-17.

[99]   Xu Q, Kanthasamy G, Reddy MB. Phytic acid protects against 6-OHDA and iron induced apoptosis in cell culture model of Parkinson's disease. The FASEB J 2006: A192.

[100]   Jagadeesh S, Banerjee PP. Inositol hexaphosphate represses telomerase activity and translocates TERT from the nucleus in mouse and human prostate cancer cells *via* the deactivation of Akt and PKCalpha. Biochem Biophys Res Com 2006; 349: 1361-7.

[101]   Solyakov L, Cain K, Tracey BM, *et al.* Regulation of casein kinase-2 (CK2) activity by inositol phosphates. J Biol Chem 2004; 279: 43403-10.

[102]   Huang C, Ma WY, Hecht SS, *et al.* Inositol hexaphosphate inhibits cell transformation and activator protein 1 activation by targeting phosphatidylinositol-3' kinase. [erratum appears in Cancer Res 1997 Nov 15; 57(22): 5198]. Cancer Res 1997; 57: 2873-8.

[103]   Vucenik I, Ramakrishna G, Tantivejkul K, *et al.* Inositol hexaphosphate (IP6) differentially modulates the expression of PKCd in MCF-7 and MDA-MB 231 cells. Proc Amer Assoc Cancer Res 1999.

[104]   Vucenik I, Ramljak D. The contradictory role of PKCdelta in cellular signaling. [comment]. Br Can Res Treat 2006; 97: 1-2.

[105]   Nickel KP, Belury MA. Inositol hexaphosphate reduces 12-O-tetradecanoylphorbol-13-acetate-induced ornithine decarboxylase independent of protein kinase C isoform expression in keratinocytes. Cancer Lett 1999; 140: 105-11.

[106]   Chen N, Ma WY, Dong Z. Inositol hexaphosphate inhibits ultraviolet B-induced signal transduction. Molecular Carcinogenesis 2001; 31: 139-44.

[107]   Hoy M, Berggren PO, Gromada J. Involvement of protein kinase C-epsilon in inositol hexakisphosphate-induced exocytosis in mouse pancreatic beta-cells. J Biol Chem 2003; 278: 35168-71.

[108]   Cocco L, Martelli AM, Barnabei O, *et al.* Nuclear inositol lipid signaling. Advan Enzyme Regul 2001; 41: 361-84.

[109]   Cocco L, Manzoli L, Barnabei O, *et al.* Significance of subnuclear localization of key players of inositol lipid cycle. Advan Enzyme Regul 2004; 44: 51-60.

[110]   Cocco L, Gilmour RS, Ognibene A, *et al.* Synthesis of polyphosphoinositides in nuclei of Friend cells. Evidence for polyphosphoinositide metabolism inside the nucleus which changes with cell differentiation. Biochem J 1987; 248: 765-70.

[111]   Cocco L, Martelli AM, Vitale M, *et al.* Inositides in the nucleus: regulation of nuclear PI-PLCbeta1. Advan Enzyme Regul 2002; 42: 181-93.

[112]   York JD, Odom AR, Murphy R, *et al.* A phospholipase C-dependent inositol polyphosphate kinase pathway required for efficient messenger RNA export. Science 1999; 285: 96-100.

[113]   Alcazar-Roman AR, Tran EJ, Guo S, *et al.* Inositol hexakisphosphate and Gle1 activate the DEAD-box protein Dbp5 for nuclear mRNA export. [see comment]. Nature Cell Biol 2006; 8: 711-16.

[114]   Weirich CS, Erzberger JP, Flick JS, *et al.* Activation of the DExD/H-box protein Dbp5 by the nuclear-pore protein Gle1 and its coactivator InsP6 is required for mRNA export. [see comment]. Nature Cell Biol 2006; 8: 668-76.

[115]   Cole CN, Scarcelli JJ. Transport of messenger RNA from the nucleus to the cytoplasm. Current Opinion in Cell Biology 2006; 18: 299-306.

[116]   Gross T, Siepmann A, Sturm D, *et al.* The DEAD-box RNA helicase Dbp5 functions in translation termination. Science 2007; 315: 646-9.

[117]   Ongusaha PP, Hughes PJ, Davey J, *et al.* Inositol hexakisphosphate in Schizosaccharomyces pombe: synthesis from Ins(1,4,5)P3 and osmotic regulation. Biochem J 1998; 335: 671-9.

[118]   Odom AR, Stahlberg A, Wente SR, *et al.* A role for nuclear inositol 1,4,5-trisphosphate kinase in transcriptional control. [see comment]. Science 2000; 287: 2026-9.

[119]   Macbeth MR, Schubert HL, Vandemark AP, *et al.* Inositol hexakisphosphate is bound in the ADAR2 core and required for RNA editing. Science 2005; 309: 1534-9.

[120]   Zhang Z, Carmichael GG. The fate of dsRNA in the nucleus: a p54(nrb)-containing complex mediates the nuclear retention of promiscuously A-to-I edited RNAs. Cell 2001; 106: 465-75.

[121]   Agarwal C, Dhanalakshmi S, Singh RP, *et al.* Inositol hexakisphosphate inhibits constitutive activation of NF- kappa B in androgen-independent human prostate carcinoma DU145 cells. Anticancer Res 2003; 23: 3855-61.

[122] Kapral M, Parfiniewicz B, Strzałka-Mrozik B, *et al.* Evaluation of the expression of transcriptional factor NF-kappaB induced by phytic acid in colon cancer cells. Acta Pol Pharm 2008: 697-702.

[123] Janus SC, Weurtz B, Ondrey FG. Inositol hexaphosphate and paclitaxel: symbiotic treatment of oral cavity squamous cell carcinoma. Laryngoscope 2007; 117: 1381-8.

[124] Thompson LU, Zhang L. Phytic acid and minerals: effect on early markers of risk for mammary and colon carcinogenesis. Carcinogenesis 1991; 12: 2041-5.

[125] Baten A, Ullah A, Tomazic VJ, *et al.* Inositol-phosphate-induced enhancement of natural killer cell activity correlates with tumor suppression. Carcinogenesis 1989; 10: 1595-8.

[126] Zhang Z, Song Y, Wang XL. Inositol hexaphosphate-induced enhancement of natural killer cell activity correlates with suppression of colon carcinogenesis in rats. World J Gastroent 2005; 11: 5044-6.

[127] Cholewa K, Parfiniewicz B, Bednarek I, *et al.* The influence of phytic acid on TNF-alpha and its receptors genes' expression in colon cancer Caco-2 cells. Acta Pol Pharm, 2008; 65: 75-9.

[128] Illies C, Gromada J, Fiume R *et al.* Requirement of inositol pyrophosphate for full exocytotic capacity in pancreatic beta cells. Science 2007; 318: 1299-302.

[129] Koldobskiy MA, Chakraborty A, Werner JK *et al.* p53-mediated apoptosis requires inositol hexakisphosphate kinase-2. Proc. Natl. Acad. Sci. U.S.A.2010; 107: 20947-51

[130] Morrison BH, Haney R, Lamarre E *et al.* Gene deletion of inositol hexakisphosphate kinase 2 predisposes to aerodigestive tract carcinoma.Oncogene. 2009; 28: 2383-92.

[131] Chakraborty A, Kim S, Snyder SH. Inositol pyrophosphates as mammalian cell signals. Sci Signal 2013; 4(188): re1. doi: 10.1126/scisignal.2001958.

[132] Luo HR, Huang YE, Chen JC *et al.* Inositol pyrophosphates mediate chemotaxis in *Dictyostelium via* pleckstrin homology domain-PtdIns(3,4,5)P3 interactions. Cell. 2003; 114: 559-72.

[133] Saiardi A, Bhandari R, Resnick AC.*et al.* Phosphorylation of proteins by inositol pyrophosphates. Science 2004; 306, 2101-5

[134] Wilson MS, Livermore TM, Saiardi A. Inositol pyrophosphates: between signalling and metabolism. Biochem J 2013; 452: 369-79.

CHAPTER 4

Analysis & Assays of Inositol and Inositol Phosphates

Abstract: While the quantitation of inositol in biological specimens and that of InsP_6 in food is rather simple and the methodologies are well-established, accurate quantitation of inositol phosphates, especially InsP_6 in biological specimens *in vivo* has not been easy, many being dogmatic about even its presence in human tissues. In spite of that, it appears that finally accurate and specific methods to quantitate and follow InsP_6 in cells, tissues, plasma, urine *etc.* may have been developed that might be useful in both research and clinical laboratories.

Keywords: Fluorescence, HPLC, gas chromatography, GC-MS, ion-exchange chromatography, LC-MS.

INTRODUCTION

Detection and accurate quantitation of any substance are critical for further research and practical applications; inositol and its phosphates are no exceptions. There are several methods for quantitation of inositol; but those for inositol phosphates have been limited, and mostly in the area of food science. Fortunately, there are recent encouraging reports of quantitative methods for detection of inositol hexaphosphate (InsP_6) in biological specimens. For, unless we can detect and accurately quantify InsP_6 and other inositol phosphates in biological samples, our progress in the field will continue to be stifled.

ANALYSIS OF INOSITOL

The determination of *myo*-inositol in biological samples and foods could be of great interest. Methods of analysis of inositol can be classified as enzymatic, high performance liquid chromatography (HPLC) and gas chromatography (GC), and HPLC/MS methods.

Enzymatic Method

An enzymatic assay for *myo*-inositol was developed by MacGregor & Matschinsky in 1984 using a *myo*-inositol dehydrogenase to oxidize *myo*-inositol [1]. The chemical reaction is that the *myo*-inositol is oxidized by NAD+-dependent *myo*-inositol dehydrogenase and coupled to re-oxidation of NADH with oxaloacetate and malate dehydrogenase, and fluorimetrically measuring the resultant malate [1]. If re-oxidation of NADH is performed with iodonitrotetrazolium chloride and diaphorase, the resultant formazan can be measured spectrophotometrically [2]. The amount of *myo*-inositol and its absorbance is linear in the range of 0.5 to 3 nmol *myo*-inositol per assay; and the assay is quantitative for *myo*-inositol in amounts ranging from 1 to 20 nmol [2-4]. Thus, quantitated amounts of *myo*-inositol in sera from apparently healthy subjects by enzymatic assay are consistent with the data determined by gas chromatographic assay [3]. Use of this rapid and simple assay in biological specimen shows that the serum and urinary *myo*-inositol concentration (mean ± SD) is significantly different in diabetic patients compared to healthy individuals without diabetes mellitus [5, 6].

High-Performance Liquid Chromatography

Many methods for the qualitative and quantitative analysis of carbohydrates using high-performance liquid chromatography (HPLC) have been developed [7]. An HPLC procedure to detect nanogram quantities of the *p*-nitrobenzoate derivatives of sorbitol and inositol following their separation on a Porasil column was first described in 1984 [8]. HPLC methods with refractive index or photometric detection, or with pulsed amperometric detection and no prior derivatization were further developed [9, 10]. The methods incorporate a pre-column derivatization reaction using aqueous extracts with benzoyl chloride as a modifying agent. The benzoylated derivatives are isolated by HPLC using reversed-phase gradient chromatography and quantified *via* absorbance detection at 231 nm. Calibration curves are linear in the range of 1.4-89 nmol for *myo*-inositol. Subsequently, the HPLC method was used to accurately measure *myo*-inositol levels in plasma and tissue as well as in the food, infant formula *etc.* [9, 11, 12].

Gas Chromatography

A rapid and more sensitive capillary gas chromatographic (GC) method was developed for the profile analysis of urinary polyols including *myo*-inositol as their trifluoroacetyl derivatives [13], and these urinary polyols were verified by gas chromatography/mass spectrometry. Compared to HPLC methods, GC method has better sensitivity but requires previous derivatization to volatile compounds so that *myo*-inositol can be determined as trifluoroacetyl derivative or hexa-O-trimethylsilyl ethers [13-15].

HPLC/Mass Spectrometry

Using mass detection for its sensitivity and selectivity, a HPLC-mass spectrometry (MS) method was developed in 2004 for the analysis of *myo*-inositol [16]. This sensitive quantitative method is applied for direct determination in urine and saliva samples that require a small sample volume (an ordinary inositol urinary determination requires 20 mL of sample), and no prior complex purification or derivatization, thus reducing the sampling time compared with GC methods.

Urine and saliva samples are only previously purified by passing through an anion-exchange resin and concentrations as low as 138 and 461 $\mu g\ L^{-1}$ in saliva and urine could be respectively quantified [16]. In order to enhance the specificity of tandem mass spectrometry, a LC-MS/MS method has been developed for quantification of *myo*-inositol in brain homogenate [17]. It was further confirmed to be a rapid and reliable quantitation assay [18], in which *myo*-inositol content is determined by using spiked calibration curves and mass spectrometry, and a novel chiral LC/MS/MS method to resolve *myo*-inositol from other endogenous inositol epimers; and confirm the selectivity of the quantitative procedure. This method shows a linear range of 0.100–100 $\mu g\ mL^{-1}$. For accurate MS-based quantification, it is important to resolve inositol from other monosaccharides and sugar alcohols that may suppress the MS signal. This method utilizes a lead-form resin based column online to a triple quadruple tandem mass spectrometer, which requires minimum sample preparation and no derivatization. It allows separation and selective detection of *myo*-inositol from other inositol stereoisomers, as well as separation of inositol from hexose monosaccharides of the same molecular weight, including glucose, galactose, mannose and fructose [19]. This method shows calibration curves in water or urine matrices is linear between 2.5 and 50 μM. Analyses of intra- and inter-assay precision indicate that the method is robust and reproducible, which have been extensively used to determine a quantitative analysis of *myo*-inositol in urine, blood and nutritional supplements [19]. Interestingly these assay results show that the concentration of *myo*-inositol in urine varied quite widely between individuals, even when normalized to creatinine, indicating variations in dietary intake of inositol as well as differential catabolism of inositol in the kidneys [19].

ANALYSIS OF INOSITOL PHOSPHATES

There appears to be a dichotomy insofar as analysis of inositol phosphates is concerned in the fields of cell biology, the biological samples, and food sciences. And until recently very little interdisciplinary communication, much less collaboration and technology transfer has been in evidence. Part of the reason being the prevailing dogma within food & nutritional science that inositol phosphates, especially $InsP_6$ does not (or cannot) exist in mammalian cells, let alone in humans. Examination of earlier reports on intracellular inositol phosphates from the disciplines of cell biology and biochemistry on the other hand reveals a lack of reference to the presence of $InsP_6$ in foods. There are methodological challenges as well: experiments in cell biology and biochemistry have the advantage of using radiolabelled inositol phosphates, a luxury that food scientists and biomedical scientists do not have. So, how do we detect non-radiolabelled inositol phosphates in biological specimens? And the numerous stereoisomers of various inositol phosphates pose additional challenge.

In Foods

In the absence of methodologies for direct measurement, the analytic methods and quantitation of $InsP_6$ have been indirect by analyzing the amount of *myo*-inositol or the phosphate groups. The first report of $InsP_6$ assay in food substances was reported by Heubner and Stadler [20] who analyzed it by extraction from ground cereal powder with hydrochloric acid and subsequent precipitation as Fe^{3+}-$InsP_6$ after stepwise addition of $FeCl_3$. $InsP_6$ content is deduced on the basis of the iron content. Another indirect approach of assaying $InsP_6$ content was to measure the phosphate content [21]. Harland and Oberleas [22] used anion exchange chromatography with more accurate results. However, the quest for finding the most precise method for determination of $InsP_6$ continues. The use of HPLC with preceding purification and concentration with AG 1 resin improved the sensitivity of the assay; however, problem persists in separating the other inositol phosphates from $InsP_6$.

Insofar as food is concerned, Phillippy and his coworkers successfully achieved good separation of different inositol phosphates with anion exchange AS3 columns [23, 24].

In Biological Specimens

As described above, while the quantitation of inositol in biological specimens is rather simple and the methodologies are well-established, that for inositol phosphates, especially InsP_6 has not only been easy, but a principal factor in tardiness in progress in the field has been the dogma that it does not exist in mammalian system! Notwithstanding this denial, a few scientists have been doggedly pursuing it, and the results have been rewarding.

Aside from the 'politics' of it, the other limiting factor in our quest to accurately quantitate and follow InsP_6 in cells, tissues, plasma, urine *etc.* has been the inability of InsP_6 to significantly absorb light from any region of the UV-visible spectrum; this makes it challenging to quantitate with spectroscopic methods.

The analyses of other inositol phosphates (InsP_{1-5}) within the cells by using various columns such as AG 1, SAX, Partisil SAX 10 columns, *etc.* have proven to be sensitive and accurate methods to differentiate radiolabelled inositol phosphates [25, 26]. Using a modification of the ion-pair RP HPLC Sulpice *et al.*, [27] could discriminate inositol phosphates such as Ins$P_{1,4,5}$ from other organic phosphates. However, the HPLC methods used in the field of cell biology have not been used in food and nutrition research. In our pursuit to detect InsP_6 in micromolar to nanomolar range, various different methodologies such as colorimetry, capillary electrophoresis, flow injection analysis, fluorescence detection, light scattering detection, chemically suppressed conductivity, nuclear magnetic resonance (NMR) spectroscopy, mass spectroscopy (MS), *etc.* have been utilized. Most if not all of these assays have been indirect and not without controversy; one of the main factors being poor recovery of InsP_6 from biological samples. Please see Schlemmer *et al.,* [28] for a review of the various analytical methods.

Grases and his colleagues have been indefatigably pursuing the methodologies to detect minute amounts of InsP_6 in biological samples. Using a "direct, sensitive, selective bioanalytical method" his laboratory reported the plasma levels of InsP_6 in rats, dogs and humans. With very straightforward ways for sample pretreatment and total chromatographic time for each sample analysis being 7 minutes, this LC-MS methodology offers a sensitive means to assay biological samples for research and clinical use [29]. The investigators also report that the mean normal plasma level in rat to be 98 ng mL^{-1} when fed an InsP_6-free diet as opposed to 340 ng mL^{-1} on normal diet; the mean values in dog and human plasma were 166.9 and 179.6

ng mL^{-1} respectively. InsP_6 in spiked human plasma stored at room temperature was stable for 3-4 h, and over three freeze/thaw cycles; the lower temperature being $-80 \pm 10\ ^{\circ}$C [29].

The importance of accurate analytical methods for detection of InsP_6 cannot be overemphasized. Hadi Alkarawi and Zotz [30] reviewed 45 published studies with information on InsP_6 content in leaves. InsP_6 was almost always detected when studies specifically tried to detect it, and accounted for up to 98% of total P. However, they argue that such extreme values, which rival findings from storage organs, are dubious and posit that these probably result from measurement errors.

CONCLUDING REMARKS

It appears that finally accurate and specific methods to quantitate and follow inositol and its phosphates (InsP_6) in cells, tissues, plasma, urine *etc.* may have been developed that will be useful in both research and clinical laboratories. However, as in all new technologies, their validation and reproducibility need to be firmly established.

REFERENCES

[1]   MacGregor LC, Matschinsky FM. An enzymatic fluorimetric assay for myo-inositol. Anal Biochem 1984; 141: 382-9.
[2]   Ashizawa N, Yoshida M, Aotsuka T. An enzymatic assay for myo-inositol in tissue samples. J Biochem Biophys Methods 2000; 44: 89-94.
[3]   Dolhofer R, Wieland OH. Enzymatic assay of myo-inositol in serum. J Clin Chem Clin Biochem 1987; 25: 733-6.
[4]   Laursen SE, Knull HR, Belknap JK. Sample preparation for inositol measurement: Sep-Pak C18 use in detergent removal. Anal Biochem 1986; 153: 387-90.
[5]   Yamakoshi M, Takahashi M, Kouzuma T *et al.* Determination of urinary myo-inositol concentration by an improved enzymatic cycling method using myo-inositol dehydrogenase from Flavobacterium sp. Clin Chim Acta 2003; 328: 163-71.
[6]   Kouzuma T, Takahashi M, Endoh T *et al.* An enzymatic cycling method for the measurement of myo-inositol in biological samples. Clin Chim Acta 2001; 312: 143-51.
[7]   Verhaar LA, Kuster BF. Liquid chromatography of sugars on silica-based stationary phases. J Chromatogr 1981; 220: 313-28.
[8]   Petchey M, Crabbe MJ. Analysis of carbohydrates in lens, erythrocytes, and plasma by high-performance liquid chromatography of nitrobenzoate derivatives. J Chromatogr 1984; 307: 180-4.
[9]   Lauro PN, Craven PA, DeRubertis FR. Two-step high-performance liquid chromatography method for the determination of myo-inositol and sorbitol. Anal Biochem 1989; 178: 331-5.
[10]   Podeschwa MA, Plettenburg O, Altenbach HJ. Stereoselective synthesis of several azido/amino- and diazido/diamino-myo-inositols and their phosphates from p-benzoquinone. Org Biomol Chem 2003; 1: 1919-29.
[11]   Indyk HE, Woollard DC. Determination of free myo-inositol in milk and infant formula by high-performance liquid chromatography. Analyst 1994; 119: 397-402.

[12]   Frieler RA, Mitteness DJ, Golovko MY *et al.* Quantitative determination of free glycerol and myo-inositol from plasma and tissue by high-performance liquid chromatography. J Chromatogr B Analyt Technol Biomed Life Sci 2009; 877: 3667-72.

[13]   Haga H, Nakajima T, Determination of polyol profiles in human urine by capillary gas chromatography. Biomed Chromatogr 1989; 3: 68-71.

[14]   March JG, Forteza R, Grases F. Determination of inositol isomers and arabitol in human urine by gas chromatography-mass spectrometry. Chromatographia 1996; 42: 329-31.

[15]   Koning AJD Determination of myo-inositol and phytic acid by gas chromatography using scyllitol as internal standard. Analyst 1994; 119: 1319-23.

[16]   Perelló J, Isern B, Costa-Bauzá A *et al.* Determination of myo-inositol in biological samples by liquid chromatography-mass spectrometry. J Chromatogr B Analyt Technol Biomed Life Sci 2004; 802: 367-70.

[17]   Kindt E, Shum Y, Badura L, *et al.* Development and validation of an LC/MS/MS procedure for the quantification of endogenous myo-inositol concentrations in rat brain tissue homogenates. Anal Chem 2004; 76: 4901-8.

[18]   Bathena SP, Huang J, Epstein AA *et al.*, Rapid and reliable quantitation of amino acids and myo-inositol in mouse brain by high performance liquid chromatography and tandem mass spectrometry. J Chromatogr B Analyt Technol Biomed Life Sci 2012; 893-894: 15-20. doi: 10.1016/j.jchromb.2012.01.035. Epub 2012 Feb 28.

[19]   Leung KY, Mills K, Burren KA *et al.* Quantitative analysis of myo-inositol in urine, blood and nutritional supplements by high-performance liquid chromatography tandem mass spectrometry. J Chromatogr B Analyt Technol Biomed Life Sci 2011; 879: 2759-63.

[20]   Heubner W, Stadler H. ber eine Titrationsmethode zur Bestimmung des Phytins. Biochem. Z 1914; 64: 422-437.

[21]   McCance RA, Widdowson EM. Phytin in human nutrition. Biochem J 1935; 29: 2694-9.

[22]   Harland BF, Oberleas D. A Modified Method for Phytate Analysis Using an Ion-Exchange Procedure: Application to Textured Vegetable Proteins. Cereal Chem 1977; 54: 827-32.

[23]   Phillippy BQ, Johnston MR. Determination of Phytic Acid in Foods by Ion Chromatography with Post-Column Derivatization. Journal of Food Science 1985; 50: 541-2.

[24]   Phillippy BQ, Bland JM, Evens TJ. Ion chromatography of phytate in roots and tubers. J Agric Food Chem 2003; 51: 350-3.

[25]   Berridge MJ, Dawson RM, Downes CP *et al.* Changes in the levels of inositol phosphates after agonist-dependent hydrolysis of membrane phosphoinositides. Biochem J 1983; 212: 473-82.

[26]   Heslop JP, Irvine RF, Tashjian AH Jr *et al.* Inositol tetrakis- and pentakisphosphates in GH4 cells. J Exp Biol 1985; 119: 395-401.

[27]   Sulpice JC, Gascard P, Journet E *et al.*, The separation of [32P]inositol phosphates by ion-pair chromatography: optimization of the method and biological applications. Anal Biochem 1989; 179: 90-7.

[28]   Schlemmer U, Frølich W, Prieto RM *et al.* Phytate in foods and significance for humans: food sources, intake, processing, bioavailability, protective role and analysis. Mol Nutr Food Res 2009; 53 Suppl 2: p. S330-75.

[29]   Tur F, Tur E, Lentheric I *et al.*, Validation of an LC-MS bioanalytical method for quantification of phytate levels in rat, dog and human plasma. J Chromatogr B Analyt Technol Biomed Life Sci 2013; 928: 146-54.

[30]   Hadi Alkarawi H, Zotz G. Phytic acid in green leaves. Plant Biol (Plant Biol (Stuttgart). 2013 Dec 16. doi: 10.1111/plb.12136. [Epub ahead of print].

CHAPTER 5

Pharmacokinetics of InsP_6

Abstract: Studies have shown that InsP_6 is rapidly absorbed from the gastrointestinal tract, distributed through the plasma to various organs including the brain, and excreted from the lungs *via* exhaled air and through the kidneys *via* urine as inositol, InsP_{1-5} and InsP_6. That it is distributed in brain indicates that it crosses the blood-brain barrier. Also challenging the dogma is that at least in mice, InsP_6 is distributed in the red blood cells as well. InsP_6 is rapidly taken up by malignant cells wherein it is precipitously dephosphorylated to various lower phosphates; depending on the cell types, the proportion of these lower inositol phosphates varies. Inositol is also rapidly absorbed by humans.

Keywords: Blood-brain barrier, GC-MS, internalization, red blood cells, visualization.

INTRODUCTION

Owing to the charged nature of inositol hexaphosphoric acid, there has been a myth that InsP_6 is not absorbed by humans and cannot be taken up by the cell. The originator(s) and the propagator(s) of this misinformation failed to take into account that a) the naturally occurring state of InsP_6 is the calcium-magnesium inositol hexaphosphate, and b) there are mechanisms within the cells such as pinocytosis to handle charged particles, even if the naturally occurring InsP_6 was indeed highly charged. And finally, early studies dating back to the mid-1990s showing the absorption and tissue distribution of InsP_6 have been ignored.

Absorption, Distribution and Metabolism

The earliest pharmacokinetic study of the InsP_6 was performed by Singsen *et al.,* in 1950 [1] who used radioactive phosphorus ^{32}P as soon as it became generally available to investigators. They fed a diet of ^{32}P-labeled 1% Ca- InsP_6 to turkeys for two weeks and found the label in bones. To determine the source of phosphorus in chicks and poults, Gillis and coworkers [2] compared the absorption and distribution of ^{32}P labeled InsP_6 and inorganic ^{32}P orthophosphate; and the role of vitamin D in utilization of both types of phosphorus. Their *in vitro* experiments demonstrated that radioactive inorganic phosphate can exchange with phosphate in the InsP_6 molecule.

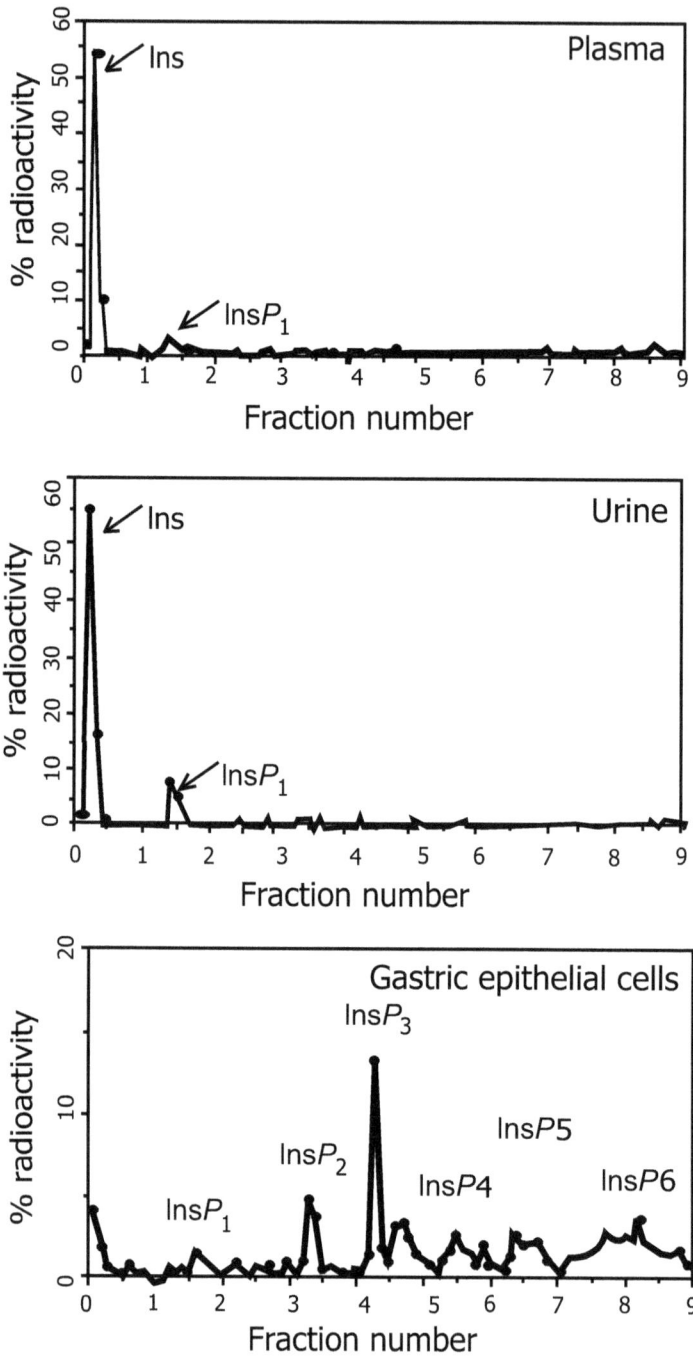

Fig. (5.1). Ion exchange chromatography of plasma (1 h), urine (12 h) and gastric epithelial cells (1 h) following intragastric administration of [^{3}H]-InsP_6 [8].

Studies in Rats

To circumvent that problem, thus in 1980 Nahapetian and Young [3] investigated the uptake of $InsP_6$ by using uniformly labeled ^{14}C (U-^{14}C) in rats given high and low calcium diets. U-^{14}C labeled *myo*-inositol was administered to wheat by stem injection 2 weeks after anthesis (milk state) at a time when $InsP_6$ synthesis is high. The seeds were allowed to mature naturally at which time they were harvested and their $InsP_6$ extracted; rats were given radioactive $InsP_6$ or inositol in distilled water. They reported that 94% of the administered $InsP_6$ was absorbed by the animals, distributed through blood to liver, kidneys, brain, and bone; 60% of the administered $InsP_6$ was oxidized to CO_2 and excreted through exhaled air. This rather elegant and painstaking landmark study was designed primarily to look into the effect of high and low calcium in diet on the absorption of $InsP_6$: its absorption and utilization is inhibited with high calcium diet (Ca/P molar ratio: 2.24), and almost quantitatively absorbed (94% of the dose) when the diet had low calcium (Ca/P molar ratio: 0.21). It also gave valuable information about the overall absorption, tissue distribution and some aspects of the metabolism of orally administered $InsP_6$ in an organism.

In the mid- to late 1980's, the anticancer action of $InsP_6$ was discovered in Shamsuddin's laboratory [4-7]. Against the backdrop of prevailing dogma that $InsP_6$ cannot be absorbed by the animals, much less distributed to various organs, prior reports by Nahapetian and Young [3], Gillis *et al.,* [2] and Singsen *et al.,* [1] notwithstanding; it was necessary to further investigate the fate of orally administered $InsP_6$ especially to understand the mechanism of this very intriguing action. Using ^3H-$InsP_6$ in rats, Shamsuddin's laboratory then embarked on this line of investigation as well. Shamsuddin and colleagues demonstrated that $InsP_6$ is rapidly absorbed from the stomach and small intestine; 11% of the radioactivity was detected within the walls of stomach ($4.4 \pm 3.7\%$) and upper small intestine ($6.6 \pm 1.9\%$), skeletal muscles ($6.5 \pm 2.6\%$) and skin ($4.0 \pm 1.5\%$) at 1 hour; smaller amounts were detected in the blood, brain, heart, lungs, spleen, testicles and thymus.

Much of the radioactivity at 24 h was in the muscles ($18.1 \pm 3.4\%$), skin ($10.1 \pm 3.3\%$), liver ($4.0 \pm 0.9\%$) and kidneys ($2.2 \pm 1.1\%$); again, proportionately smaller amounts, but higher than those at 1 hour were detected in the blood, brain, heart, lungs, spleen, testicles and thymus [8]. The radioactivity in the gastrointestinal tract (both in the contents and in the wall) was high at 1 hour, but as expected, low at 24 hour indicating absorption into the body. Light microscopic autoradiography of stomach, small intestine, liver and kidneys gave visual confirmation of the presence of grains in those organs. Autoradiographic grains were seen in the submucosal capillaries, lymphatics and venules of the stomach, small intestine and colon; in the hepatic central veins and surrounding sinusoidal spaces; and in the venules of renal medulla. Interestingly very few or no grains were detected over the epithelial cells of the stomach, jejunum, colon and, kidney tubules and mesenchymal cell; however small number of grains were seen in the goblet cells of small intestine and colon [8].

Ion exchange chromatography of the plasma and urine showed that almost all the radioactivity was in the inositol or $InsP_1$ form. On the other hand, analysis of gastric epithelial cells showed little inositol, but substantial amounts of $InsP_{1-3}$ and lesser amounts of $InsP_{4-6}$ (Fig. **5.1**).

The striking presence of autoradiographic grains in the lumen of gastrointestinal tract and in mucosal and submucosal capillaries and lymphatics and their absence in the majority of the cells (except goblet cells) indicate that $[^3H]\text{-}InsP_6$ is very quickly transported from the lumen to vascular channels. However, the presence of $InsP_{1-6}$ in gastric epithelial cells and the absence of $InsP_6$ and $InsP_{2-5}$ with a preponderance of inositol and $InsP_1$ in plasma as early as 1 hour following ingestion indicate that $InsP_6$ is very rapidly dephosphorylated within the gastric cells in rats.

Detection of radioactivity in brain indicates crossing the 'blood-brain' barrier' by $InsP_6$; further confirmation was provided by Grases and his coworkers (*vide infra* [9]).

Using gas chromatography-mass detection analysis of HPLC chromatographic fractions, Grases *et al.,* [9] measured unlabeled total $InsP_3$ and $InsP_6$ as they occur within cell culture, tissues, and plasma, and their changes depending on the presence of exogenous $InsP_6$. When rats were fed on a purified diet in which $InsP_6$ was undetectable the levels of $InsP_6$ in brain were 3.35 ± 0.57 µmol kg^{-1} and in plasma 0.023 ± 0.008 µmol L^{-1}. The presence of 1% $InsP_6$ in diet dramatically influenced its levels in brain and in plasma. When rats were given an $InsP_6$-sufficient diet, the levels of $InsP_6$ were about 100-fold higher in brain tissues (36.8 ± 1.8 µmol kg^{-1}) than in plasma (0.29 ± 0.02 µmol kg^{-1}); $InsP_6$ concentrations were 8.5-fold higher than total $InsP_3$ concentrations in plasma (0.033 ± 0.012 µmol L^{-1}) and brain (4.21 ± 0.55 µmol kg^{-1}). When animals were given an $InsP_6$-poor diet there was a 90% decrease in $InsP_6$ content in both brain tissue and plasma; however, there was no change in the level of total $InsP_3$. These results indicate that exogenous $InsP_6$ directly affects its physiological levels in plasma and brain of normal rats without changes on the total $InsP_3$ levels.

Studies in Mice

Eiseman *et al.,* [10] investigated the pharmacokinetics of orally and intravenously administered $[^{14}C]$-$InsP_6$ in SCID mice containing xenotransplanted human breast cancer cells MDA-MB-231. Following intravenous injection, plasma $InsP_6$ concentration peaked at 5 minutes and remained detectable up to 45 minutes. Liver $InsP_6$ concentrations were more than 10-fold higher than plasma concentrations, whereas concentrations in other normal tissues were similar to plasma. Chromatographic analysis showed the presence of only inositol in the xenografts. Since all the prior *in vivo* experiments were performed by using oral administration of $InsP_6$ and experiments at Shamsuddin laboratory and at the National Cancer Institute (Frederick, MD) demonstrated that intravenous $InsP_6$ is lethal, it was not clear why the investigators chose to do so anyway. Be that as it may, their experiments have yielded the information hence available in public domain that intravenously, 25 mg kg^{-1} of $InsP_6$ is non-lethal in mice.

Interestingly, very important information also obtained is the concentration of $InsP_6$ in red blood cells at 5 minutes after intravenous administration; it was 14 fold higher than the concomitant plasma concentration, challenging another age-old dogma that exogenous $InsP_6$ cannot cross red blood cell membrane! This should have some practical relevance insofar as the use of $InsP_6$ in red blood cells in sickle cell disease (Chapter 14).

In the xenografts, $InsP_6$ was not detectable; however inositol concentration slowly increased to peak at 6 hours. Following oral administration, $InsP_6$ was detected in liver; but only inositol was detectable in other tissues. As in the rats [8], in mice too, within 10 minutes inositol was detected in plasma after $InsP_6$ was given orally.

Their data from both intravenous and oral administration in mice are similar to those in rats by Sakamoto *et al.,* [8] that exogenous $InsP_6$ is rapidly absorbed and dephosphorylated once inside the organism.

Human Studies

Seven healthy volunteers (3 males and 4 females) were placed on an $InsP_6$-poor diet for 15 days following which they were place on $InsP_6$-normal diet for 16 days. Volunteers on an $InsP_6$-poor diet became deficient in $InsP_6$ - the basal level in plasma being 0.07 ± 0.01 mg L^{-1} or 0.106 ± 0.015 μmol L^{-1}. These volunteers became deficient in $InsP_6$ if they consumed $InsP_6$-poor diet, in as little as 2 weeks. Consuming an $InsP_6$-normal diet resulted in the plasma level to raise 3-5 folds to 0.26 ± 0.03 mg L^{-1} or 0.393 ± 0.045 μmol L^{-1}. The maximum concentration of $InsP_6$ in plasma after a single bolus of $InsP_6$ was achieved after 4 hours [11]. The volunteers ingested different doses of $InsP_6$ and two different salt forms: Ca-Mg-$InsP_6$ or Na- $InsP_6$. There was no difference in excretion pattern with the different salts or doses. These results also suggest that in humans too, $InsP_6$ absorption takes place mostly from the stomach, and rather quickly.

In Vitro Studies

Contrary to the dogma that owing to the highly negative charges, $InsP_6$ cannot be internalized within cells, based on the ability of mammalian cells to internalize complex macromolecules by endocytosis, pinocytosis *etc.*, Shamsuddin [12] however postulated that $InsP_6$ could similarly gain access within the cells. YAC-1 (mouse T cell leukemia), K562 (human erythroleukemia) and HT-29 (human colon adenocarcinoma) cell lines were incubated with $[^3H]$-$InsP_6$ [13]. After 1 hour, $31.3 \pm 3.1\%$ of administered radioactivity was taken up by YAC-1 cells as opposed to $6.2 \pm 0.9\%$ by human erythroleukemia K562 cells and $6.6 \pm 3.8\%$ by human colon cancer HT-29 cells. There was a difference in the rate at which various cells internalize $InsP_6$. Kinetic study showed that the uptake was most rapid by the YAC-1 murine leukemia cells, being linear at 1 minute and reaching a plateau at 10 minute. The uptake by K562 cells was less and did not reach a steady state level even after 12 hours of incubation; HT 29 cells showed the least uptake of radioactivity. Differential centrifugation and high resolution subcellular fractionation of cell homogenates demonstrated that within the various cellular compartments, 80% (HT-29) to 97% (YAC-1) of the total radioactivity was in the cytosol. Kinetic study showed that the peak of the total absorption was obtained after 30 minutes of cell exposure to radiolabeled $InsP_6$, after which a plateau was reached. Analysis of the radioactivity accumulated within the cells showed variable proportions of *myo*-inositol and $InsP_{1-6}$, with a preponderance of $InsP_1$ and $InsP_2$. The presence of $[^3H]$-*myo*-inositol and $[^3H]$-$InsP_{1-6}$ suggests that $InsP_6$ may, in some cells at least, be absorbed as such and that a variable degree of dephosphorylation of $InsP_6$ takes place both extra- and intracellularly.

Using GC-MS analysis of HPLC fraction to determine non-radiolabelled $InsP_6$ as may be naturally found within the cell, Grases *et al.,* [9] found that $InsP_6$ contents were 16.2 ± 9.1 µmol kg^{-1} in MDA-MB 231 human breast cancer cells and 15.6 ± 2.7 µmol kg^{-1} in human K562 erythroleukemia cells. These values were approximately 3-fold higher than those of InsP_3 (4.8 ± 0.5 µmol kg^{-1} and 6.9 ± 0.1 for MDA-MB 231 and K562 cells respectively). Treatment of malignant cells with InsP_6 resulted in a 2-fold increase in the intracellular concentrations of total InsP_3 (9. ± 1.3 and 10.8 ± 1.0 µmol kg^{-1} for MDA-MB 231 and K562 cells respectively), without changes in InsP_6 levels. Thus, as opposed to the normal healthy rats, following InsP_6 treatment, increased intracellular levels of total InsP_3 was observed in human malignant cell lines. These data are similar to earlier report of Shamsuddin *et al.*, [14] who observed a 41% increase in intracellular InsP_3 following InsP_6 treatment of K562 cells.

How is InsP_6 Internalized within the Cells?

Ferry *et al.*, [15] treated human uterine cancer HeLa cells with InsP_6 at 1 mM concentration and induced apoptosis. InsP_6 caused mitochondrial permeabilization, followed by cytochrome c release, which later caused activation of the apoptotic machinery, caspase 9, caspase 3 and poly (ADP-ribose) polymerase. When InsP_6 was applied together with histone, the effective concentration to induce apoptosis was approximately 10-fold lower. These investigators also used [^{3}H]-InsP_6 to follow internalization within HeLa cells and they too showed that InsP_6 is dephosphorylated; in HeLa cells to InsP_3, InsP_4 and InsP_5. However, pre-incubation of HeLa cells with colchicine - a pinocytosis inhibitor did not produce the dephosphorylated forms of InsP_6 supporting the view that the lower inositol phosphates are the products of intracellular dephosphorylation of InsP_6.

Visualization

Visual confirmation of the internalization of InsP_6 by endocytosis/pinocytosis is provided by Riley *et al.*, [16]. The investigators synthesized FAM-InsP_5, a fluorescent conjugate of InsP_5 that allows direct visualization of its interaction with cells. FAM-InsP_5 was internalized by bronchial carcinoma cell line H1229 and the process of FAM-InsP_5 uptake is non-receptor-mediated endocytosis which is blocked at 4 °C and probably involved interaction of the ligand with the glycocalyx. These investigators also showed a difference in uptake of InsP_6 by different cell types. However, since the antiproliferative actions of InsP_5 and InsP_6 are seen following their direct application to the cancer cells *in vitro*, these findings are enigmatic because internalized FAM- InsP_5 appeared in lysosomes and apparently did not enter the cytoplasm [16]. How would the compartmentalized $InsP_5$ and/or $InsP_6$ interact with cytosolic molecules remains an interesting puzzle.

In any event, using a synthesized new receptor tetranaphthoimidazolium as a fluorescent chemosensor, Lee *et al.,* also provide visual documentation of $InsP_6$'s entry into the living cells, both normal (WI38 VA13 subclone 2RA and HeLa adenocarcinoma [17]. Their confocal fluorescence images clearly show internalization and intracytoplasmic distribution of $InsP_6$ thereby irrevocably destroying the myth that it cannot be transported inside the cell.

PHARMACOKINETICS OF *MYO*-INOSITOL

Inositol has been supplemented in infant formulas since the late 1990s at approximately 44 mg/100 kcal (350 mg/l) and there have been clinical trials of it in preterm infants with respiratory distress syndrome. Early fetal serum inositol levels are 2-10 times higher *in utero* than adult levels which decrease gradually towards term [18]. If an infant is not receiving enteral milk feedings, serum levels fall to levels substantially below those that would have been present *in utero* [18].

Hallman and colleagues investigated the pharmacokinetics of *myo*-inositol in preterm infants (mean birth weight 1365 g, gestational age 30.1 weeks) between 48 hours and 10 days of age at a dose 40 mg/kg every 6 hours (equivalent to preterm human milk feedings) [19]. Serum inositol concentration increased between days 2 and 3 from a mean of 566 μmol L^{-1} to 823 μmol L^{-1} in the infants given supplement, whereas it fell from 451 μmol L^{-1} to 292 μmol L^{-1} in the controls. On day 16, serum inositol values remained higher in the infants given supplement than in those given placebo (mean 334 μmol L^{-1} *v* 146 μmol L^{-1}, $P = 0.014$). The infants who developed bronchopulmonary dysplasia (a form of lung pathology in premature infants) had significantly higher renal inositol clearance, lower inositol intake, and lower serum inositol concentrations. Inositol supplementation increased the saturated phosphatidylcholine/sphingomyelin ratio in tracheal aspirates. These results show that supplementation with inositol (in preterm infants) leads to a rise in serum inositol concentration concomitant to an improvement in the surfactant phospholipids [19].

Pharmacokinetic studies on inositol stereoisomers in mice (male ICR mice) following oral administration of 1 g/kg body weight of inositol stereoisomers *myo*-inositol, D-*chiro*-inositol and *scyllo*-inositol demonstrate that all of the three stereoisomers are rapidly absorbed with elevated levels (2.5 - 6.5 mM) in blood plasma; none of the three stereoisomers was seen in untreated samples. Interestingly, plasma of *scyllo*-inositol-administered animals contained substantial amount of *myo*-inositol, suggesting a possible metabolic conversion of *scyllo*-inositol to *myo*-inositol in mice [20].

CONCLUDING REMARKS

The preceding demonstrates beyond any reasonable doubt that $InsP_6$ is rapidly absorbed and distributed widely throughout the mammalian system including the brain and red blood cells thereby challenging several myths about it. $InsP_6$ is also absorbed by the various malignant cells almost immediately after exposure *in vitro*. Additional research into the methodology to follow these fascinating molecules and their derivatives, as well as the application of these methodologies for research and practical application for human and animal health are needed.

REFERENCES

[1]   Singsen EP, Matterson LD, Kozeff A: Phosphorus in poultry nutrition IV. Radioactive phosphorus as a tracer in studying the metabolism of phytin by the turkey poult. Poultry Sci 1950; 29: 635-9.
[2]   Gillis MB, Keane KW, Collins RA. Comparative metabolism of phytate and inorganic P32 by chicks and poults. J Nutr 1957; 62: 13-26.
[3]   Nahapetian A, Young VR: Metabolism of 14C-phytate in rats: effect of low and high dietary calcium intakes. J Nutr 1980; 110: 1458-72.
[4]   Elsayed A, Chakravarthy A, Shamsuddin A. Inositol hexaphosphate from corn decreased the frequency of colorectal cancer in azoxymethane-treated rats. Laboratory Investigation 1987; 56: 21A.
[5]   Shamsuddin AM, Elsayed AM, Ullah A. Suppression of large intestinal cancer in F-344 rats by inositol hexaphosphate. Carcinogenesis 1988; 9: 577-80.
[6]   Shamsuddin AM, Ullah A. Inositol hexaphosphate inhibits large intestinal cancer in F344 rats 5 months following induction by azoxymethane. Carcinogenesis 1989; 10: 625-6.
[7]   Shamsuddin, AM, Ullah A, Chakravarthy AK. Inositol and inositol hexaphosphate suppress cell proliferation and tumor formation in CD-1 mice. Carcinogenesis 1989; 10: 1461-3.
[8]   Sakamoto K, Vucenik I and Shamsuddin AM: ³[H]-phytic acid (inositol hexaphosphate) is absorbed and distributed to various tissues in rats. Journal of Nutrition 1993; 123: 713-20.
[9]   Grases F, Simonet BM, Vucenik I *et al.* Effects of exogenous inositol hexakisphosphate (InsP₆) on the levels of InsP₆ and of inositol trisphosphate (InsP₃) in malignant cells, tissues and biological fluids. Life Sciences 2002; 71: 1535-46.
[10]   Eiseman J, Lan J, Guo J *et al.* Pharmacokinetics and tissue distribution of inositol hexaphosphate in C.B17 SCID mice bearing human breast cancer xenografts. Metabolism 2011; 60: 1465-74, 2011. doi: 10.1016/j.metabol.2011.02.015. Epub 2011 Apr 12.
[11]   Grases F, Simonet BM, Vucenik I *et al.* Absorption and excretion of orally administered inositol hexaphosphate (IP₆ or phytate) in humans. Biofactors 2001; 15: 53-61.
[12]   Shamsuddin AM: Metabolism and cellular functions of IP₆: A review. Anticancer Res 1999; 19: 3733-6.
[13]   Vucenik I, Shamsuddin AM. [3H]Inositol hexaphosphate (phytic acid) is rapidly absorbed and metabolized by murine and human malignant cells *in vitro*. J Nutr 1994; 124: 861-8.
[14]   Shamsuddin AM, Baten A, Lalwani ND: Effects of inositol hexaphosphate on growth and differentiation in K-562 erythroleukemia cell line. Cancer Lett 1992; 64: 195-202.

[15]   Ferry S, Matsuda M, Yoshida H *et al.* Inositol hexakisphosphate blocks tumor cell growth by activating apoptotic machinery as well as by inhibiting the Akt/NFkappaB-mediated cell survival pathway. [Erratum appears in Carcinogenesis. 2003 Jan; 24(1): 149]. Carcinogenesis 2002; 23: 2031-41.

[16]   Riley AM, Windhorst S, Lin HY *et al.* Cellular internalisation of an inositol phosphate visualised by using fluorescent InsP$_5$. Chembiochem. 2014; 15: 57-67. doi: 10.1002/cbic.201300583. [Epub 2013 Dec 6]

[17]   Lee M, Moon JH, Jun EJ *et al.* A tetranaphthoimidazolium receptor as a fluorescent chemosensor for phytate. Chem Commun (Camb). 2014 Apr 23. [Epub ahead of print] DOI: 10.1039/c4cc02036g

[18]   Phelps DL, Ward RM, Williams RL *et al.* Pharmacokinetics and safety of a single intravenous dose of myo-inositol in pre-term infants of 23-29 week. Pediatr Res 2013; 74: 721-9.

[19]   Hallman M, Arjomaa P, Hoppu K. Inositol supplementation in respiratory distress syndrome: relationship between serum concentration, renal excretion and lung effluent phospholipids. J Pediatr 1987; 110: 604-10.

[20]   Yasmashita Y, Yamaoka M, Hasunuma T *et al.* Detection of orally administered inositol stereoisomers in mouse blood plasma and their effects on translocation of glucose transporter 4 in skeletal muscle cells. J Agr Food Chem 2013; 61: 4850-4, doi: 10.1021/jf305322t. Epub 2013 May 13.

CHAPTER 6

Experimental Cancer Prevention by InsP_6 & Inositol

Abstract: *Myo*-inositol and InsP_6 have independent anticancer action against divergent types of malignancies in a consistent and reproducible manner. The anti-cancer action is dose-dependent and inositol acts synergistically with InsP_6 to enhance the anticancer action of the latter. Since inositol and InsP_6 are common constituents of cereals and legumes, comparative study of high bran diet with equivalent amount of InsP_6 showed that while the 'high fiber' diet failed to reduce the incidence of experimental cancer, InsP_6 alone at equivalent dose significantly inhibited mammary cancer formation. Combined with the other studies, these data strongly suggest that supplementation with pure inositol + InsP_6 may be better than gobbling a large amount of fiber.

Keywords: AOM, CD, colitis, Crohn's disease, DMBA, DMH, pre-initiation, post-initiation, tumor promoter, UC, ulcerative colitis.

INTRODUCTION

Cancer - A Global Problem

In 2012, 14.1 million adults were diagnosed with cancer and 8.2 million people died of the disease worldwide. The number of new victims is estimated to go up by 69% in 2030; Asia, Latin America and Africa leading the increases by 75%, 86% and 87% respectively. Of the 12.7 million new cases, at 6.09 million Asia had the lion-share, followed by Europe (3.2 million), North America (1.6 million), Latin America and the Caribbean (906,008), and Africa (681,094) [1]. In the United States alone, an estimated 560,000 people succumb to the disease annually; in addition, almost 1.4 million new cases are diagnosed every year. It is therefore no exaggeration to state that cancer is a major public health problem.

The earliest known cases of cancer were described around 2625 B.C.E. by the ancient Egyptian physician Imhotep; the 45 of his teaching cases were about "bulging tumors of the breast" and as to the treatment he offered "none" [2, 3]. In contrast to Imhotep, 4600+ years later, though modern physicians and surgeons are offering some treatment, we are still a long way from curing the disease. Over 40 years ago in December 1971 US President Richard M. Nixon declared "war on cancer" and since that time, an enormous amount of money (over US$90 billion) has been spent by the US government, let alone the private sector in an endeavor to learn about the biology, treatment and prevention of this disease[1]. From time to time, the 'generals' leading the war – those bosses engaged in the field of cancer research – report exciting new gains they achieved in the battlefield. But overall, in spite of continued efforts and a great deal of optimism, we are simply not yet in a position to declare victory; we have not won the war against this disease, unfortunately not yet!

Research and research-funding organizations, such as the *National Cancer Institute* and the *American Cancer Society* – just to name two - often make decision as to the next moves in the ongoing battle against cancer. It is at such organizations that decision-makers set the new targets and provide support for the scientific community - all in an effort to develop "silver bullets" that might serve to attack cancer cells. Usually through the "peer review" process scientists working in the field generally approve of this approach to fund their research and follow the leaders not unlike the foot-soldiers in the battlefield following their commanders.

"A common <u>illusion</u> is that strategic objectives are necessary to discover the cure for cancer and AIDS and that groups of sufficient size need to be mobilized for wars and crusades against these enemies. <u>Nothing could be more misguided</u> [underlines added]. In the history of triumphs in medical research such wars and crusades have invariably failed because they lacked the necessary weapons – the essential knowledge of basic life processes. Instead, some of the major advances – X-rays, penicillin, polio vaccine, and genetic engineering – have come from the efforts of individual scientists to understand Nature..." opines Kornberg [4].

Before we discuss about the role of inositol and InsP_6 let's review some of the basic facts about cancer with apologies to those who might find these to be too elementary.

Cancer Development

The etymology of the term "cancer" can likely be traced back to the Latin word for "crab". This is a fitting analogy because cancer literally "adheres to any part that it seizes upon in an obstinate manner like the crab". Cancer does not just appear overnight. In order for a cancer to develop, a whole series of complex interactions involving genetic, hereditary, and environmental factors has to come

---

[1]Apparently the phrase "war on cancer" has been the product of news reports, there was never an official declaration (Marshall E *Science* 331: 1540, 2011).

into play. For instance, harmful environmental influences might include an exposure to toxic substances, such as chemicals, radiation from the sun or, from anthropogenic activity (x-rays), and even viruses. Furthermore, we know that diet plays a crucial role - it can greatly influence the actual outcome that may result from interplay of the various contributing factors.

Before a cancer can establish itself and ultimately overpower the body's immune system, an entire sequence of steps must occur. In situations where a cancer is triggered by an exposure to a chemical substance (either as experimental carcinogenesis, or environmental exposure), we can distinguish two main phases: initiation and promotion. A certain chemical may act as the initiator inducing changes in our cells that heighten the likelihood they will become cancerous. The process of cancer initiation by itself however does not produce cancer. A specific chemical may indeed cause lasting damage to a cell's DNA, but a so-called promoter is then necessary in order to turn that damaged cell into one that is cancerous. A cancer promoter might consist of another chemical, radiation such as x-rays or UV rays from sun exposure, or it could also be a virus.

Just as a cancer initiator cannot cause the disease on its own, the same principle applies to the promoter; it does not cause cancer by itself. It is only when an initiator works in conjunction with a promoter that a cancer can be produced. However a promoter does not have to be present immediately following a cancer initiation. In fact, a cancer-promoting factor may appear much, much later, thus explaining why some cancers take years to develop. A prominent example of this is asbestos, which can lead to lung cancer. Asbestos is the cancer initiator, and smoking can act as the promoter.

It is well known that even when a person has been exposed to an initiator and is subsequently also exposed to a promoter, cancer will not necessarily result in every instance. This can be explained by the strength (or health) of the host's immune system amongst others. There are enzymes to repair the damaged DNA, thus enabling damaged cells to heal. Repair of double-strand breaks in DNA is essential for maintaining the stability of the genome, failure to repair may result in loss of genetic information, chromosomal translocation, and even cell death.

CANCER PREVENTIVE STUDIES

For more than half-a-century numerous substances with potential anticancer function have been tested by various investigators worldwide. It is neither the mission, nor the intention of the eBook to delve into those. We will discuss only experiments that are germane to the main theme of the eBook, *viz.* inositol and inositol phosphates. While we will refer to the important and elegant work of other investigators, we have taken the liberty of giving a historical background as to how the pioneering experiments were conceived and executed for the benefit of the inquisitive readers.

Dietary Fiber

The role of dietary fiber in prevention of cancer, initially of colon cancer has reached public awareness over the last 60+ years. Our dietary fiber comes from the plants we eat and is not truly digested in our bodies; neither our digestive enzymes nor the enzymes produced by our intestinal flora are able to fully decompose it. Fiber can be classified according to its resistance to digestion: *Insoluble fiber* that is highly resistant to digestion and, *soluble fiber* that is comparatively less resistant. Fiber's main components are polysaccharides consisting of cellulose, hemicellulose, pectins, gums, mucilages, and lignin.

Soluble fiber is found in fruits and vegetables, and in some grains (in oats, for instance). Cereals - notably wholegrain cereals represent the major source of insoluble fiber. Fruits and vegetables also contain some insoluble fiber. The richest source of insoluble fiber is the outer coating of cereal grains, also known as bran.

Fiber and Colon Cancer: Epidemiological Connection

In India as in most large countries, eating habits vary greatly from region to region; there are also differences between the various religious groups. SL Malhotra conducted an extensive study of these dietary habits, comparing the foods eaten in India's North with those eaten in the South. He conclusively demonstrated that people who eat a diet high in fiber have a lower risk of developing certain diseases including colon cancer; colon cancer was significantly less prevalent in Northern India than in the South. He noted that "[W] while the North Indian diets are rich in roughage, cellulose, and vegetable fibers, these are almost completely lacking in the South Indian diets" [5].

How does fiber inhibit colon cancer? Fiber by absorbing cancer-causing or cancer-promoting chemicals (carcinogens) in the gut lumen and by speeding up the stool's transit time can accomplish this as the carcinogens have less time to interact with the cells in the intestines. Quite interestingly, around the same time it is believed that George Oettle of South Africa have made similar observations though unpublished, as SL Malhotra across the Indian Ocean; however, the credit for having advocated the benefits of dietary fiber and its role in preventing colon cancer is generally given to Denis Burkett [6].

Types of Dietary Fibers

Large-scale population studies of the relationship between fiber and the incidence or prevalence of cancer however suggest that not all fibers are related to a lower incidence of cancer. A review of several studies found negative correlation, no correlation or even positive correlation with dietary fiber intake and colon cancer [7]. Indeed, while most animal experiments have shown a suppression of colon cancer by high dietary fiber, there has been enhancement of colon carcinogenesis in rats fed wheat bran [8]! This raises the issue that not all fibers are same and that there are other factors which might be responsible; and epidemiological studies from Scandina *via* point to that [9, 10]. Only high fiber diets based specifically on cereals grains have been shown to be strongly and consistently linked to a lower frequency of colon cancer [11].

How Does Fiber Prevent Cancer?

Dietary Fiber – InsP$_6$

Based on the Scandinavian studies showing a low prevalence of colorectal and breast cancer correlating with an increased consumption of cereal grains in Finland as opposed to Denmark with high total fiber intake and double the prevalence of the cancers, one is poised to ask: what is that the cereal grains have that protects the Finnish population from cancers of the breast and colon [11]? Graf & Eaton hypothesized that the Finnish cereal-rich diet had more InsP$_6$. In contrast, while the Danes had a total fiber intake double that of the Finns, yet their diet contained less InsP$_6$; and the presumed benefit is owing to the antioxidant property of InsP$_6$ [11].

Inhibition of Carcinogenesis by InsP$_6$ and Inositol

Though Graf & Eaton [11] had proposed that the antioxidant property of InsP$_6$ is the mechanism of protection of the colon from cancer, Shamsuddin however hypothesized that InsP$_6$ enters the intracellular inositol phosphate pool and affects the signal transduction process perhaps *via* InsP$_3$ the ubiquitous second messenger. Therefore, the putative anticancer action of $InsP_6$ should be observed not just in colon, but also in other organs; *i.e.,* it could be a broad-spectrum anticancer agent [12-16]. Since there are enzymes to phosphorylate and dephosphorylate $InsP_6$ and inositol, it is considered plausible that addition of inositol to $InsP_6$ might lead to increased production of the second messenger $InsP_3$ which in turn could potentiate the action of $InsP_6$ [14]. While the antioxidant mechanism was not discounted, it could however, not even hypothetically prevent cancer in organs remote from the gastrointestinal tract as at the time $InsP_6$ was not considered to be absorbed into the body. Contrary to the prevailing dogma, Shamsuddin's hypothesis obviously included the absorption of $InsP_6$ from the gastrointestinal tract.

Pre-Initiation Study in Colon Cancer Model

Since there was no indication whether $InsP_6$ would work at all, rats were given $InsP_6$ in drinking water one week prior to the administration of the initiating carcinogen azoxymethane (AOM) - pre-initiation phase; it was designed to test if $InsP_6$ could prevent the earliest process of cancer formation (Fig. **6.1**).

After 6 months, the average number of tumors developed by the control rats was 4.6 per animal as opposed to 3 tumors per animal ($p < 0.01$) in the rats treated with $InsP_6$. Interestingly the tumor volume of $InsP_6$-treated rates was approximately two thirds smaller. The rate of cell division in the colon away from the tumor of $InsP_6$-treated animals was very similar to the rate of cell division found in the healthy control animals and significantly ($p < 0.001$) less than that in the carcinogen-treated animals, pointing to regulatory function that serves to normalize the abnormal increase in cell division induced by the carcinogens. Animals that had received $InsP_6$ but not the carcinogens showed normal rates of cell division [12, 17].

Post-Initiation Study in Colon Cancer Model

That the tumors in the animals treated with $InsP_6$ were smaller suggested that $InsP_6$ might be effective even after cancer is already established. Thus animals were given $InsP_6$ as early as two weeks after, and as late as five months after the initiation with carcinogens (post-initiation). $InsP_6$ treatment significantly inhibited the development of cancer in these rats even late in the disease process; and the rate of cellular mitosis in non-cancerous colon in $InsP_6$-treated animals was lower than the untreated control (Table **6.1**). Once again, interestingly enough, the tumor volume was smaller in the $InsP_6$-treated animals; therefore raising the prospect of therapeutic applications for existing cancers [13, 18].

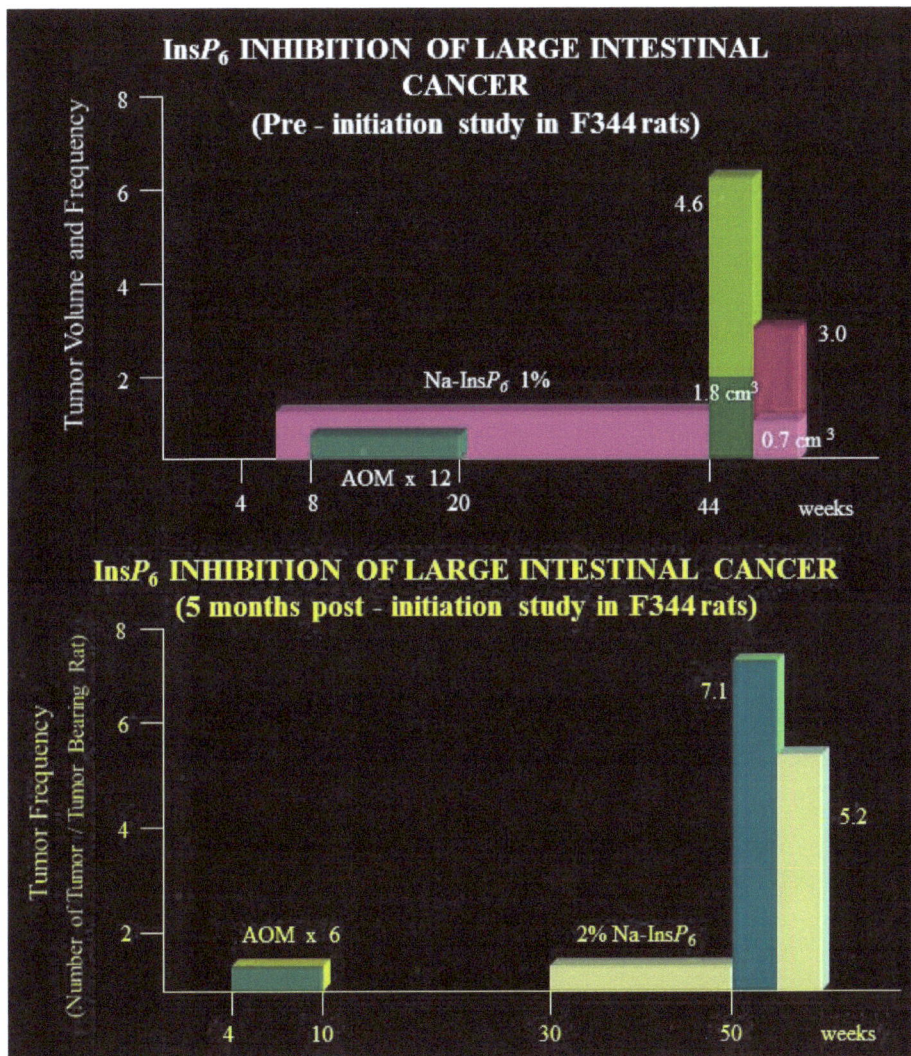

Fig. (6.1). Schematic representation of the experimental design of Pre-Initiation (upper) and Post-Initiation experiments, and results thereof. In the Pre-Initiation study, the carcinogen AOM (azoxymethane) was injected 1 week after the rats were started on 1% Na-$InsP_6$. At 44 weeks, the control animals had 4.6 tumors per rat whereas there were 3 tumors per $InsP_6$-treated rat ($p <$ 0.01). The average tumor size in the $InsP_6$-treated rats was 0.7 cm³ compared to 1.5 cm³ in control. In the Post-Initiation study, oral administration of 2% Na-$InsP_6$ in drinking water was started 5 months after the carcinogen AOM. At 50 weeks, the $InsP_6$-treated rats had 5.2 tumor/rat as opposed to 7.1 tumor/rat in the control animals ($p < 0.02$).

Table 6.1. Experimental inhibition of colon cancer after 5 months [13].

Treatment	Tumors/Rat	% of Mitotic Rate in Non-Tumorous Area
Carcinogen only Azoxymethane (AOM)	7.1	2.3
AOM + InsP$_6$	5.2	1.0

Further experiments on dose-dependent relationship with 0.1% and 1% InsP$_6$ showed a dose-related reduction in the prevalence of tumors [19]. Two weeks following the beginning of InsP$_6$ supplementation, rats were given six injections of AOM at a dose of 8 mg/kg body weight/week and were killed 30 weeks following the last injection. Compared to the untreated control rats injected with AOM, 1% InsP$_6$ reduced the tumor prevalence by 52.2% ($p < 0.01$), tumor frequency by 55.8% ($p = 0.001$) and tumor size by 62.3% ($p = 0.001$); 0.1% InsP$_6$ showed a lesser reduction in tumor prevalence (21%) but a greater reduction in tumor size 71% ($p = 0.001$) [19].

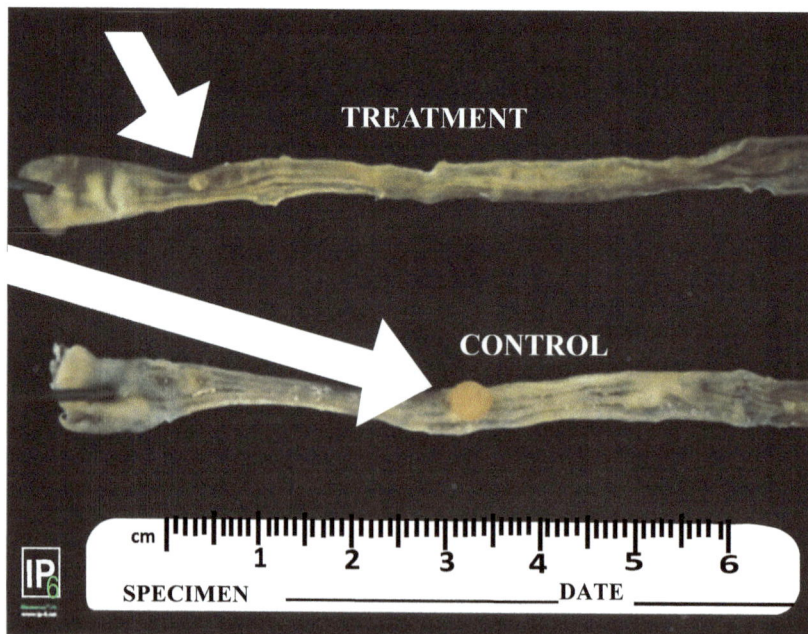

Fig. (6.2). Two representative tumors of rat colon. Note that the tumor in the InsP$_6$-treated rat is much smaller - approximately 2/3rd smaller on the average than the untreated carcinogen-control (**$p < 0.01$**) [13].

Inositol + InsP₆ in Colon Cancer Models

As discussed in the previous chapters and alluded to earlier, $InsP_6$ is subject to dephosphorylation resulting in formation of inositol and $InsP_{1-5}$; and that $InsP_6$ could enter into the intracellular inositol phosphate pool and through $InsP_3$ or other lower inositol phosphates mediate the effect. To address this issue, using a different species of animals (CD-1 mice) and a different carcinogen 1,2-dimethylhydrazine (DMH 15 mg/kg/week x13)), a combination of 1% *myo*-inositol + 1% $InsP_6$ was given [14].

Table 6.2. Effect of inositol+InsP₆ on DMH-induced colon cancer in CD-1 mice.

Experimental Group	Tumors/Mice %	Tumor/10 Mice	Tumors/Tumor Bearing Animal
DMH	63[a]	11.6	1.83
DMH+InsP₆	47[b]	6.2	1.30
DMH+Inositol	30[c]	4.5	1.50
DMH+InsP₆ +Inositol	25[d]	2.5	1.00

Data represent total cancers – microscopic and macroscopic. [a] *versus* [d] significant at $p < 0.001$, [a] *versus* [c] significant at $p < 0.001$, [b] *versus* [d] significant at $p<0.005$, [b] *versus* [c] not significant [14].

As can be seen in Table **6.2**, the combination of 1% $InsP_6$ + 1% inositol resulted in a significantly lesser prevalence of tumors than 1% $InsP_6$ alone ($p = 0.005$). Quite surprisingly, while *myo*-inositol is an *in vitro* growth promoting factor used in all cell and tissue culture medium as an essential ingredient, caused a significant decrease ($p = 0.001$) in tumor prevalence. As in the rat-AOM studies discussed above, there was also a concomitant reduction in the rate of mitosis in the non-tumorous areas of the colon in the $InsP_6 \pm$ inositol treated mice; being within the range of normal rate in the control animals [14]. Thus the results of this experiment show that inositol potentiates the anticancer action of $InsP_6$.

Studies of the natural killer (NK) cells from these groups of mice showed a similar boosting of the carcinogen-induced depressed NK activity by $InsP_6 \pm$ inositol; and $InsP_6$ + inositol treatment gave the best results [20, 21]. This is discussed further in Chapter 17.

Anticancer Action of Inositol & InsP₆ in Other Experimental Models

Since the hypothesis for the anticancer cancer action of $InsP_6 \pm$ inositol was due to the lower inositol phosphates important in cellular signal transduction common to all the cells, it was therefore logical to expect to see similar results in other cancers as well. Thus experiments were conducted in other models with different carcinogens and different species (the experiments on human cancer cell lines will be discussed in the next chapter).

Mammary Tumor Inhibition

Mammary tumors were induced in rats with 7,12-dimethylbenz(*a*)anthracene (DMBA). Starting a week prior to induction with DMBA, the drinking water of female Sprague-Dawley rats was supplemented with either 15 mM InsP_6, 15 mM *myo*-inositol, or 15 mM InsP_6 + 15 mM *myo*-inositol; a control group received no inositol compounds. Animals (55-day-old) were given a single dose of DMBA (20 mg) in 1 ml of sesame oil by oral intubation. Four additional groups did not receive DMBA, but were drinking tap water, InsP_6, inositol, or InsP_6+inositol of the same molarity as experimental groups; they were observed for the duration of the study to monitor for any putative toxicity following this long-term treatment. As opposed to the DMBA-only group, rats treated with InsP_6 ± inositol showed a 48% reduction in the number of tumors/tumor bearing animal (tumor multiplicity) and a 40% reduction in the number of tumors/animal. In contrast to 20% rats in DMBA-only group, only 0-8% animals in the treatment groups had 5 or more tumors. The tumor incidence was reduced by 19% in InsP_6 ± inositol group as compared to control untreated animals. The tumors in the treated groups were also 16% smaller in size [22].

Following this pilot study, a second experiment of rat mammary carcinogenesis was conducted. Starting two weeks prior to induction with DMBA, the drinking water of female Sprague-Dawley rats was supplemented with either: 15 mM InsP_6, 15 mM *myo*-inositol, or 15 mM InsP_6 + 15 mM *myo*-inositol; a control group received no inositol compounds. After 45 weeks of treatment, the animals in all the three treatment regimens showed a significant reduction ($p < 0.05$) in tumor incidence. Tumor number, multiplicity and tumor burden were also significantly ($p < 0.05$) reduced by InsP_6 ± inositol. When all the parameters were taken into consideration, the best results were obtained by the combination treatment of InsP_6 + inositol. Thus, it was reproducibly demonstrated that InsP_6 ± inositol inhibited experimental mammary carcinoma [23].

Additional experiments of rat mammary carcinogenesis were performed at Shamsuddin laboratory that conclusively demonstrated that it is the InsP_6 in the bran part of cereals that is responsible for cancer inhibition (Table **6.3**) [24].

Table 6.3. Comparison of the effects of bran *v* pure IP$_6$ in suppression of mammary cancer.

Treatment/Diet Group	Tumor Incidence	Tumors per Rat	Rats with $\geq$3 Tumors
DMBA	79.0%	2.8	47%
DMBA + 20% bran	70.0%	3.1	36%
DMBA + 0.4% InsP_6	52.6%	2.2	15%

DMBA = 7,12-dimethyl-benz[a] anthracene. Source: Adapted from Ref [24].

In comparison to the animals subjected to the carcinogen DMBA, the rats receiving 0.4% InsP_6 (a dose level comparable to the calculated amount of InsP_6 in the bran) in their drinking water were the only animals to show a statistically significant reduction of tumor incidence. The rats given 0.4% InsP_6 in drinking water, equivalent to that in 20% bran, had a 33.5% reduction in tumor incidence ($p < 0.02$) and 48.8% fewer tumors ($p < 0.03$). Supplemental dietary fiber in the form of bran exhibited a very modest, statistically non-significant inhibitory effect, which was also not dose dependent. In contrast, animals given InsP_6 showed significant reduction in tumor number, incidence, and multiplicity. The decrease in the percentage of rats showing three or more tumors per animal was statistically significant. Thus InsP_6 an active substance responsible for cereal's beneficial anticancer effect, was clearly more effective than 20% bran in the diet [24]. In practical terms, intake of InsP_6 may be a more pragmatic approach than guzzling enormous quantities of fiber for cancer prophylaxis.

Aside from that, interestingly, 4 control animals not subjected to a treatment with carcinogens (and also not given any bran or InsP_6), still developed some palpable tumors; the tumors they produced were however non-cancerous (benign – fibroadenomas). Spontaneously appearing tumors were also discovered in 20% of the animals given the high fiber diet; however, no such tumors were found in the InsP_6 group. Thus, InsP_6 had a protective effect, preventing the development of spontaneously appearing mammary tumors and not just the carcinogen-induced tumors [24].

Additional Studies on Prevention of Experimental Cancers

There have been many studies, albeit not as many as it should have been, on the effect of InsP_6 in various organs systems. Following the original studies of the efficacy, these later studies were mostly designed to investigate the mechanisms as to how inositol phosphates, especially InsP_6 exert their action. To avoid redundancy, some of these studies have been discussed in the chapter on Signal

Transduction (Chapter 3), others appropriately in the chapters on Mechanisms (Chapters16-20). Here are the studies that were more focused on the efficacy:

Parenteral supplementation of iron was found to augment tumor yield ($p = 0.012$) and oral iron was found to augment tumor incidence in a rat-DMH model of colon carcinogenesis [25]. Treatment of the animals with $InsP_6$ reversed the augmenting effect of oral iron on tumor yield and incidence ($p = 0.09$ for both) [25].

Jariwalla *et al.*, used a rat fibrosarcoma model to investigate whether $InsP_6$ administered in diet as 12.6% dipotassium pentamagnesium salt would inhibit the growth of the transplanted tumors promoted by a special diet containing 5% saturated fatty acid and 1.2% magnesium oxide. Supplementation of the diet by $InsP_6$ reduced the tumor incidence and growth rate [26]. It is of note that the tumor inhibition observed by Jariwalla *et al.*, was achieved by addition of a very high amount (12.6%) of $InsP_6$ administered with food; whereas similar reduction was observed with a much lower dose (1-2%, and as low as 0.1%) of $InsP_6$ in drinking water [13-24]. This suggested that mixing with diet may not be the most efficient strategy to administer $InsP_6$.

That adding $InsP_6$ with the diet may not be the best route is also demonstrated rather serendipitously in a rat wide-spectrum organ carcinogenesis model. Animals were initiated with two intraperitoneal injections of 1000 mg/kg body weight 2,2'-dihydroxy-di-*n*-propylnitrosamine (DHPN) followed by two intragastric administrations of 1500 mg/kg body weight N-ethyl-N-hydroxy-ethylnitrosamine (EHEN), and then three subcutaneous injections of 75 mg/kg body weight 3,2'-dimethyl-4-aminobiphenyl (DMAB) during the first 3 weeks [27]. Starting 1 week after the last injection, groups of rats received diet containing 2% $InsP_6$ besides other test substances, or basal diet alone for 32 weeks. Histological examination revealed an enhancement of the incidence of urinary bladder papillomas (benign lesions that are precursor of cancer) by $InsP_6$. $InsP_6$ showed a modest, statistically non-significant inhibition of the development of liver and pancreatic tumors [27]. While this earlier study suggested a positive correlation between ingestion of $InsP_6$ and the incidence of urinary bladder papillomas, subsequent studies by the same group of investigators showed that the sodium salt, but not the potassium or the magnesium salt of $InsP_6$, could encourage development of pre-neoplastic lesions of the urinary bladder [28]. Thus factors other than $InsP_6$ may affect the experimental and clinical outcomes.

Arnold and colleagues from the Chemoprevention Branch of the Division of Cancer Prevention and Control of the US National Cancer Institute, Bethesda,

Maryland reported on the finding of their evaluation of 99 different chemopreventive agents of interest in a Rat Tracheal Epithelial Transformation assay. $InsP_6$ (IHP in their abbreviation) showed a 78% inhibition (Table **2** page 541 of their paper [29]). However, no further details were available.

CANCER PREVENTION BY *MYO*-INOSITOL

Lung Carcinogenesis

As presented earlier, that *myo*-inositol inhibits colon and mammary carcinogenesis had been demonstrated by Shamsuddin and co-workers [14, 23]. Subsequently, Estensen & Wattenberg demonstrated its efficacy in benzo[*a*]pyrene (BP) induced lung and forestomach carcinogenesis model in mice [30]. A diet containing 3% *myo*-inositol fed beginning 1 week after BP administration (post-initiation) reduced the number of pulmonary adenomas by 40% but did not prevent forestomach tumors. Feeding a diet containing both *myo*-inositol and dexamethasone resulted in an additive effect on the inhibition of pulmonary adenoma formation. The combination of *myo*-inositol plus dexamethasone produced almost identical inhibition of forestomach tumor formation to that of dexamethasone alone [30].

The investigators then used a dual pre-initiation and post-initiation schedule wherein in the pre-initiation study the animals were given *myo-i*nositol and or dexamethasone 2 weeks prior to the first dose of carcinogen and continued for the duration of the experiment. They reported reductions in lung tumor formation as follows: *myo*-inositol, 64%; dexamethasone, 56%; and both together, 86% ($P < 0.001$ for all three) [31].

Subsequent experiments on the stages of progression demonstrated that dietary *myo*-inositol failed to suppress progression from adenoma to carcinoma when started 12 weeks post-carcinogen [32].

Colitis-Induced Cancer Prevention

Chronic inflammation of the colon and rectum – ulcerative colitis (UC) and Crohn's disease (CD), together known as inflammatory bowel diseases (IBD) predispose to cancer formation in the long run. The dextran sulfate sodium (DSS)-induced ulcerative colitis (UC) in rodents is a good experimental model for investigating colitis. In this model, feeding of 2-fold iron diet to mice subjected to low dose, cyclic, long-term DSS treatment (to mimic flare-up and flare down activity of UC in human) increased colorectal carcinoma incidence from approximately 19% in mice with normal diet to approximately 88% in 2-fold iron diet [33-35].

Administration of *myo*-inositol in the drinking water to DSS colitis mice significantly inhibited colitis-induced cancer development as evaluated by reduction of tumor incidence, multiplicity, and volume [36]. Chronic active colitis was frequently observed in non-tumor colorectal mucosa and exhibited as ulceration, reactive hyperplasic epithelial change, glandular distortion and basal lymphoplasmacytosis. According to established criteria for UC indices [33-35], as shown in Table **6.4**, the UC indices, ulcer formation, and the area of inflammation in mice administered with *myo*-inositol were significantly inhibited in comparison to the colon of mice treated with DSS.

Mac3-positive macrophages were frequently observed in the non-inflamed mucosa of the colon in mice treated with DSS. They were significantly reduced in non-inflamed and inflamed mucosa in mice treated with inositol. The mean number of Mac3-positive cells in non-inflamed mucosa was $541 \pm 59/mm^2$ in DSS-treated mice v $302 \pm 65/mm^2$ in inositol treated mice ($p < 0.01$).

Table 6.4. **Inhibition of DSS-induced colorectal inflammation by 1% *myo*-inositol[*].**

Groups	Histological Score				
	Severity	Ulceration	Hyperplasia	Area Involved	UC Index
DSS control	1.64 ± 0.03	0.61 ± 0.10	1.70 ± 0.16	3.28 ± 0.13	7.22 ± 0.30
Inositol	1.48 ± 0.06	0.36 ± 0.17^a	1.23 ± 0.14	1.38 ± 0.06^a	4.45 ± 0.13^a

*Data are expressed as the mean ± SE. The mice were subjected to 15 cycles of DSS treatment. UC was scored in non-cancerous mucosa of the colon. Three H&E-stained slides per mouse from 6 mice per group were analyzed. The histological scores for each animal were the average of the scores for the three slides. The UC index was the sum of the individual scores for disease severity, ulceration, glandular hyperplasia, and area of inflammatory involvement. [a]p<= 0.05 (Student *t*-Test).

Nitrotyrosine accumulation, a biomarker of nitro-oxidative stress, was analyzed immunohistochemically to determine the effect of inositol compounds on inflammation-caused nitro-oxidative damage. Nitrotyrosine immunostaining showed strong reactivity in the macrophages in non-inflamed and inflamed mucosa of the colon in DSS-induced UC mice. Inositol markedly reduced nitrotyrosine staining intensity and nitrotyrosine positive cell numbers in both non-inflamed and inflamed areas of the colon. The mean number of nitrotyrosine-positive cells per area of mucosa was $638 \pm 87/mm^2$ in the colon in DSS-treated

UC mice v 389 ± 71/mm^2 in inositol treated mice. Statistically significant inhibition of cell proliferation in non-inflamed and dysplastic epithelia was observed in DSS-treated mice administered with inositol [36]. These results strongly indicate that inositol's prevention of colitis-induced carcinogenesis is possibly *via* blocking inflammation-induced cell proliferation and nitro-oxidative stress.

CONCLUDING REMARKS

Experiments in different species of laboratory animals of both sexes and divergent organ systems showed cancer inhibition by InsP_6 with or without *myo*-inositol, and the latter have shown synergistic effect with the former. Additional experiments on the efficacy of InsP_6 and *myo*-inositol as potential therapeutic agents will be discussed in the next chapter.

REFERENCES

[1]   O'Callaghan T. The prevention agenda. Nature 471: S2-S4, 24 March 2011.
[2]   Mukherjee S.: The Emperor of All Maladies, A Biography of Cancer. Scribner, New York, 2010.
[3]   Pederson T. On cancer and people. Science 2011; 332: 423.
[4]   Kornberg A. The NIH did it. Science 1997; 278: 1863.
[5]   Malhotra SL. Geographical distribution of gastrointestinal cancers in India with special reference to causation. Gut 1968; 8: 361-72.
[6]   Burkitt DP. Epidemiology of cancer of the colon and rectum. Cancer 1971; 28: 3-13.
[7]   Zaridge DG. Environmental etiology of large-bowel cancer. J Natl Cancer Inst 1983; 70: 389-400.
[8]   Jacobs LR. Enhancement of rat colon carcinogenesis by wheat bran consumption during the stage of 1,2-dimethyl hydrazine administration. Cancer Res 1983; 43: 4057-61.
[9]   Jensen OM, MacLennan R. Dietary factors and colorectal cancer in Scandinavia. Isr J Med Sci 1975; 15: 329-34.
[10]   MacLennan R, Jensen OM, Mosbech J *et al*. Diet, transit time, stool weight, and cancer in two Scandinavian populations. Am J Clin Nutr 1978; 31: S239-S242.
[11]   Graf E, Eaton JW. Dietary suppression of colonic cancer: Fiber or phytate? Cancer 1985; 56: 717-8.
[12]   Shamsuddin AM, Elsayed AM, Ullah A. Suppression of large intestinal cancer in F-344 rats by inositol hexaphosphate. Carcinogenesis 1988; 9: 577-80.
[13]   Shamsuddin AM, Ullah A. Inositol hexaphosphate inhibits large intestinal cancer in F344 rats 5 months following induction by azoxymethane. Carcinogenesis 1989; 10: 625-6.
[14]   Shamsuddin AM, Ullah A, Chakravarthy AK. Inositol and inositol hexaphosphate suppress cell proliferation and tumor formation in CD-1 mice. Carcinogenesis 1989; 10: 1461-3.
[15]   Shamsuddin AM, Vucenik I, Cole KE. IP$_6$: a novel anti-cancer agent. Life Sciences 1997; 61: 343-54.
[16]   Shamsuddin AM. Metabolism and cellular functions of IP$_6$: a review. Anticancer Research 1999; 19: 3733-6.
[17]   Elsayed A, Chakravarthy A, Shamsuddin A. Inositol hexaphosphate from corn decreased the frequency of colorectal cancer in azoxymethane-treated rats. Laboratory Investigation 1987; 56: 21A.
[18]   Elsayed A, Ullah A, Shamsuddin A. Post- initiation dietary supplementation with corn derived inositol hexaphosphate (IP$_6$) inhibits large intestinal carcinogenesis in F-344 rats. Federation Proceedings 1987; 46: 585.
[19]   Ullah A, Shamsuddin AM. Dose-dependent inhibition of large intestinal cancer by inositol hexaphosphate in F-344 rats. Carcinogenesis 1990; 11: 2219-22.

[20]   Baten A, Ullah A, Tomazic VJ, *et al*. Inositol phosphate induced enhancement of natural killer cell activity correlates with tumor suppression. Carcinogenesis 1989; 10: 1595-8.

[21]   Shamsuddin AM.: Reduction of Cell Proliferation and Enhancement of NK-Cell Activity, US5082833 (1992).

[22]   Vucenik I, Sakamoto K, Bansal M *et al*. Inhibition of rat mammary carcinogenesis by inositol hexaphosphate (phytic acid). A pilot study. Cancer Letters 1993; 75: 95-102.

[23]   Vucenik I, Yang GY, Shamsuddin AM. Inositol hexaphosphate and inositol inhibit DMBA-induced rat mammary cancer. Carcinogenesis 1995; 16: 1055-8.

[24]   Vucenik I, Yang GY, Shamsuddin AM. Comparison of pure inositol hexaphosphate and high-bran diet in prevention of DMBA-induced rat mammary carcinogenesis. ncer. Nutr Cancer 1997; 28: 7-13.

[25]   Nelson RU, Yoo SJ, Tanure JC *et al*. The effect of iron in experimental colorectal carcinogenesis. Anticancer Res 1989; 9: 1777-82.

[26]   Jariwalla RJ, Sabin R, Lawson S *et al*. Effects of dietary phytic acid (phytate) on the incidence and growth rate of tumors promoted on Fischer rats by a magnesium supplement. Nutrition Res 1988; 8: 813-7.

[27]   Hirose M, Ozaki K, Takaba K *et al*. Modifying effects of the naturally occurring antioxidants γ-oryzanol, phytic acid, tannic acid and n-tritriacontane-16, 18-dione in a rat wide-spectrum organ carcinogenesis model. Carcinogenesis 1991; 12: 1917-21.

[28]   Takaba K, Hirose M, Ogawa K *et al*. Modification of *N*-butyl-*N*-(4-hydroxylbutl) nitrosamine-initiated urinary bladder carcinogenesis in rats by phytic acid and salts. Food Chem Toxicol 1994; 32: 499-503.

[29]   Arnold JT, Wilkinson BP, Sharma S *et al*. Evaluation of chemopreventive agents in different mechanistic classes using a rat tracheal epithelial cell culture transformation assay. Cancer Research 1995; 55: 537-43.

[30]   Estensen RD, Wattenberg LW. Studies of chemopreventive effects of myo-inositol on benzo[a]pyrene-induced neoplasia of the lung and fore stomach of female A/J mice. Carcinogenesis 1993; 14: 1975-7.

[31]   Wattenberg LW, Estensen RD. Chemopreventive effects of myo-inositol and dexamethasone on benzo[a]pyrene   and   4-(methylnitrosoamino)-1-(3-pyridyl)-1-butanone-induced   pulmonary carcinogenesis in female A/J mice. Cancer Res 1996; 56: 5132-5

[32]   Estensen RD, Jordan MM, Wiedmann TS *et al*. Effect of chemopreventive agents on separate stages of progression of benzo[α]pyrene induced lung tumors in A/J mice. Carcinogenesis 2004; 25: 197-201.

[33]   Yang GY, Taboada S, Liao J. Inflammatory bowel disease: a model of chronic inflammation-induced cancer. Meth Molec Biol (Clifton, N.J.). 2009; 511: 193-233.doi: 10.1007/978-1-59745-447-6_9. PMID: 19347299 ISSN: 10643745

[34]   Seril DN, Liao J, West AB *et al*. High-iron diet: Foe or feat in ulcerative colitis and ulcerative colitis-associated carcinogenesis. J Clin Gastroenterol 2006; 40: 391-7.doi: 10.1097/00004836-200605000-00006. PMID: 16721219 ISSN: 01920790

[35]   Seril DN, Liao J, Ho K-LK *et al*. Dietary iron supplementation enhances DSS-induced colitis and associated colorectal carcinoma development in mice. Dig Dis Sci 2002; 47: 1266-78.doi: 10.1023/A: 1015362228659. PMID: 12064801

[36]   Liao J, Seril DN, Yang AL *et al*. Inhibition of chronic ulcerative colitis associated adenocarcinoma development in mice by inositol compounds. Carcinogenesis 2007; 28: 446-454.doi: 10.1093/carcin/bgl154. PMID: 16973672 ISSN: 01433334

CHAPTER 7

Experimental Cancer Regression by InsP_6 & Inositol

Abstract: Towards the eventual goal of translating the cancer inhibitory effect of inositol & InsP_6 seen in experimental animals *in vivo* to cancer prevention and therapy in humans, studies of human cancer cells lines were performed *in vitro*; and human cancer cells xenotransplanted in nude mice were used to test the therapeutic potential. A consistent and reproducible broad-spectrum anticancer action of InsP_6 ± inositol was observed in various models; in human breast cancer model InsP_6 showed synergism with standard chemotherapeutic agents Adriamycin or Tamoxifen. Suppression of cancer growth was seen in xenotransplanted rhabdomyosarcoma model and regression of preexisting cancer was seen in hepatoma model. InsP_6 ± inositol also inhibited various steps in cancer metastasis, inositol potentiated the action. These findings are compelling reasons for application of InsP_6+inositol for human cancer control.

Keywords: Adriamycin, CEA, chemotherapeutic agent, collagen, fibronectin, haptotactic migration, HepG2, hepatoma, hepatocellular carcinoma, laminin, MCF-7, MDA MB 231, metastasis, MMP, nude mice, Tamoxifen, TRAMP, synergism, xenotransplantation.

INTRODUCTION

In the previous chapter, the cancer preventive studies were discussed in experimental animals *in vivo*. In the post-initiation studies, administration of InsP_6 2 weeks or even 5 months after carcinogen administration (a time when 100% of the animals have multiple tumors) there was not only a reduction in the incidence of tumors concomitant to a reduction in cell division in the non-tumorous tissue, but the tumors in the InsP_6 ± inositol group were also approximately two-thirds smaller, strongly suggesting that InsP_6 ± inositol may have therapeutic efficacy as well [1, 2]. Thus, Shamsuddin laboratory embarked on the studies of experimental therapy; first on testing the efficacy against human cancer cell lines *in vitro*, and human cancer cells xenotransplanted in athymic nude mice. Whether inositol and InsP_6 affect the metastasis process was investigated in transplanted and metastatic cancers models *in vitro* and *in vivo*. Once again, the following descriptions are largely from the experiments from Shamsuddin laboratory as they were the earliest and dealt with the effects; later studies from Shamsuddin and others which mostly address mechanism are discussed in Chapters 17-21.

STUDIES OF CANCER CELL LINES *IN VITRO*

Cancer cell lines consist of populations of cancer cells that are removed from actual tumors and then cultured, maintained and propagated in a laboratory setting. *In vitro* studies were carried out on such cancer cell lines - originating both from humans and from rodents. $InsP_6$ consistently lowered the cell propagation rate of all the cell lines tested.

Hematopoietic Cells

The first studies were done on K-562 human erythroleukemia cell line with either continuous or non-continuous treatment with $InsP_6$. For continuous treatment (0.05% Na-$InsP_6$), cells were re-suspended in the media and left undisturbed until the next change of media every 3rd or 4th day at which time cells, both treated and control, were counted. Non-continuous treatments were done with 0.1% Na-$InsP_6$; after 1 hour of incubation, the cells were washed twice and cultured; the procedure being repeated every 3rd or 4th day [3].

Irrespective of whether it was a short-term non-continuous or continuous treatment, $InsP_6$ caused a 19-36% reduction in cell population ($p < 0.001$, Fig. **7.1**) with concomitant increased differentiation as evidenced by ultrastructural

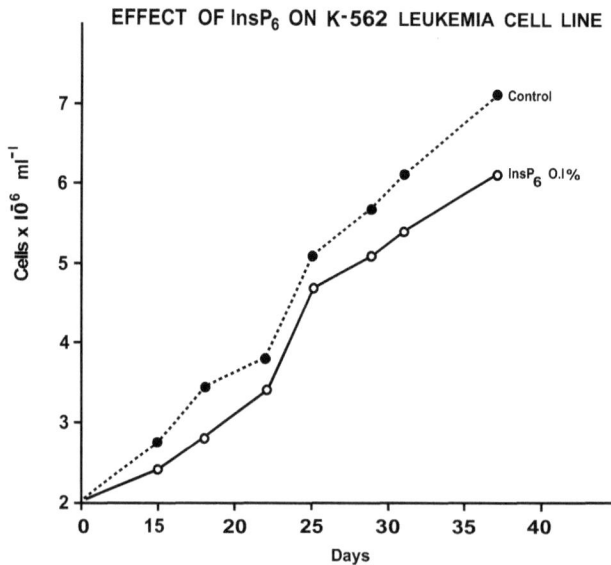

Fig. (7.1). Growth inhibition of K-562 human erythroleukemia cells with 0.1% Na-$InsP_6$.

morphology and increased hemoglobin content [3]. Following treatment with $InsP_6$, the concentration of intracellular Ca^{2+} ($[Ca^{2+}]_i$) was increased by 57% ($p <$ 0.02). Likewise, a 41% increase ($p < 0.05$) in $InsP_3$ and a 26% decrease ($p < 0.02$) in $InsP_2$ were noted 1 hour following treatment with $InsP_6$. Contrary to the dogma that cell division is associated with increased $[Ca^{2+}]_i$; paradoxically, there was reduced cell growth and enhanced differentiation that was associated with increased $[Ca^{2+}]_i$ and increased $InsP_3$ in the presence of $InsP_6$ [3]. Similar studies of growth inhibition were conducted in YAC-1 mouse lymphoma, and the HL-60 human leukemia lines with near identical results (unpublished observation).

Colon Cancer

$InsP_6$ inhibited the growth of HT-29 colon cancer cells at 0.33-20 mM concentrations in a dose-dependent manner ($p < 001$).

Inositol or inositol hexasulfate used as controls or media without $InsP_6$ did not show any suppressive effect. The expression of the epithelial differentiation marker cytokeratin, and 'tumor marker' carcinoembryonic antigen (CEA) were both augmented by either $InsP_6$ or inositol at all concentrations tested, although the degree of augmentation was milder with inositol than with $InsP_6$ (Figs. **7.2**, **7.3**). The expression of the tumor marker D-galactose-ß-[1→3]-N acetyl-D-galactosamine (Gal-GalNac) was augmented (100.7% increase) by low dose (0.66 mM) of $InsP_6$ but was subsequently suppressed with higher concentrations of it (Fig. **7.4**).

The combination of $InsP_6$+inositol both at 0.66 mM concentrations resulted in augmentation ($p < 0.001$) of cytokeratin expression, while that of CEA remained unchanged [4, 5].

While in this experiment the inhibitory effect of $InsP_6$ on cell proliferation was not altered by combination with additional inositol at any concentrations, Schröterová *et al.,* [6] reported otherwise. The investigators tested three different cell lines derived from human colon cancers of different stages of malignancy (HT-29, SW-480 and SW-620) with $InsP_6$ ± *myo*-inositol *in vitro* for inhibition of cell growth and induction of apoptosis. Their data showed an enhancement of proapoptotic effect of $InsP_6$ by inositol in all the cells lines studied [6]. $InsP_6$ and inositol at concentrations of 0.2, 1 and 5 mM were tested for 24, 48 and 72 hours.

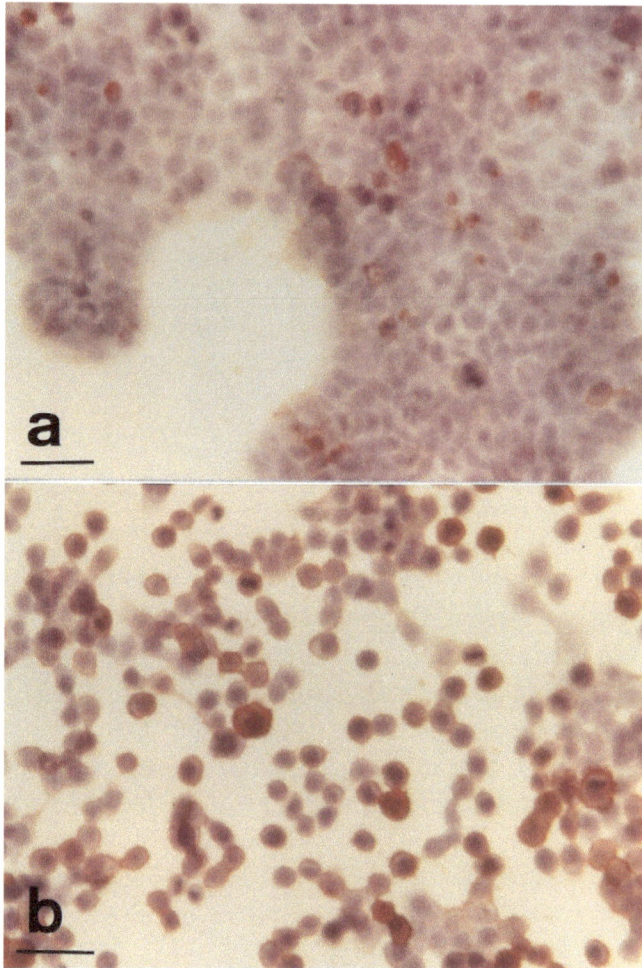

Fig. (7.2). Effect of InsP$_6$ on CEA expression in HT-29 human colon cancer cells. Immunocytochemical staining demonstrates a remarkable increase in CEA positive cells as indicated by brown coloration. (**a**) Control, (**b**) 1.0 mM InsP$_6$ for 48 hours; note the fewer number of cells as a result of InsP$_6$ treatment; scale bar 20 μm.

At all concentrations tested, InsP$_6$ ± inositol decreased proliferation of the cell lines, with the maximum decrease being observed in HT-29 cells. Their findings are quite intriguing as they contradict those of Shamsuddin and coworkers in line with the fact that inositol is essential for the growth of cells in artificial media and therefore was not expected to yield a synergistic reaction *in vitro* as seen *in vivo*. Further studies are needed to understand this enigmatic result.

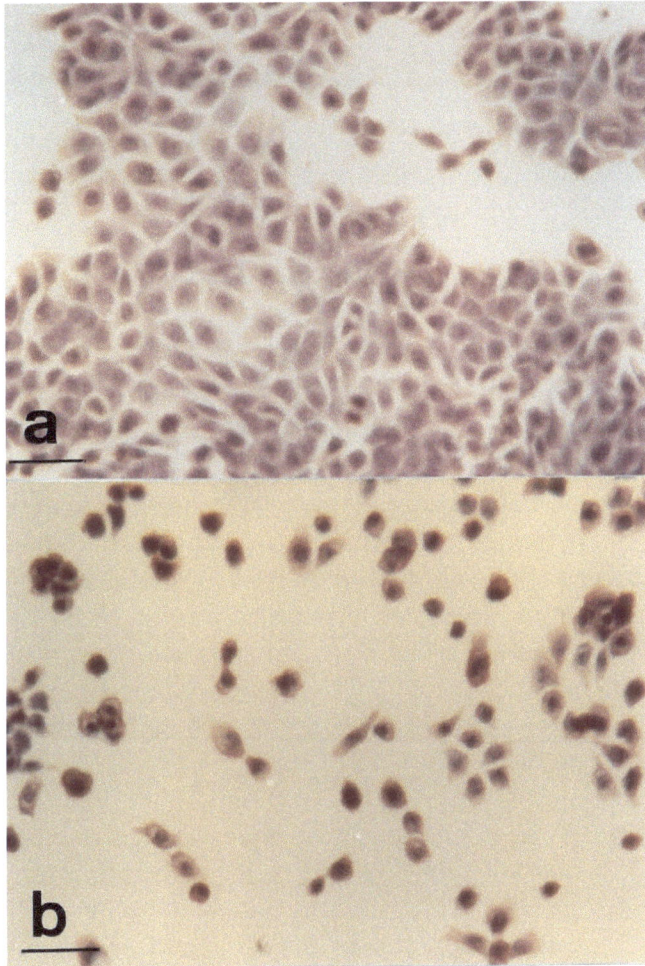

Fig. (7.3). Effect of $InsP_6$ on cytokeratin expression in HT-29 human colon cancer cells. Immunocytochemical staining demonstrates a marked increase in cytokeratin positive cells (brown color) (**a**) Control, (**b**) 1.0 mM $InsP_6$ for 48 hours; note the fewer number of cell which are also more rounded as a result of $InsP_6$ treatment; scale bar 20 μm.

DNA-synthesis was also suppressed by $InsP_6$ as early as 6 hour after treatment with 1 mM concentration ($p < 0.05$) and continued to 48 hour ($p < 0.01$). Measurement of intracellular inositol and its phosphates showed a 84-98% decrease of inositol, $InsP_1$ and $InsP_2$; $InsP_3$ was reduced by 39% and $InsP_4$ and $InsP_5$ by 21% and 13% respectively, whereas intracellular $InsP_6$ was increased by 24.6% at 5 min following ^{3}H-$InsP_6$ [5]. Alkaline phosphatase activity (brush border enzyme, associated with absorptive cell differentiation), increased following 1 and 5 mM $InsP_6$ treatment for 1-6 days. The expression of a mucin antigen associated with goblet cell differentiation and defined by the monoclonal antibody CMU10 was augmented ($p < 0.0001$) by $InsP_6$ [5].

The effects on differentiation, apoptosis *etc.*, are further discussed in the Chapter 19. Herein we will discuss the practical application of the differentiating feature insofar as tumor markers are concerned.

Modulation of Colon Cancer Markers by InsP$_6$: Practical Application

The disaccharide tumor marker Gal-GalNAc detected by the galactose oxidase-Schiff sequential reaction is not expressed by normal epithelial cells, but by cancerous and precancerous cells of colon, lung, breast *etc.* Studies by Sakamoto *et al.,* [4] and, Yang & Shamsuddin [5] reproducibly demonstrated that concomitant to reduced cell proliferation and enhanced differentiation of HT-29 colon cancer cells, the treated cells stopped producing Gal-GalNAc although the progression of mucin accumulation within the cells continued uninterrupted; *i.e.,* there was reversion of the cancer cells to normal phenotype by $InsP_6$. Because a) the tumor marker Gal-GalNAc is easily detected in rectal mucin of patients with colonic cancer and precancer with high sensitivity and specificity, and b) Gal-GalNAc is suppressed by $InsP_6$ treatment, it can be used to monitor the efficacy of chemoprevention by $InsP_6$.

Gal-GalNAc is expressed in the mucus samples from the colon (obtained by digital rectal examination) or bronchus (through coughed-up sputum) of a person afflicted with cancer. Periodic (re)testing for the marker Gal-GalNAc in a person with previous history of colon or lung cancer could give guidance as to the efficacy of therapy and relapse of the cancer. For instance, a person who showed a prior positive result for Gal-GalNAc, a negative result at a later time would be a good indication that he/she became free of the cancer. If a treatment with inositol+$InsP_6$ is indeed able to reduce the risk of cancer and to prevent its recurrence, it would be advantageous if $InsP_6$+inositol could at the same time suppress the expression of Gal-GalNAc. Support for such a strategy has been given by Vucenik *et al.,* [7].

One hundred and thirty seven rectal mucus samples of randomly selected patients with colorectal cancers were investigated for the expression of Gal-GalNAc by a sequential reaction of galactose oxidase and Schiff's reagent (GOS). Aside from a sensitivity of 100% and specificity of 96.8% for colorectal cancer, the marker was expressed as GOS positive test in 60% (32 of 53) of the samples collected from

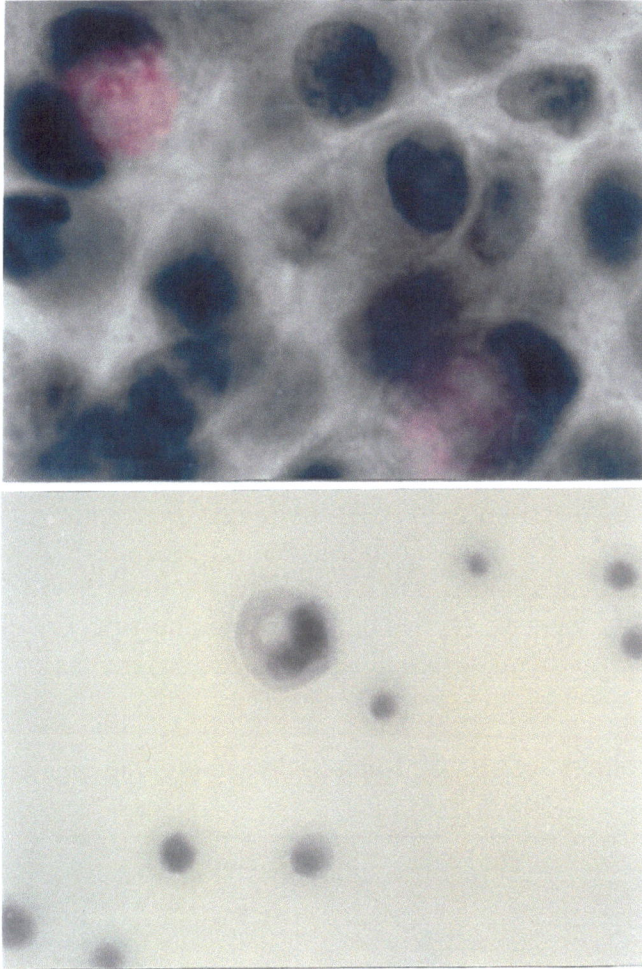

Fig. (7.4). HT-29 human colon cancer cells. Untreated control cells in the upper panel are stained with sequential galactose oxidase-Schiff reaction to identify the cancer marker Gal-GalNAc as seen by magenta colored mucin within the cancer cells [18]. $InsP_6$-treated cells in the lower panel show fewer cells (inhibition of cell growth) with concomitant reduction in cell size and production of mucin (vacuole in the center) yet, the mucin does not express the cancer marker Gal-GalNAc.

patients after tumor resection, showing the persistence of the biochemical changes even though the malignant tumors were removed. Five patients out of these 32 (16%) postoperative cases with positive GOS test had a tumor recurrence within a year $(0.05 < p < 0.10)$, suggesting that a persistently positive GOS test in this population may serve as a predictor of tumor recurrence. Gal-GalNAc thus is an early and intermediate biomarker, suitable not only for the detection of malignancy in its inception, but also for monitoring of people at high risk for cancer, and the efficacy of the cancer therapy as well as secondary prevention by this technology [7].

Prostate Cancer

In Vitro

PC-3 cells from prostate carcinoma require the presence of testosterone for their development. As in breast cancer, prostate cancer can also be treated by cutting off its source of hormones - for instance, by surgically removing the testes (orchiectomy), or by drugs which eliminate the male hormones.

A significant dose- and time-dependent growth inhibition of PC-3 cells was observed ($p < 0.05$) 24 hours after 1 mM $InsP_6$ treatment *in vitro* (Fig. **7.5**) [8]. DNA synthesis as determined by [^{3}H]-thymidine incorporation assay was also suppressed by $InsP_6$ in a dose-dependent manner, occurring as early as 3 hours after treatment and continuing up to 48 hours ($p < 0.01$ at 1 mM $InsP_6$; Fig. **7.6**). A nine to ten-fold increase ($p < 0.01$) in expression of HLA class I molecule associated with tumor immunosurveillance and cell differentiation was induced by $InsP_6$ (Fig. **7.7**).

Growth Inhibition of PC-3 Human Prostate Cancer Cells by InsP

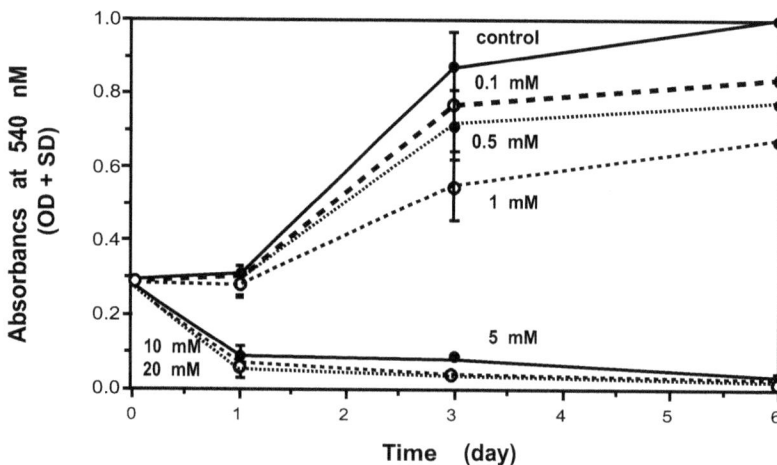

Fig. (7.5). Time- and dose-dependent growth inhibition of PC-3 prostate cancer cells tested by MTT incorporation assay. Note a significant ($p < 0.001$) inhibition with 0.1 mM $InsP_6$ after 3 days of treatment [8].

The marker for prostatic cell differentiation, prostate acid phosphatase, was significantly ($p < 0.05$) increased after 48 h treatment at 0.5-5 mM InsP_6 (Fig. **7.8**).

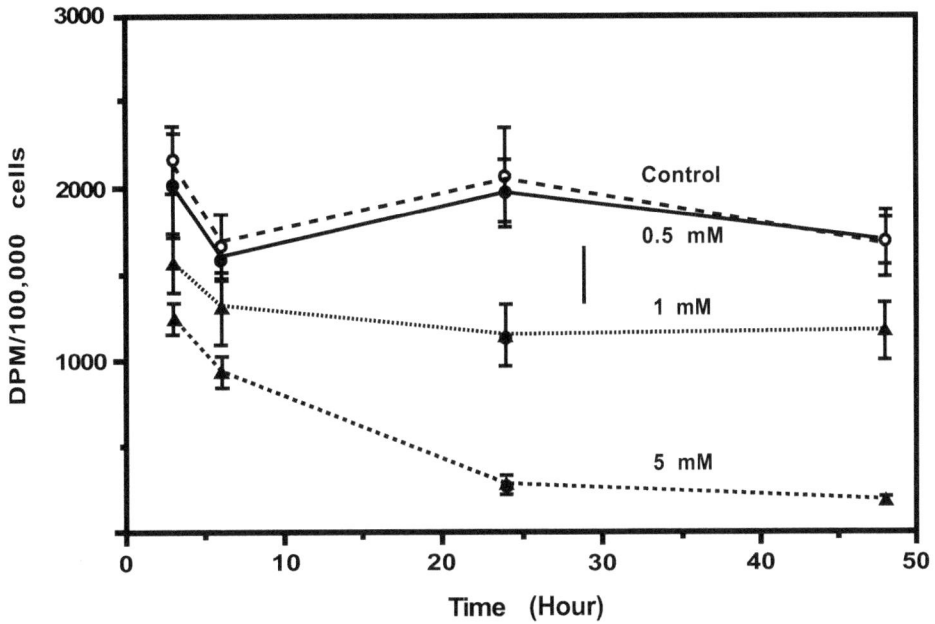

Fig. (7.6). DNA synthesis of PC-3 cells as measured by [^{3}H]thymidine incorporation assay. Significant inhibition started at 1 mM InsP_6 as early as 3 hours after treatment and continued for up to 48 hours [8].

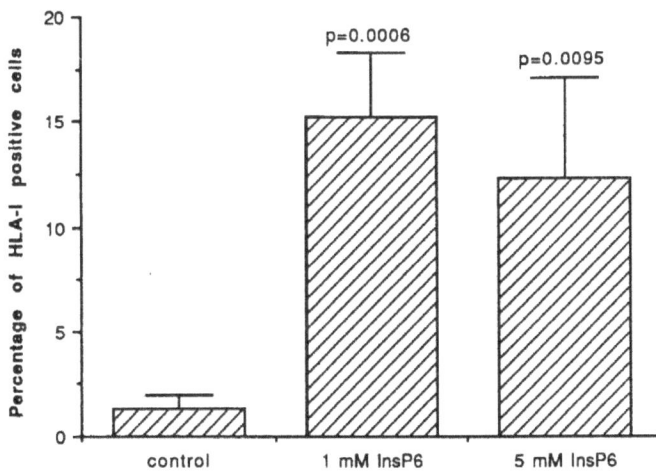

Fig. (7.7). Significant induction (10-fold increase $p < 0.001$) of HLA class I antigen is noted with InsP_6 treatment after 48 hours [8].

Of note is that compared to HT-29 colon cancer cells [4, 5], a more effective growth inhibition of prostate cancer cells was observed with $InsP_6$ treatment. While a significant growth inhibition of HT-29 cells was noted at ≥ 1 mM $InsP_6$, the same was achieved at 0.1 mM for PC-3 cells [8]. Additionally, dose- or time-dependent de-phosphorylation of $InsP_6$ by the enzyme phytase showed no statistically significant difference in growth, pointing to the fact that it is $InsP_6$ and not the dephosphorylated products (lower inositol phosphates) that caused significant growth inhibition of PC-3 cells, though there appeared to be trend of decreased cell growth with time (Fig. **7.9**).

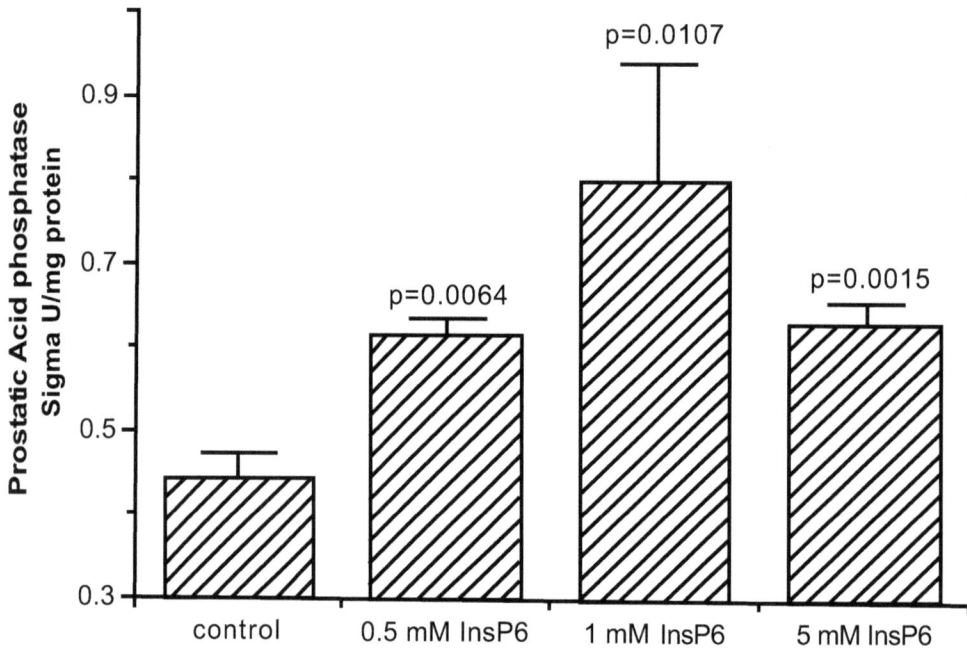

Fig. (7.8). Prostatic acid phosphatase activity in PC-3 cells is significantly increased after $InsP_6$ treatment [8].

Androgen-independent prostate cancers are harder to treat. Diallo *et al.*, had shown that treatment of PC-3 cells with $InsP_6$ induces the transcription of a subset of nuclear factor-κB (NF-κB)-responsive and pro-apoptotic BCL-2 family genes [9]. They then showed that although NF-κB subunits p50/p65 translocate to the nucleus of PC3 cells in response to $InsP_6$, inhibition of NF-κB-mediated transcription does not modulate $InsP_6$ sensitivity. There was enhanced killing of androgen-independent prostate cancer cells when $InsP_6$ is combined with proteasome inhibitors [10]. These investigators also studied whether the pH of

Time-dependent effect of Phytase Induced dephosphorylation
on InsP6 Inhibition of PC-3 cells

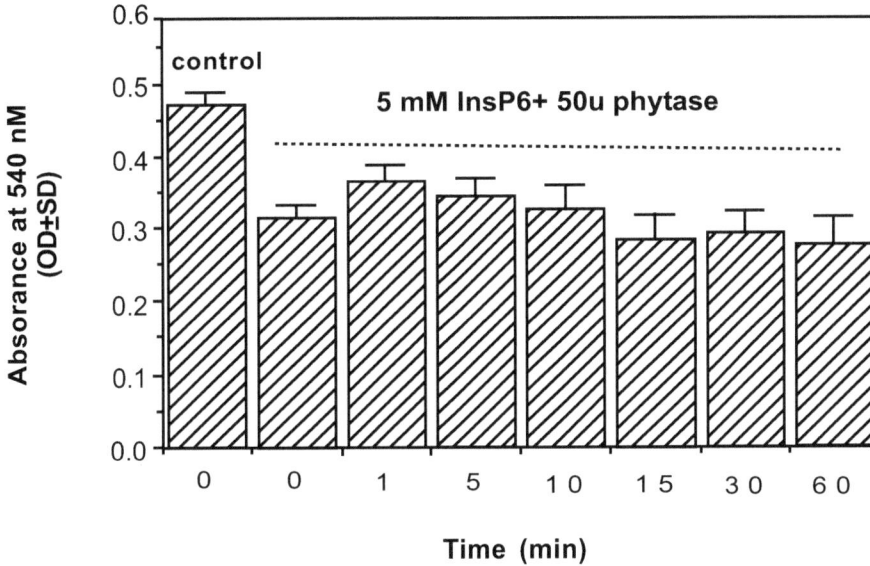

Fig. (7.9). Time-dependent effect of phytase-induced dephosphorylation of $InsP_6$ on PC-3 cell growth. Difference between control v $InsP_6$+phytase (50 U) at 0 time is significant at $p < 0.01$; however, reaction with phytase for 1-60 minutes did not significantly ($p > 0.05$) inhibit the cell growth though there appears to be a trend towards decreased cell growth [8].

$InsP_6$ could influence its anti-cancer activity *in vitro*. PC-3 cells were exposed to $InsP_6$ at pH 5, pH 7, and pH 12. $InsP_6$ at pH 5 and pH 12 were more potent at lowering the metabolic activity of PC-3 cells than at pH 7. Treatment with $InsP_6$ at pH 12 also caused the greatest inhibition in cellular proliferation, accumulation of cells at sub-G_1 and a reduction in phospho-AKT and phospho-PDK1 and up-regulated phospho-ERK. These data strongly suggest that the pH of $InsP_6$ modulates its anti-cancer activity. However, although $InsP_6$ at any pH reduced the metabolic rate and the proliferation of hormone-refractory PC-3 cells, it was at pH 5 and pH 12 that it offered the most significant inhibition on the metabolic rate of PC-3 cells; and the most significant *in vitro* reduced cellular proliferation was achieved at pH 12 [11]. The explanation for this is speculative at this time; clearly further investigations are needed to fully understand its significance insofar as the practical application of $InsP_6$ in administering to patients is concerned.

Extensive and ongoing investigation by Agarwal *et al.,* [12-16] has helped elucidate the molecular mechanisms of $InsP_6$'s anticancer action; these are discussed in Chapters 19-21. In here we describe the data germane to eventual clinical application.

Agarwal and coworkers showed that another androgen-dependent prostate cancer cell line LNCaP is also inhibited by $InsP_6$ [12] While PC-3 cells are androgen-dependent, DU 145 cells are androgen-independent; *in vitro*, $InsP_6$ inhibited both the anchorage-dependent and anchorage-independent growth of these cells [13]. Studies in the transgenic adenocarcinoma of mouse prostate (TRAMP) derived TRAMP-C1 cells showed that $InsP_6$ treatment of cells at 0.5-4.0 mM concentration for 24-72 hours inhibited the cell growth by 17-76%; and resulted in 6-35% cell death, in a dose- and time-dependent manner. $InsP_6$ treatment resulted in up to 92% cells in G_0-G_1 phase as compared to controls and induced a moderate to strong (up to 14-fold over control) apoptotic cell death [14].

In Vivo

In their effort towards therapeutic application of $InsP_6$ in humans, studies were extended to invasive human prostate cancer PC-3 and C4-2B cells, and PC-3 tumor xenograft growth in nude mice; $InsP_6$ treatment (2% in drinking water) inhibited tumor growth and reduced tumor weight by 52% to 59% ($p < 0.001$) [15]. Immunohistochemical analysis of xenografts showed that $InsP_6$ treatment significantly reduced the expression of molecules associated with cell survival/proliferation and angiogenesis together with an increase in apoptotic markers [15].

The investigators then employed anatomical and dynamic contrast-enhanced (DCE) magnetic resonance imaging (MRI) to investigate the efficacy of $InsP_6$ *in vivo* in a non-invasive manner in the TRAMP model. Male TRAMP mice, beginning at 4 weeks of age, were given 1%, 2%, or 4% (w/v) $InsP_6$ in drinking water and monitored using MRI [16]. Longitudinal assessment of prostate volumes by conventional MRI and tumor vascularity by gadolinium-based DCE-MRI showed a profound reduction in tumor size, partly due to the anti-angiogenic effects of $InsP_6$ treatment [16]. Interestingly, $InsP_6$ decreased the expression of glucose transporter GLUT-4 protein together with an increase in the levels of phospho-AMP-activated kinase in prostate tissues of mice. $InsP_6$ also significantly decreased glucose metabolism and membrane phospholipid synthesis, in addition to causing an increase in *myo*-inositol levels in the prostate [16].

In summary, InsP$_6$ has been demonstrated to reproducibly inhibit cell growth *in vitro* and block growth and angiogenesis of prostate cancer in different animal models *in vivo*.

Modulation of Prostate Cancer Markers by InsP$_6$

As shown in Figs. (**7.7**, **7.8**), there was an increased expression of the prostatic cell differentiation marker prostatic acid phosphatase as well as an increased expression of HLA class I antigen; the expression of the common prostatic cancer marker prostate specific antigen (PSA) was not tested. However, clinical data from patients indicate a reduction of PSA level following oral InsP$_6$+inositol (unpublished observation).

Breast Cancer

In the previous chapter (Chapter 6) we have discussed the results of *in vivo* studies of mammary carcinogenesis. Of direct relevance to the day to day activity in humans is the study comparing the efficiency of supplemental dietary fiber in the form of bran which exhibited a very modest, statistically non-significant inhibitory effect. In contrast, animals given equivalent amount of pure InsP$_6$ showed significant reduction in tumor number, incidence, and multiplicity [17]. This study clearly showed that pure InsP$_6$ is more effective than a high fiber diet in preventing experimental mammary tumors. Thus, for cancer prevention, prophylactic intake of InsP$_6$ may not only be more effective, but also more practical than gorging on large quantities of fiber.

Human Mammary Cancer Cell Lines In Vitro

As in prostate cancer and androgens, estrogen is a hormone believed to be of great significance in the genesis and treatment of breast cancer. Some breast cancers show a growth response when specific hormones are present. In order for a tumor to respond to a certain hormone, it must have a receptor that is specific to that hormone. Breast cancers are commonly tested for the presence (or absence) of estrogen or progesterone receptors (ER or PR). For instance, MCF-7 breast cancer cells are estrogen receptor-positive. In contrast, MDA MB-231 breast cancer cells are receptor-negative. InsP$_6$ has been shown to be equally successful in curtailing the growth of *both* of these cell lines.

A dose-dependent growth inhibition was observed in both MCF-7 (ER+) and MDA-MB-231 (ER-) by InsP$_6$. Statistically significant growth inhibition ($p <$

0.05) was observed starting at 1 mM $InsP_6$ as early as after the first day of treatment and continued up to 6 days for both the cell lines [18]. DNA synthesis in both the cell lines was suppressed by $InsP_6$ occurring as early as 3 hours after the beginning of treatment and continued up to 48 hours; significant inhibition ($p < 0.05$) started at 1 mM $InsP_6$ after 6 hours of treatment [18]. Compared to untreated cells, a 5-fold ($p < 0.05$) and 22-fold ($p < 0.01$) increase in expression of lactalbumin, associated with luminal cell differentiation was identified by immunocytochemistry after 48 hours of treatment with 1 and 5 mM $InsP_6$ [18]. This has been discussed additionally in Chapter 19.

Synergistic Action of $InsP_6$ with Standard Chemotherapeutic Agents

Combination of different chemotherapeutic agents is common in treatment of cancer of breast and other organs. The effect of $InsP_6$ on the standard anticancer chemotherapeutic drugs Tamoxifen and Adriamycin was tested on three human breast cancer cell lines: estrogen receptor (ER) α-positive MCF-7, ER α-negative MDA-MB 231 and Adriamycin-resistant MCF-7 (MCF-7/Adr) [19]. Compared to MCF-7 cells, MCF-7/Adr cells required a much lower concentration of $InsP_6$ to show growth inhibition; IC_{50} for MCF-7/Adr cells being 1.26 mM compared to 4.18 mM for MCF-7 cells. The ER-negative MDA-MB 231 cells were also highly sensitive to $InsP_6$ with IC_{50} being 1.32 mM. Growth suppression was markedly increased when $InsP_6$ was administered prior to the addition of Adriamycin, especially against MCF-7 cells ($p < 0.0001$) [19].

Synergism was also observed when $InsP_6$ was administered after Tamoxifen in all three cell lines studied (Fig. **7.10**). These data not only reconfirm that

$InsP_6$ alone inhibits the growth of breast cancer cells; but it also acts synergistically with Adriamycin or Tamoxifen, being particularly effective against ER α-negative cells and Adriamycin-resistant cell lines, thus offering itself as an adjuvant agent along with standard chemotherapeutic drugs.

Rhabdomyosarcoma

Rhabdomyosarcomas are highly aggressive cancers of muscle; they consist of cells that appear similar to primitive skeletal muscle-forming cells. In children and in young adults, these tumors are the most prevalent soft tissue sarcomas; typically they appear before the age of 20. Due to their aggressive nature, these

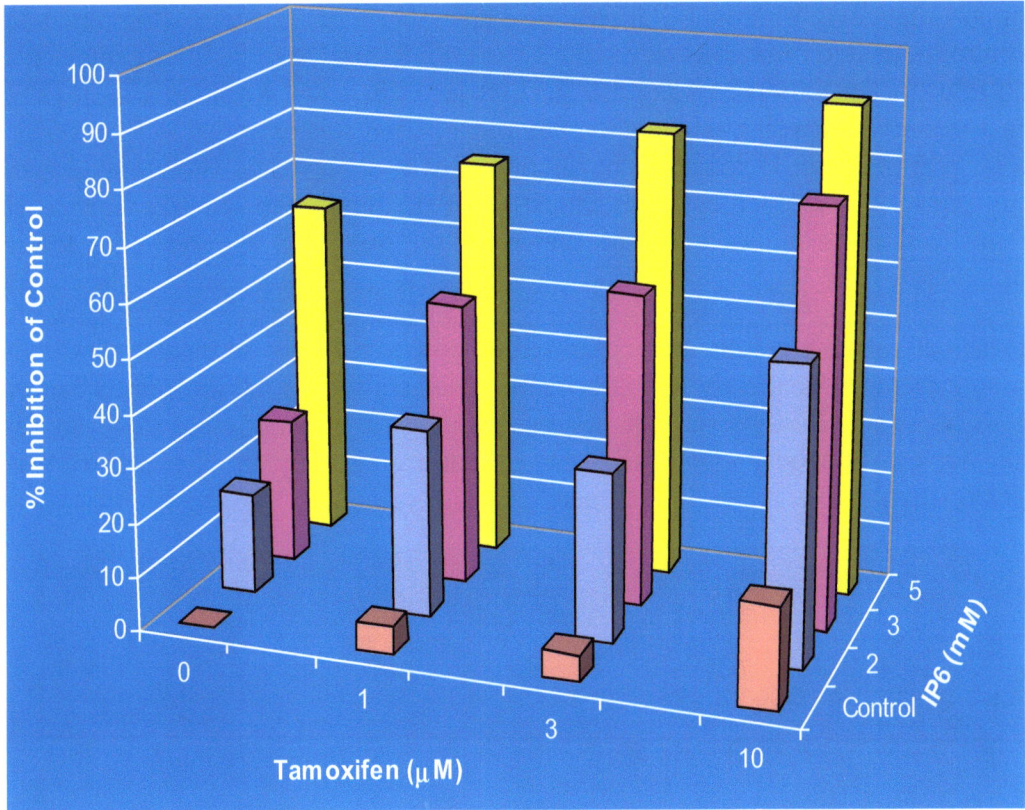

Fig. (7.10). Simultaneous treatment of $InsP_6$+Tamoxifen yielded synergism in MDA-MB 231 cells. Data points represent the mean of triplicates [19].

cancers are generally treated by combining radiation and chemotherapy with surgical approaches; however, in circumstances where the cancer has already metastasized, these conventional forms of therapy remain largely ineffective.

$InsP_6$ suppressed growth of human rhabdomyosarcoma cell line (RD) *in vitro* in a dose-dependent fashion. A 50% inhibition of cell growth (IC_{50}) was induced by < 1.0 mM $InsP_6$. Exposure of RD cells to $InsP_6$ led to differentiation; the cells became larger with abundant cytoplasm and expressing higher levels of muscle-specific actin. Consistent with *in vitro* observation, $InsP_6$ suppressed RD cell growth *in vivo*, in a xenografted nude mice model [20].

Six weeks old NIH athymic nude mice were inoculated with 10^7 RD cells subcutaneously in the dorsal region. Two days after inoculation of the RD cells, the mice received 40 mg/kg body weight of $InsP_6$ in 0.1 mL PBS around the site of inoculation (peritumoral) every other day for 2 or 5 weeks. Compared to control untreated mice receiving only 0.1 mL PBS, InsP_6-treated mice produced a 25 fold smaller tumors ($p = 0.008$), as observed after a two weeks treatment [20]. In a second experiment, wherein the treatment period was extended to five weeks, a 49 fold ($p = 0.001$) reduction in tumor size was observed in mice treated with InsP_6 (Fig. **7.11**). Histological examination of the tumor showed no evidence of tumor cell necrosis, however, the tumors of the treated group showed a lower mitotic rate [20].

Inasmuch as these results are exciting, they however do not address the issue of cancer therapy completely as InsP_6 treatment was given to the inoculated cancer cells and not fully developed tumor masses. Thus the next studies done on yet another cancer model - hepatocellular carcinoma whereby InsP_6 treatment was given after the tumors were developed.

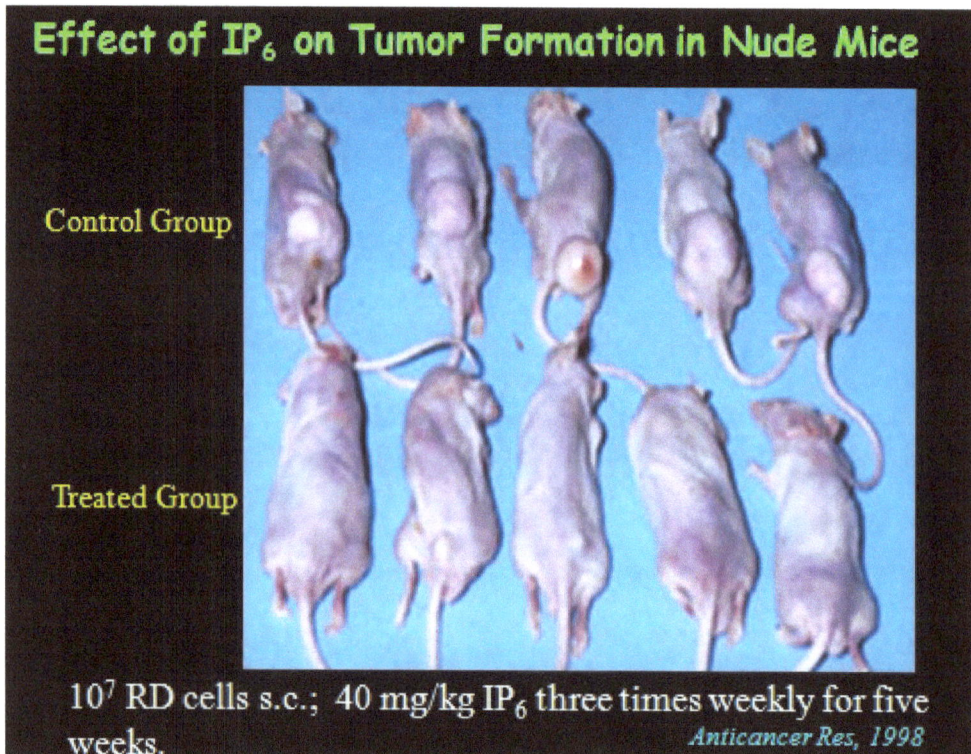

Fig. (7.11). Nude mice 40 days after subcutaneous inoculation of 10^7 RD cells and injected with 40 mg/kg InsP_6 thrice weekly for 5 weeks or only PBS as control. The tumors in the treated animals (lower row) are 49-fold smaller in aggregate than the control animals in upper row ($p = 0.001$) [20].

Liver Cancer

Liver cancer, also known as hepatocellular carcinoma (HCC) is common in areas where hepatitis B virus is widespread. In North America the incidence of liver cancer ranges between 3 - 7 cases per 100,000 people and is predominantly associated with liver cirrhosis - caused by alcohol, excessive iron, and other toxins. Irrespective of its cause, liver cancer is a malignant disease with only a very minimal chance for a positive outcome - death often occurs within months of the cancer's diagnosis. While numerous therapeutic approaches have been proposed, their effectiveness remains tentative.

Compared to most other cancer cell lines except PC-3 prostate cancer cells, human hepatocellular carcinoma HepG2 cells were quite sensitive to $InsP_6$, IC_{50} being 0.338 mM. Treatment with $InsP_6$ decreased the ability of HepG2 cells to form colonies, as assessed in the plating efficiency assay. Morphological changes induced by $InsP_6$ were consistent with differentiation of HepG2 cells.

Exposure of HepG2 cells to $InsP_6$ drastically decreased the rate of production of α-fetoprotein (AFP), a tumor marker of HCC in a dose-dependent manner (Table **7.1**). Furthermore, $InsP_6$ treatment caused a decreased expression of mutant *p53* protein in HepG2 cells, with no significant change in the expression of wild-type *p53*. The expression of *p21*WAF1 protein was increased by 1.5 fold [21].

In subsequent experiments, HepG2 cells were treated with a single exposure to 5.0 mM $InsP_6$ *in vitro* and after 48 hours they were inoculated into athymic nude mice (10^7 cells per mouse) subcutaneously. No tumor was found in mice which had received HepG2 cells pretreated with $InsP_6$ whereas 71% of mice receiving the same number of control untreated HepG2 cells developed solid tumors at the transplantation site ($p < 0.03$). For a tumor suppression/regression study, when the transplanted tumors reached 8-10 mm in diameter, intra-tumoral injection of $InsP_6$ (40 mg/kg) was given for 12 consecutive days, after which the animals were sacrificed. At autopsy, the tumor weight in $InsP_6$ -treated mice was 86% to 1180% (340% average) less than that in control mice (0.33 +/- 0.12 g *versus* 1.13 +/- 0.25 g, $p = 0.016$) [22]. Two mice, a control untreated and a mouse treated with $InsP_6$ revealing tumor regression can be seen in Fig. (**7.12**).

This study not only showed unequivocal regression of preexisting human cancers xenotransplanted in nude mice, but also demonstrated that a single $InsP_6$ treatment totally suppressed the tumorigenicity of the cancerous cells.

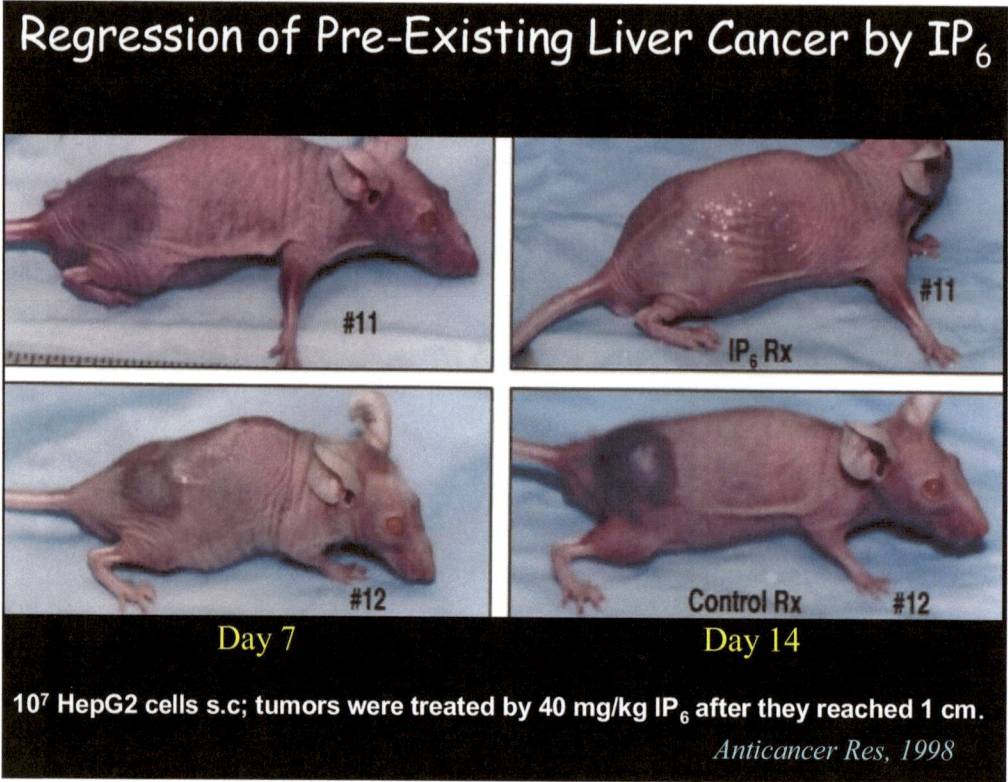

Fig. (7.12). The tumor in mouse #11 has totally regressed in a week during the period its sibling #12 has 2-3 fold larger tumor [22].

Monitoring the Efficacy of InsP$_6$ Treatment by Tumor Marker

The tumor marker α-fetoprotein (AFP) is in standard use to monitor liver cancer - increased amount of it in the blood is a sign of increased tumor load, and *vice versa*. HepG2 cells as in liver cancer in a patient produce AFP. As noted above and can be seen in Table **7.1**, the suppression of AFP secretion by HepG2 cells

Table 7.1. Effect of InsP_6 treatment on α-fetoprotein (AFP) secretion in HepG2 cells.

Treatment	AFP in Media ng/mL	AFP in Cells pg/Cell
0 InsP_6 (Control)	1,3790 ± 65.7	27.6 ± 1.3
0.5 mM InsP_6	63.0 ± 24.0	5.3 ± 2.0
1.0 mM InsP_6	1.9 ± 0.2	0.3 ± 0.0

Difference between control *v* Treatment groups is significant at $p < 0.0001$ [21].

following InsP_6 treatment was rather dramatic; there was a virtual shut-down of AFP production. Thus, the efficacy of InsP_6 treatment can be monitored by routine periodic follow-up of AFP.

Miscellaneous Cancers

The broad-spectrum anti-cancer action of InsP_6 ± Inositol has been further proven by the demonstration of growth inhibition of skin cancer melanoma [23, 24], brain tumor glioblastoma multiforme [25], oral cavity squamous cell carcinoma [26], pancreas [27] *etc.*, *in vitro*.

Experimental Models of Metastatic Cancer

The process of cancer cell metastasis involves the detachment of cells, invasion into adjacent blood vessels and lymphatic channels, and establishing a new home at the distant site. Once they are in the capillaries at a distant site, they may leave the vasculature and infiltrate the target organ. This complex process depends on the ability of cancer cells to adhere, migrate, and invade. The adhesion of cancer cell surface integrin receptors to the extracellular matrix (ECM) results in integrin clustering and subsequent intracellular signal transduction, which leads to cytoskeletal rearrangements necessary for cellular motility. Therefore, adhesion of cells to the ECM is essential for migration besides cell growth, maturation *etc.*

Cancer cell migration depends on the ability of cells to interact with the ECM for traction. Migrating cells form special lamellipodia structures at the leading edges. Lamellipodia contain orthogonal cross-weave patterns of actin filaments with membrane ruffling that are usually present at the leading edge of motile cells. The traction and contraction of actin filaments move the cells along the ECM. However, for cells to invade through the basement membrane, they need to secrete proteolytic enzymes, such as the matrix metalloproteinases (MMPs), to digest through the surrounding matrices.

InsP_6's role in new blood vessel development (angiogenesis) is described in Chapter 21. Discussed here are the early and late events and the outcome of InsP_6 treatment *in vitro* and *in vivo*.

In Vitro

The potential of InsP_6 to inhibit cell adhesion, migration and invasion - the key steps in cancer metastasis were studied on the highly invasive estrogen receptor negative MDA-MB231 human mammary cancer cell line [28].

InsP_6 treatment caused a 65% reduction of cell adhesion to fibronectin ($p = 0.002$) and a 37% reduction to collagen ($p = 0.005$). To determine whether a decrease in cell adhesion leads to a decrease in cell motility, migration assays were performed; InsP_6 decreased both the number of migrating cells and the distance of cell migration into the denuded area by 72% ($p < 0.001$). Haptotatic cell migration (directional motility) in a modified Boyden chambers was also reduced in a dose-dependent manner [28]. While cell migration on fibronectin was inhibited by 65% ($p < 0.001$, Fig. **7.13**), migration on collagen and laminin was decreased by 32% ($p < 0.01$) and 13% ($p < 0.05$), respectively. Immunocytochemistry revealed the absence of lamellipodia structure in InsP_6-treated cells as compared to untreated cells, corresponding to a diminished ability of cancer cells to form cellular network as determined by Matrigel outgrowth assay. Likewise, cell invasion also was decreased (by 72% after InsP_6 treatment, $p = 0.001$) in a dose-dependent fashion. InsP_6 significantly ($p = 0.006$) inhibited the secretion of matrix metalloproteinase (MMP) MMP- 9 as assessed by zymography [28].

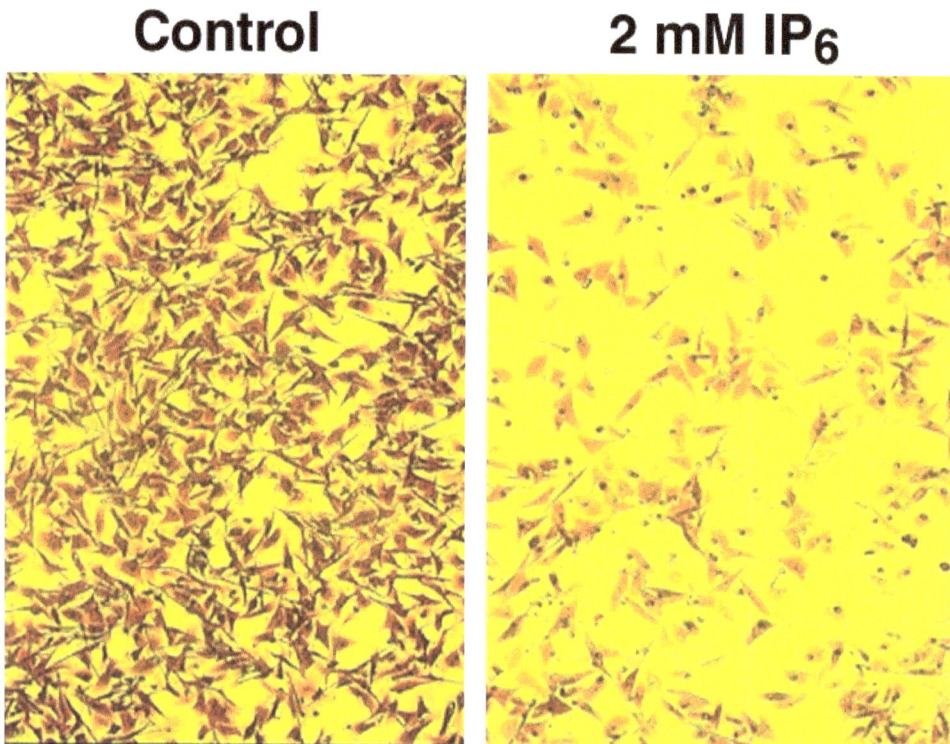

Control   **2 mM IP$_6$**

Fig. (7.13). Haptotactic migration of MD-MB231 on the proteins of extracellular matrix; InsP_6 treatment inhibits the migration of cancer cells on fibronectin matrix by 65% ($p < 0.001$, right panel) [28].

InsP_6 treatment also caused a significant ($p < 0.005$) decrease in the expression of integrin heterodimers α2β1 collagen receptor, α5β1 fibronectin receptor and αvβ3 vitronectin receptor. There was a dramatic 82% decrease in the expression of α5β1 on α2β1-treated cells ($p < 0.0001$), indicating a decrease in cell surface expression of the heterodimers [28].

No effect was seen when inositol hexasulfate, an analogue of InsP_6 was used as a control. Immunocytochemistry showed a lack of clustering of paxillin; tyrosine-phosphorylated proteins in InsP_6-treated cells were discontinuous and scattered around the cell periphery, whereas the patterns were more dense and localized in control cells. Consistent with these observations, focal adhesion kinase (FAK) autophosphorylation at tyrosine-397 residue was suppressed, albeit modestly, by InsP_6 treatment, suggesting a down-regulation in the integrin-mediated signaling pathway [29]. These results show that InsP_6 inhibits the metastasis of human breast cancer cells *in vitro* by affecting cancer cell adhesion, migration and invasion.

In Vivo

In the previous studies of cancer regression, InsP_6 was injected peritumorally. A yet different route of administration was tested in a transplanted and metastatic tumor model. Tumors were developed in the animals by injecting them subcutaneously with mouse fibrosarcoma (FSA-1) cells - cancers of mesenchymal (non-epithelial) cell origin, as opposed to the common cancers of lung, colon, breast, stomach, ovary *etc.*, which are of epithelial origin. FSA-1 cells also having the ability to metastasize served as a model to expand the study towards the eventual goal of human therapeutics. There were two parts to this pilot study: first the effect of InsP_6 treatment on subcutaneously transplanted tumor cells and the survival of mice; the second investigation was the effect of InsP_6 ± inositol on metastatic cancer [30].

Transplanted Tumor and Mouse Survival Study

C3H/JSed mice were injected subcutaneously with 5×10^6 FSA-1 cells. Animals were treated with 1 mL injections of 0.25% Na-InsP_6 intraperitoneally every other day, and the size of subcutaneous tumors was also measured every other day. A statistically significant inhibition of tumor sizes ($p < 0.001$), as well as an improvement of the animals' survival time ($p < 0.05$) in the InsP_6-treated mice were observed in comparison to those animals that had not received any InsP_6 [30].

Metastatic Tumor Study

For metastasis study, 10^5 FSA-1 cells were injected intravenously, which developed into metastatic lesions in the mice's lungs. The animals were treated daily with $InsP_6$, inositol or $InsP_6$ + inositol *via* intraperitoneal injection and were sacrificed on day 28.

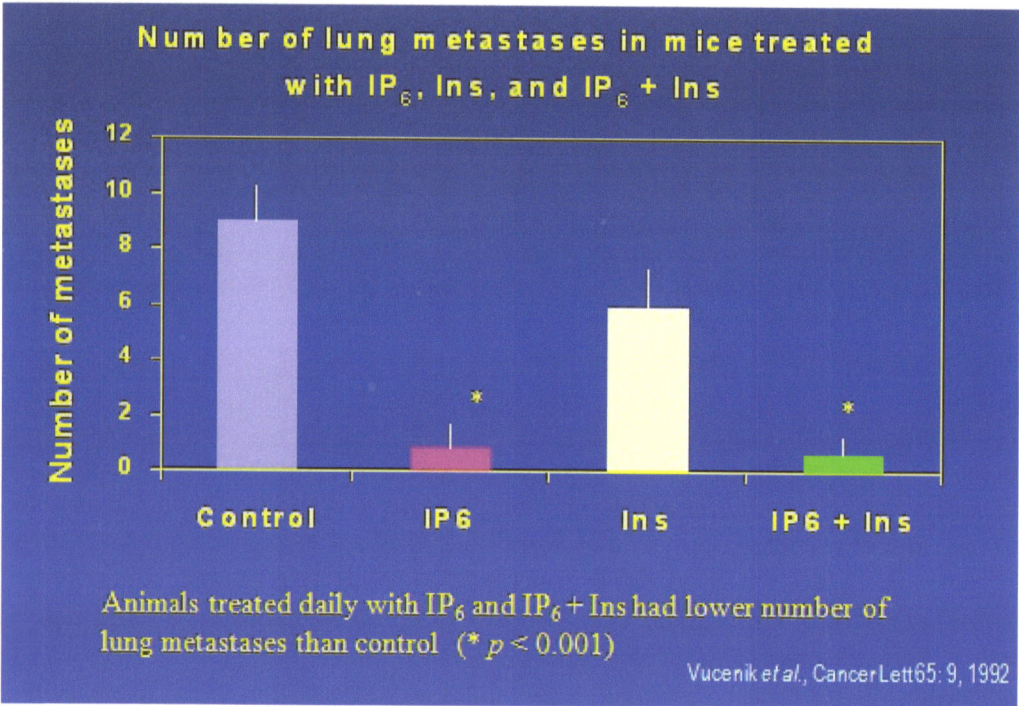

Number of lung metastases in mice treated with IP_6, Ins, and IP_6 + Ins

Animals treated daily with IP_6 and IP_6 + Ins had lower number of lung metastases than control ($* p < 0.001$)

Vucenik *et al.*, CancerLett 65: 9, 1992

Fig. (7.14). Bar graph showing the numbers of pulmonary metastasis in various groups of treated mice and untreated controls [30].

Animals treated with $InsP_6$ or $InsP_6$+inositol had significantly lower number of pulmonary metastasis than the saline-treated control ($p < 0.001$); inositol treatment did not result in a statistically significant reduction (Fig. **7.14**). Likewise, though the combination of $InsP_6$ + inositol yielded the best results, the difference with $InsP_6$-alone treatment was not significant [30].

CONCLUDING REMARKS

Consistent and reproducible broad-spectrum anticancer action of $InsP_6$ ± inositol was observed in various *in vitro* and *in vivo* models of human cancers.

Suppression of cancer growth as well as regression of preexisting cancer was seen in rhabdomyosarcoma and hepatoma models. InsP_6 ± inositol also inhibited various steps in cancer metastasis and concomitant to an inhibition of cancer cell proliferation caused a reduction of cancer-associated tumor marker expression; the latter offers the potential for following up patients with periodic marker evaluation for efficacy of therapy. The data strongly establish InsP_6 ± inositol as broad-spectrum anticancer agents, both for prevention and therapy of human cancers.

REFERENCES

[1]   Shamsuddin AM, Ullah A. Inositol hexaphosphate inhibits large intestinal cancer in F344 rats 5 months following induction by azoxymethane. Carcinogenesis 1989; 10: 625-6.
[2]   Elsayed A, Ullah A, Shamsuddin A. Post- initiation dietary supplementation with corn derived inositol hexaphosphate (IP$_6$) inhibits large intestinal carcinogenesis in F-344 rats. Federation Proceedings 1987; 46: 585.
[3]   Shamsuddin AM, Baten A, Lalwani ND. Effects of inositol hexaphosphate on growth and differentiation in K-562 erythroleukemia cell line. Cancer Letters 1992; 64: 195-202.
[4]   Sakamoto K, Venkatraman G, Shamsuddin AM. Growth inhibition and differentiation of HT-29 cells *in vitro* by inositol hexaphosphate (phytic acid). Carcinogenesis 1993; 14: 1815-9.
[5]   Yang GY, Shamsuddin AM. IP$_6$-induced growth inhibition and differentiation of HT-29 human colon cancer cells: involvement of intracellular inositol phosphates. Anticancer Res 1995; 15: 2479-87.
[6]   Schröterová L, Hasková P, Rudolf E, *et al*. Effect of phytic acid and inositol on the proliferation and apoptosis of cells derived from colorectal carcinoma. Oncology Reports 2010; 23: 787-93.
[7]   Vucenik I, Gotovac J, Družijanić N *et al*. Usefulness of galactose oxidase-Schiff test in rectal mucus for screening of colorectal malignancy. Anticancer Res 2001; 21: 1247-56.
[8].   Shamsuddin A M, Yang GY. Inositol hexaphosphate inhibits growth and induces differentiation of PC-3 human prostate cancer cells. Carcinogenesis 1995; 16: 1975-9.
[9]   Diallo JS, Péant B, Lessard L *et al*. An androgen-independent androgen receptor function protects from inositol hexaphosphate toxicity in PC3/PC3AR prostate cancer cell line. Prostate 2006; 66: 1245-56.
[10]   Diallo JS, Betton B, Parent N *et al*. Enhanced killing of androgen-independent prostate cancer cells using inositol hexakisphosphate in combination with proteasome inhibitors. Brit J Cancer 2008; 99: 1613-22.
[11]   Betton B, Gannon PO, Koumakpayi IH *et al*. Influence of pH on the cytotoxic ability of inositol hexakisphosphate (IP6) on prostate cancer. Fron Oncol Oct 31; 1: 40. doi: 10.3389/fonc.2011.00040. eCollection 2011.
[12]   Agarwal C, Dhanalakshmi S, Singh RP *et al*. Inositol hexaphosphate inhibits growth and induces G1 arrest and apoptotic death of androgen-dependent human prostate carcinoma LNCaP cells. Neoplasia 2004; 6: 646-59.
[13]   Zi X, Singh RP, Agarwal R. Impairment of erbB1 receptor and fluid-phase endocytosis and associated mitogenic signaling by inositol hexaphosphate in human prostate carcinoma DU145 cells. Carcinogenesis 21: 2225-2235, 2000.
[14]   Sharma G, Singh RP, Agarwal R. Growth inhibitory and apoptotic effects of inositol hexaphosphate in transgenic adenocarcinoma of mouse prostate (TRAMP-C1) cells. Int J Oncol 2003; 23: 1413-8.
[15]   Gu M, Roy S, Raina K *et al*. Inositol hexaphosphate suppresses growth and induces apoptosis in prostate carcinoma cells in culture and nude mouse xenograft: PI3K-Akt pathway as potential target. Cancer Res 2009; 69: 9465-72. doi: 10.1158/0008-5472.CAN-09-2805.
[16]   Raina K, Ravichandran K, Rajamanickam S *et al*. Inositol hexaphosphate inhibits tumor growth, vascularity, and metabolism in TRAMP mice: a multiparametric magnetic resonance study. Cancer Prev Res 2013 Jan; 6(1): 40-50. doi: 10.1158/1940-6207.CAPR-12-0387. Epub 2012 Dec 4

[17]   Vucenik I, Yang GY, Shamsuddin AM. Comparison of pure inositol hexaphosphate and high-bran diet in prevention of DMBA-induced rat mammary carcinogenesis. Nutr Cancer 1997; 28: 7-13.

[18]   Shamsuddin AM, Yang G-Y, Vucenik I. Novel anti-cancer functions of IP$_6$: Growth inhibition and differentiation of human mammary cancer cell lines *in vitro*. Anticancer Res 1996; 16: 3287-92.

[19]   Tantivejkul K, Vucenik I, Eiseman J *et al*. Inositol hexaphosphate (IP$_6$) enhances the anti-proliferative effects of adriamycin and tamoxifen in breast cancer. Breast Cancer Res Treat 2003; 79: 301-12.

[20]   Vucenik I, Kalebic T, Tantivejkul K *et al*. Novel anticancer function of inositol hexaphosphate: inhibition of human rhabdomyosarcoma *in vitro* and *in vivo*. Anticancer Res 1998; 18: 1377-84.

[21]   Vucenik I, Tantivejkul K, Zhang ZS *et al*. IP$_6$ in treatment of liver cancer. I. IP6 inhibits growth and reverses transformed phenotype in HepG2 human liver cancer cell line. Anticancer Res 1998; 18: 4083-90.

[22]   Vucenik I, Zhang ZS, Shamsuddin AM. IP$_6$ in treatment of liver cancer. II. Intratumoral injection of IP$_6$ regresses pre-existing human liver cancer xenotransplanted in nude mice. Anticancer Res 1998; 18: 4091-6.

[23]   Rizvi I, Riggs DR, Jackson BJ *et al*. Inositol hexaphosphate (IP6) inhibits cellular proliferation in melanoma. Journal of Surgical Research 2006; 133: 3-6.

[24]   Schneider JG, Alosi JA, McDonald DE *et al*. Effect of pterostilbene on melanoma alone and in synergy with inositol hexaphosphate. Am J Surg 2009; 198: 679-84.

[25]   Karmakar S, Banik NL, Ray SK. Molecular mechanism of inositol hexaphosphate-mediated apoptosis in human malignant glioblastoma T98G cells. Neurochem Res 2007; 32: 2094-102.

[26]   Janus SC, Weurtz, B, Ondrey FG. Inositol hexaphosphate and paclitaxel: symbiotic treatment of oral cavity squamous cell carcinoma. Laryngoscope 2007; 117: 1381-8.

[27]   Somasundar P, Riggs DR, Jackson BJ *et al*. Inositol hexaphosphate (IP6): a novel treatment for pancreatic cancer. J Surgical Res 2005; 126: 199-203.

[28]   Tantivejkul K, Vucenik I, Shamsuddin AM. Inositol hexaphosphate (IP$_6$) inhibits key events of cancer metastasis: I. *In vitro* studies of adhesion, migration and invasion of MDA-MB 231 human breast cancer cells. Anticancer Res 2003; 23: 3671-9.

[29]   Tantivejkul K, Vucenik I, Shamsuddin AM. Inositol hexaphosphate (IP$_6$) inhibits key events of cancer metastasis: II. Effects on integrins and focal adhesions. Anticancer Res 2003; 23: 3681-9

[30]   Vucenik I, Tomazic VJ, Fabian D *et al*. Antitumor activity of phytic acid (inositol hexaphosphate) in murine transplanted and metastatic fibrosarcoma, a pilot study. Cancer Letters 1992; 65: 9-13.

CHAPTER 8

Clinical Studies of Human Cancer Treatment

Abstract: Despite compelling experimental evidence of the efficacy of $InsP_6$ + inositol as an anticancer cocktail, there has been very limited enthusiasm for clinical studies of cancer prevention and/or therapy. Notwithstanding this lack of interest, results from the handful of clinical studies are extremely encouraging. *Myo*-inositol with or without $InsP_6$ has shown to be effective in preventing precancerous lesions of lung cancer. Experimentally, $InsP_6$ acts synergistically with standard chemotherapeutic drugs such as doxorubicin and tamoxifen; emerging clinical data support that. $InsP_6$ + inositol available as dietary supplement has been reported to cause abrogation of breast, lung and colorectal cancers, with or without standard therapies; reduce chemotherapy-induced negative side-effects and provide a better quality of life.

Keywords: Alopecia, beta-glucan, breast cancer, chemotherapy, colon cancer, lung cancer, polychemotherapy, quality of life, radiotherapy.

INTRODUCTION

The preceding chapters have shown that the combination of $InsP_6$ + inositol has consistent and reproducible anticancer action against a variety of malignancies. This has been independently demonstrated by investigators worldwide, the lack of funding support and, more enthusiastic and productive research notwithstanding. The enthusiasm of clinical investigators to translate these most encouraging laboratory data to the advantage of patients in the clinic has also been virtually non-existent; once again, the lack of interest on the part of funding agencies to make $InsP_6$ + inositol a priority may be the critical factor. In any event, the combination of $InsP_6$ + inositol available as a dietary supplement since 1998 is being used worldwide with very positive outcome as evidenced by patient self-reporting on the Internet. But that does not establish it as a scientific fact. Described herein are the results of the limited clinical studies of $InsP_6$ + inositol in cancer patients.

WHAT SHOULD BE THE PROPERTIES OF GOOD ANTICANCER AGENTS?

"A good anticancer agent needs to be selective: it should only affect malignant cells and spare normal cells and tissues." This property was shown for $InsP_6$. When the fresh CD34$^+$ cells from bone marrow was treated with different doses of $InsP_6$, an inhibition of growth was observed that was specific to leukemic

A.K.M. Shamsuddin and Guang-Yu Yang

progenitors from chronic myelogenous leukemia patients, but no cytotoxic or cytostatic effect was observed on normal bone marrow progenitor cells under the same conditions [1]. Likewise, $InsP_6$ inhibited the colony formation of Kaposi Sarcoma cell lines, KS Y-1 (AIDS-related KS cell line) and KS SLK (Iatrogenic KS) and CCRF-CEM (human adult T lymphoma) cells in a dose-dependent manner; however, in striking contrast to the anti-cancer drug taxol used as a control, $InsP_6$ did not affect the ability of normal cells (peripheral blood mononuclear cells and colony-forming T cells) to form colonies in a semisolid methylcellulose medium [2].

"Malignant and normal cells are known to have different metabolisms, growth rate, expression of receptors, *etc.*, but the mechanism for this differential selectivity of $InsP_6$ for normal and malignant cells needs to be further investigated.

Another important aspect of cancer treatment is overcoming acquired drug resistance." Studies in Shamsuddin laboratory demonstrate that *in vitro* $InsP_6$ acts synergistically with doxorubicin and tamoxifen, being particularly effective against estrogen receptor-negative and doxorubicin-resistant tumor cell lines, both conditions that are challenging to treat [3]. These data are particularly important, as tamoxifen is usually given as a chemopreventive agent in the post-treatment period, and doxorubicin has enormous cardiotoxicity and its use is associated with doxorubicin resistance. Emerging clinical data (*vide infra*) support that $InsP_6$ + inositol when used in combination with standard chemotherapeutic agents augment the latter's ability to control tumor growth.

CLINICAL STUDIES OF InsP_6 & INOSITOL

Myo-Inositol

Myo-Inositol has been considered to be safe and essential enough to be included not only as a member of B-vitamin family, but is a necessary component of baby formulas as well as growth media for keeping the cells alive (culturing) in the laboratory. It has already been in therapeutic use, in humans in large dosage (up to 12 g/day) for certain neuropsychiatric disorders (Chapter 13).

Lam and colleagues from the British Columbia Cancer Agency, Vancouver, Canada, and Universities of British Columbia and Minnesota, and the US National Cancer Institute in Bethesda, Maryland report on a Phase I, open-label, multiple dose, dose-escalation clinical study of *myo*-inositol [4]. They assessed the safety, tolerability, maximum tolerated dose, and potential chemopreventive effect of *myo*-inositol in smokers with bronchial dysplasia - a precursor of lung cancer. A dose escalation study ranging from 12 to 30 g/day of *myo*-inositol for a month was first conducted in 16 subjects to determine the maximum tolerated dose. Ten new subjects were then enrolled to take the maximum tolerated dose for 3 months. The researchers also investigated the potential chemopreventive effect of *myo*-inositol by repeat auto-fluorescence bronchoscopy and biopsy. The maximum tolerated dose was found to be 18 g/day. Side effects, when present, were mild and mainly gastrointestinal in nature. A significant increase in the rate of regression of preexisting dysplastic lesions was observed (91% *versus* 48%; P = 0.014). As side benefit, statistically significant reduction in the systolic and diastolic blood pressures by an average of 10 mm Hg was observed after taking 18 g/d of *myo*-inositol for a month or more [4].

Gutafson and colleagues at the Boston University Medical Center, University of Utah in Salt Lake City, Vanderbilt-Ingram Comprehensive Cancer Center in Nashville TN, and the National Cancer Institute in Bethesda, MD in USA and British Columbia Cancer Agency in Vancouver, Canada report a significant increase in the genomic signature of phosphatidylinositol 3-kinase (PI3K) pathway activation in the cytologically normal bronchial cells of smokers with lung cancer and precancerous lesions; *myo*-inositol (9 g twice daily for 2-3 months) decreased the PI3 kinase activity in the airway of high-risk smokers with a concomitant significant regression of precancerous dysplasia. Using human lung adenocarcinoma H1299 cancer cell line and treating with varying doses of *myo*-inositol *in vitro*, the investigators also quantitated the PI3K activity; a dose-dependent decrease in PI3K activity was noted after *myo*-inositol treatment [5].

While the laboratory data have shown consistent and reproducible anti-cancer and other health benefits of $InsP_6$ and Inositol since the late 1980's, clinical studies of $InsP_6$ + inositol have been few and far between. Here are the published reports:

InsP_6 with *β-Glucan*

"β-Glucan is a long chain polymer of glucose derived from the fungal cell wall; it has been shown to have a number of immunomodulatory properties as well as effects on hematopoiesis and as a radiation protectant." Weitberg reported the results of a phase I/II clinical trial of β -(1,3)/(1,6) D-glucan + $InsP_6$ in the treatment of patients with advanced malignancies receiving chemotherapy [6]. Twenty patients with advanced malignancies receiving chemotherapy were given a β-glucan preparation plus InsP_6 and monitored for tolerability and effect on hematopoiesis. β-Glucan + InsP_6 were well-tolerated in cancer patients receiving chemotherapy and may have a beneficial effect on hematopoiesis in these patients, especially in those with chronic lymphocytic leukemia and lymphoma [6]. However owing to the mixture of β-glucan with InsP_6, both seeming to have similar effect it was not clear from the report as to how much of the reported results can be attributed to either InsP_6 or β-glucan or the combination thereof. Attempts to obtain clarification from Dr Weitberg were unsuccessful.

InsP_6 + Inositol and Colorectal Cancer

The very first study of the use of InsP_6 + inositol in cancer patients was reported at the 17th Annual Meeting of the European Association for Cancer Research in Grenada, Spain (8-11 June, 2002) by Družijanić and colleagues from the Clinical Hospital Split in Split, Croatia. They reported the results of a pilot study on six patients (age 51-70 years) with advanced colorectal cancer (Dukes C and D) receiving standard chemotherapy and InsP_6 + Inositol. Chemotherapy related side effects (nausea, vomiting, alopecia, diarrhea, paresthesis, *etc.*) were minimal and patients were able to perform their daily activities [7]. One patient with liver metastasis abstained from chemotherapy after first treatment; and she was being treated with InsP_6 + Inositol (6000 mg per day. orally) alone. "…ultrasound and abdominal CT 14 months postoperatively have shown significantly reduced tumor growth rate..." [7].

Družijanić and colleagues further reported an enhanced anticancer activity without compromising the patient's quality of life in a pilot clinical trial involving 22 patients with advanced colorectal cancer (Duke's B2, C and D) with multiple liver and lung metastasis [8]. The age range of the patients was 42-71 years. During the chemo- and radiotherapy the patients were monitored with CBC (complete blood count) with differentials, creatinine, electrolytes, various enzyme markers and the tumor markers CEA and CA 19-9.

The patients were surgically operated and subjected to adjuvant polychemotherapy according to Mayo protocol. InsP_6 + Inositol (1530 mg 4 times/day orally) were given as an adjuvant to chemotherapy during and after the chemotherapy for a one year period. An overall reduced tumor growth rate was noticed, and in some cases a regression of lesions was noted. Additionally, when InsP_6 + Inositol was given in combination with chemotherapy, side effects of chemotherapy, such as drop in leukocyte and platelet counts, nausea, vomiting, alopecia, were diminished and patients were able to perform their daily activities.

Fig. (**8.1**) shows the abdominal CT scan of a 63 years old man with inoperable Stage IV colorectal cancer with metastases in liver and lungs, before and 4 months after initiation of $InsP_6$ + Inositol treatment.

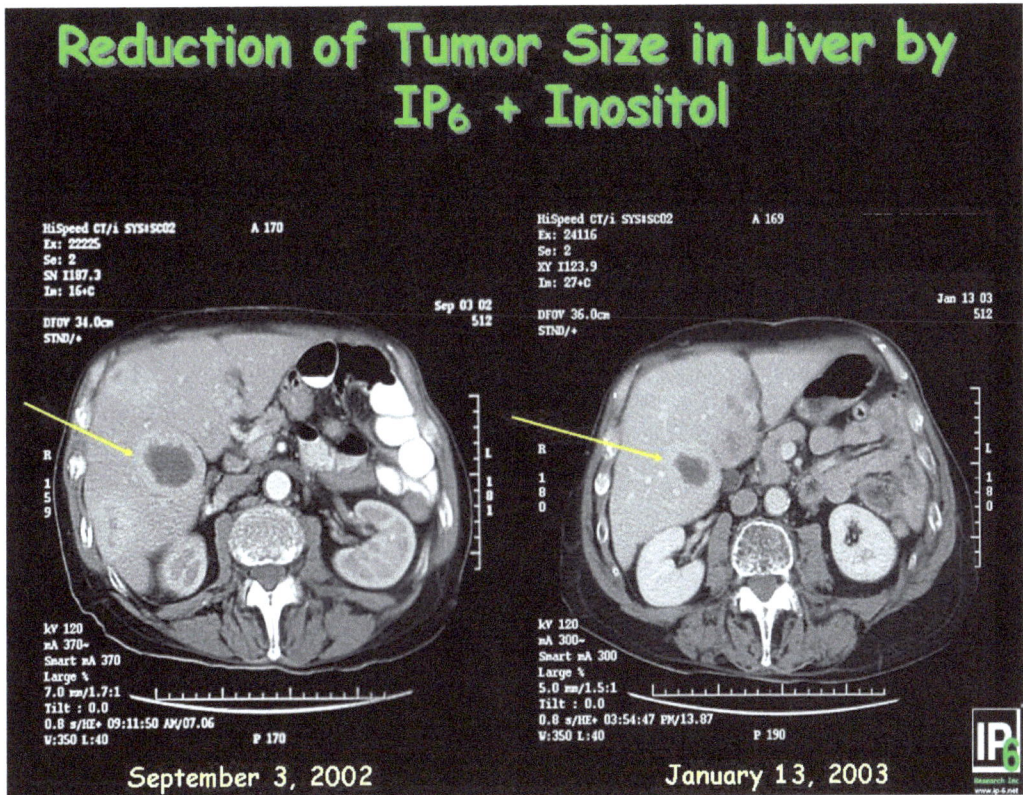

Fig. (8.1). The arrows point to the metastatic liver cancer before (left panel) and 4 months after initiation of IP_6 + Inositol treatment.

InsP_6 + Inositol and Breast Cancer

Sakamoto & Suzuki from Sakamoto Clinic of Gastroenterology and Takasaki National Hospital, Gunma, Japan presented a case of a 79 years old patient with metastatic breast cancer treated with $InsP_6$ + Inositol orally. The patient had developed pleural effusion secondary to the metastases at which time she was started on $InsP_6$ + Inositol in April 2002. Four months later, the tumor markers had dropped to a normal level. An additional 2 months later, the pleural effusion decreased in volume. However it began to increase in volume in May 2003. The tumor markers also increased continuously thereafter, reaching 35- to 40- fold higher than normal level in July 2004. While the signs of metastatic disease showed progression, her general condition was not impaired and the quality of life status remained favorable, reported the authors. "[T] the reason for the patient's favorable condition, in spite of a large burden of the relapsed tumor mass, may be partly attributed to the fact that she has taken IP6 + Inositol every day for more than two years". They conclude "[I] in combination with an oral intake of chemotherapeutic agents (anti-estrogen agent and tegafur-uracil mixture), IP6 + Inositol may contribute to improved quality of life and prolonged survival of patients with metastatic recurrence of breast cancer" [9].

Družijanić *et al.* also reported on a pilot study on 4 patients with ductal invasive breast cancer in stage II and IIIa, all treated with chemotherapy and radiotherapy, and $InsP_6$ + Inositol 2040 mg three times/day. They too report that the patients' quality of life was superior and the side-effects secondary to chemo- and radiation-therapy such as nausea, vomiting, loss of body weight, hair fall out, insomnia, depression, disorders in blood counts were minimal [10].

In a prospective randomized pilot clinical study of 14 breast cancer (invasive ductal carcinoma) patients treated with $InsP_6$ + Inositol receiving standard chemotherapy along with $InsP_6$ + Inositol (*IP6 Gold* by IP_6 International, Melbourne, Florida, USA) did not have any drop in leukocyte and platelet counts as opposed to controls. Patients who were given 3000 mg $InsP_6$ + Inositol twice a day for 6 months had significantly better quality of life (statistically significant at $p = 0.05$) (Table **8.1**) and functional status (statistically highly significant at $p = 0.0003$) (Table **8.2**) and were able to perform their daily activities [11].

All of the patients enrolled in the study had filled the questionnaires QLQ-30 and QLQ-BR23 from European organization for testing the treatment of cancer (EORTC). The questions in questionnaire were divided into the functional and symptomatic scales. The Functional scale contained questions about the physical, emotional, cognitive, social and sexual functions. Each group had a range of responses matching from 0-100 wherein a score of 100 represented the maximum compatibility with the offered answers; and 0 represented the complete lack of compatibility.

Table 8.1. Quality of Life Assessment [11].

Patients	Quality of Life	
	Mean ± SD	*p* Value
Placebo Group (7)	48.43 ± 28.96	0.05
InsP_6 + Inositol Group (7)	78.33 ± 21.60	

Based on patients' own personal assessment.

Table 8.2. Functional Status Assessment [11].

Patients	Functional Status	
	Mean ± SD	*p* Value
Placebo Group	56.29 ± 15.32	0.0003
InsP_6 + Inositol Group	87.94 ± 6.94	

Based on patients' own personal assessment.

The Symptomatic scale contained questions about side-effects of treatment, such as the general bad condition, nausea, vomiting, diarrhea, constipation, pain, insomnia, and loss of appetite, loss of body weight, hair loss, increase in body temperature and the operating complications of treatment. Replies from symptomatic scale were evaluated with the scale from 0-100, where 100 represented maximal positive personal experience with total quality of life; and 0 represented the most negative personal experience of the quality of life.

Patients' self-assessments of chemotherapy-induced symptoms are presented in the following Table **8.3**.

Table 8.3. Chemotherapy-induced side-effects: symptomatic [11].

Patients	Clinical Symptoms of Side-Effects	
	Mean ± SD	*p* Value
Placebo Group	33.81 ± 18.12	0.04
InsP_6 + Inositol Group	13.51 ± 9.98	

Based on patients' own personal assessment.

Of particular note is that owing to the unacceptable side-effects of chemotherapy, a substantial number of patients drop-out of the treatment regimen; none taking

InsP_6 + Inositol did, nor did they interrupt their standard treatment protocol due to the side-effects.

While the patient self-reporting is subjective, it is imperative that objective evidence in the form of changes on white blood cells (WBC), platelets counts and other laboratory parameters are evaluated. Decrease in these parameters result in lowered immunity (due to markedly reduced number of WBC) with resultant increased risk of infection and frequent and serious bleeding owing to very low platelet count. Table **8.4** presents the data on the WBC and platelets before and after treatment with and without InsP_6 + Inositol. While the placebo control group showed statistically significant decrease in both platelet and white blood cell count, patients taking InsP_6 + inositol did not show such drop; on the contrary these parameters seemed slightly better, albeit not statistically significant.

Table 8.4. Chemotherapy Side-Effects on Blood Cell Counts [11].

Blood Cell		Placebo Mean ± SD	InsP_6 + Inositol Mean ± SD
WBC (x10^9/L)	Before Rx	7.53 ± 1.50	6.66 ± 0.96
	After Rx	4.36 ± 1.80	6.92 ± 2.12
	p value	0.01	0.75
Platelets (x10^9/L)	Before Rx	272.71 ± 114.86	229.57 ± 31.81
	After Rx	205.00 ± 90.56	231.86 ± 47.33
	p value	0.05	0.92

Note that the decrease in WBC counts in placebo group is statistically significant whereas those in InsP_6 + Inositol treated group are not; there was no decrease in platelet counts with InsP_6 + Inositol.

"IP$_6$ + Inositol as an adjunctive therapy is valuable help in ameliorating the side effects and preserving quality of life among the patients treated with chemotherapy" concluded the authors [11].

InsP_6 + Inositol and Lung Cancer

Kosaku Sakamoto also reported of long-term survival of a patient with non-small cell carcinoma of the lung [12]. A 59-years old female with a smoking history of 30 years was diagnosed with adenocarcinoma of her left lung (T2N3M0, Stage IIIB) at the age of 49. She was undergoing chemo-radiotherapy which was discontinued in midcourse owing to intractable side-effects. She was started on

InsP_6 + Inositol 8 months following cessation of standard therapy. Of particular note is that the patient had not been taking any other medications during that period [12].

Four years and 4 months hence, she has been enjoying a "completely healthy life without any signs of relapse. Periodic check-up of her chest and abdomen by computed tomography (CT) and, recently, by multi-detector CT, revealed no evidence of tumor regression...Serum CEA level drastically decreased to normal when the CRT [chemo-radiotherapy] was terminated, and continues to be normal" [12].

CONCLUDING REMARKS

Clinical studies albeit limited, of InsP_6 + inositol have shown very encouraging results in cancer patients. As will be discussed in detail in the next chapter on radiation damage (Chapter 9), patients undergoing radiotherapy suffer from undesirable side-effects, unacceptable to many. Data show that InsP_6 + inositol can act synergistically with standard therapies, reduce the unacceptable side-effects of chemo- and radiation-therapy; and at the least it provides better quality of life.

CONFLICT OF INTEREST

Professor Shamsuddin is the inventor of several patents on InsP_6.

REFERENCES

[1]   Deliliers GL, Servida F, Fracchiolla NS *et al.* Effect of inositol hexaphosphate (IP$_6$) on human normal and leukaemic haematopoietic cells. Brit J Haemat 2002; 117: 577-87.
[2]   Tran HC, Brooks J, Agarwal S, *et al.* Effect of Inositol hexaphosphate [IP6] on AIDS Neoplastic Kaposi's Sarcoma, Iatrogenic Kaposi's Sarcoma and Lymphoma. Proceedings of the American Association for Cancer Research 2003; 44: 577-8 (#2536).
[3]   Tantivejkul K, Vucenik I, Eiseman J *et al.* Inositol hexaphosphate (IP$_6$) enhances the anti-proliferative effects of adriamycin and tamoxifen in breast cancer. Breast Can Res Treat 2003; 79: 301-12.
[4]   Lam S, McWilliams A, LeRiche J *et al.* A phase I study of *myo*-inositol for lung cancer chemoprevention. Cancer Epidemiol Biomarker Prev 2006; 15: 1526-31.
[5]   Gustafson AM, Soldi R, Anderlind C *et al.* Airway PI3K pathway activation is an early and reversible event in lung cancer development. Science Translational Medicine 2010; 2(26): 26ra25, 7 April 2010. doi: 10.1126/scitranslmed.3000251.
[6]   Weitberg AB. A phase I/II trial of beta-(1,3)/(1,6) D-glucan in the treatment of patients with advanced malignancies receiving chemotherapy. J Exper Clin Cancer Research 2008; 19: 27-40.
[7]   Družijanić N, Juricic J, Perko Z *et al.* IP-6 & Inositol: adjuvant to chemotherapy of colon cancer. A pilot clinical trial. Rev Oncología 2002; 4: (Suppl 1), 171.

[8]   Družijanić N, Juricic J, Perko Z *et al*. IP6 + Inositol as adjuvant to chemotherapy of colon cancer: Our clinical experience. Anticancer Res 2004; 24: 3474.

[9]   Sakamoto K, Suzuki Y. IP6 plus Inositol treatment after surgery and post-operative radiotherapy. Report of a case: Breast cancer. Anticancer Res 2004; 24: 3617.

[10]   Juricic J, Družijanić N, Perko Z *et al*. IP6 + Inositol in treatment of ductal invasive breast carcinoma: Our clinical experience. Anticancer Res 2004; 24: 3475.

[11]   Bacić I, Družijanić N, Karlo R *et al*. Efficacy of IP6 + Inositol in the treatment of breast cancer patients receiving chemotherapy: prospective, randomized, pilot clinical study. J Exp Clin Cancer Res 2010; 29: 12. doi: 10.1186/1756-9966-29-12.

[12]   Sakamoto K. Long-term survival of a patient with advanced non-small cell lung cancer treated with Inositol Hexaphosphate (IP6) plus Inositol treatment combined with chemo-radiotherapy. Report of a case. Anticancer Res 2004; 24: 3618.

CHAPTER 9

Radiation Protection by Inositol & its Phosphates

Abstract: Radiation is a common environmental carcinogen; humans are most often exposed to it from ultraviolet (UV) rays from the sun during sunbathing. There are other forms of radiation that we are exposed to owing to occupational or medical necessity. Radiation damage to health can be acute or chronic, the latter culminating in cancer. Uranium used in nuclear technology is also an important source of radiation. InsP_6 has shown to be a much more efficient chelator of uranium than the currently available therapy for uranium poisoning as in nuclear disasters. Inositol and InsP_6 have been proven to be protective against ultraviolet (UV) light-induced acute and chronic effects, including skin cancer.

Keywords: Alopecia, AP-1, apoptosis, NF-κB, radiation damage, skin cancer, Ultraviolet, uranium, UVB.

INTRODUCTION

Radiation is energy distributed across the electromagnetic spectrum, interacting with matter in a way that may be described as waves (having long wavelength and low frequency) and/or particles (having short wavelength and high frequency). Approximately 80% of all radiation encountered normally by mammals is from naturally-occurring sources. Radiation, particularly ionizing radiation, has an adverse effect on cells and tissues, primarily through cytotoxic effects. In humans, exposure to ionizing radiation occurs primarily through therapeutic procedures (such as anticancer radiotherapy), through occupational and/or environmental exposure to human-derived radiation sources, or through occupational exposure to naturally-occurring radiation sources as in the case of aircraft flight personnel.

Ionizing radiation is characterized as having short wavelength and high frequency; typical ionizing radiation forms include ultraviolet light, X-rays, gamma rays and cosmic radiation. Forms of ionizing radiation also include radioactive emissions of particles such as α particles or neutrons. In accord with its particulate nature, ionizing radiation causes vibration and rotation of atoms in biological molecules resulting in the ejection of electrons and the creation of free radicals. As discussed before, free radicals have a single unpaired electron in an outer orbit. This unstable configuration favors the release of energy through interactions with neighboring molecules, both inorganic and organic. In biological

A.K.M. Shamsuddin and Guang-Yu Yang

molecules, this release results in physical alteration of molecules through the creation of aberrant chemical bonds.

Table **9.1** characterizes radiation forms according to their frequency and selected biological effects.

Table 9.1. Ionizing and non-ionizing electromagnetic radiation [1].

Radiation	Frequency (Hz)	Biological Effects
Electrical power	1-50	? Increased cancer risk
Radio waves and radar	106-1,011	Thermal effects, cataract
Microwaves	109-1,010	Lens opacities
Infrared	1,011-1,014	Cataracts
Visible lights	1,015	Retinal burns (lasers)
Ultraviolet light	1,015-1,018	Skin burns, Skin cancer
X-rays and gamma rays	1,018-1,020	Acute and delayed injury, cancer
Cosmic radiation	1,027	Cataract, brain damage, cancer

Properties of radiation forms closer to the low-frequency end of the spectrum are better described as wavelike. Radiation forms closer to the high-frequency end of the spectrum have the most energy and tend to interact with matter as particles. The hazardous effects of radiation exposure are typically associated with the particulate characteristics of the radiation type.

RADIATION DAMAGE

Free radical species may also be created through chemical, enzymatic, and catalytic means by way of an intermediate reactive substance, but ionizing radiation can create free radicals directly, for example by directly hydrolyzing water into hydroxyl ($\cdot$OH) and hydrogen ($\cdot$H) free radicals. For example, when tissues are exposed to gamma radiation, much of the energy deposited in the cells is absorbed by water and results in scission of the oxygen-hydrogen covalent bonds in water, leaving a single electron on hydrogen and another on oxygen, thus creating the two radicals. The hydroxyl radical ($\cdot$OH) is the most reactive radical known in chemistry.

As mentioned earlier and further discussed in Chapter 16, free radical species react with the purine or pyrimidine bases of nucleic acids, proteins, lipids, and other biological macromolecules to produce damage to cells and tissues, and they can set off intra- and extra-cellular chain reactions, particularly in the critically ill patient. For example, reactive free-radical oxygen species initiate the activation of transcription factors through signal transduction from the cell surfaces, resulting in inflammation and tumor promotion.

DNA is the crucial target for the cytotoxic effects of ionizing radiation. Ionizing radiation is capable of damaging or altering DNA directly, causing double-stranded (ds) breaks and the formation of cross-linked pyrimidine bases, such as thymidine dimers (TT), a particularly important byproduct. Carbon-centered radicals formed directly by ionizing radiation on the deoxyribose moiety of DNA are thought to be the precursors of strand breaks. Cells undergoing extensive irreparable DNA damage generally enter into apoptosis (programmed cell death), and surviving cells bear the hallmarks of radiation damage in the form of mutations, chromosomal abnormalities and genetic instability.

Rapidly dividing cells, such as the blood forming cells (hematopoietic) in bone marrow, germ cells in testes and ovary, mucosal lining cells of the gastrointestinal tract, airway, *etc.*, are most susceptible to injury from ionizing radiation. Cells in the G_2 and mitotic phases of the cell cycle are the most likely to be damaged.

Radiation exposure from any source can be classified as acute (a single large exposure) or chronic (a series of small low-level, or continuous low-level exposures spread over time). Radiation doses from selected sources are shown in Table **9.2**; the standard reporting of radiation dosage is in millirem. Radiation sickness generally results from an acute exposure of a sufficient dose, and presents with a characteristic set of symptoms that appear in an orderly fashion, including hair loss, weakness, vomiting, diarrhea, skin burns and bleeding from the gastrointestinal tract and mucous membranes.

A sufficiently large acute dose of ionizing radiation, for example 500,000 to over 1 million millirem (equivalent to 5-10 Gy), may kill a subject immediately. Doses in the hundreds of thousands of millirem may kill within 7 to 21 days from a condition called "acute radiation poisoning." An acute total body exposure of 125,000 millirem may cause radiation sickness. Localized delivery of high doses as in radiotherapy may not cause radiation sickness, but may result in the damage or death of exposed normal cells.

Table 9.2. Radiation doses from selective sources [1].

Source	Dose in Millirem
Television	< 1/yr
Gamma Rays, Jet flight across USA	1
Mountain Vacation - 2 week	3
Atomic Test Fallout	5
U.S. Water, Food & Air (Average)	30/yr
Wood	50/yr
Concrete	50/yr
Brick	75/yr
Chest X-Ray	100
Cosmic Radiation	40/yr (+ 1 millirem/100 ft. elevation)
Natural Background San Francisco	120/yr
Natural Background Denver	50/yr
Atomic Energy Commission Limit for Workers	5,000/yr
Complete Dental X-Ray	5,000
Natural Background at Pocos de Caldras, Brazil	7,000/yr
Whole Body Diagnostic X-Ray	100,000
Cancer Therapy	500,000 (localized)
Radiation Sickness-Nagasaki	125,000 (single doses)
LD_{50} Nagasaki & Hiroshima single dose	400,000 - 500,000

An acute total body radiation dose of 100,000 - 125,000 millirem (equivalent to 1 Gy) received in less than one week would result in observable effects such as skin burns or rashes, mucosal and gastrointestinal bleeding, nausea, diarrhea and/or excessive fatigue. Longer term cytotoxic and genetic effects such as hematopoietic and immune cell destruction, hair loss (alopecia), gastrointestinal, and oral mucosal sloughing, veno-occlusive disease of the liver and chronic vascular hyperplasia of cerebral vessels, cataracts, pneumonitis, skin changes, and an increased incidence of cancer may also manifest over time. Acute doses of less than 10,000 millirem (equivalent to 0.1 Gy) typically do not result in immediately observable biologic effects, although long-term cytotoxic or genetic effects may occur [1].

Chronic exposure is usually associated with delayed medical problems such as cancer and premature aging. Chronic radiation exposure is a low level (*i.e.*, 100-5,000 millirem) incremental or continuous radiation dose received over time. Examples of chronic doses include a whole body dose of about 5,000 millirem per year, which is the dose typically received by an adult human working at a nuclear power plant. By contrast, the US Atomic Energy Commission recommends that members of the general public should receive no more than 100 millirem per year. Chronic doses may cause long-term cytotoxic and genetic effects, such as an increased risk of a radiation-induced cancer developing later in life.

Chronic doses of greater than 5,000 millirem per year (0.05 Gy per year) may result in long-term cytotoxic or genetic effects similar to those described for persons receiving acute doses. Some adverse cytotoxic or genetic effects may also occur at chronic doses less than 5,000 millirem per year. For radiation protection purposes, it is assumed that any dose above zero can increase the risk of radiation-induced cancer (*i.e.*, that there is no threshold). Epidemiologic studies have found that the estimated lifetime risk of dying from cancer is greater by about 0.04% per rem of radiation dose to the whole body.

A major source of acute exposure to ionizing radiation is the administration of human-derived therapeutic radiation in the treatment of cancer or other diseases. Subjects exposed to therapeutic doses of ionizing radiation typically receive between 0.1 and 2 Gy per treatment, and can receive as high as 5 Gy per treatment. Depending on the course of treatment, multiple doses may be received by a subject over several weeks to several months [1].

Exposure to ionizing radiation from human-derived sources can also occur in the occupational setting. Occupational doses of ionizing radiation may be received by persons whose job involves exposure (or potential exposure) to radiation, for example in the nuclear power and nuclear weapons industries. Occupational exposure may also occur in rescue and emergency personnel called in to deal with catastrophic events involving a nuclear reactor or radioactive materials. Other sources of occupational exposure may be from machine parts, plastics, solvents left over from the manufacture of radioactive medical products, smoke alarms, emergency signs, and other consumer goods. Exposure may also occur in military or civilian persons who serve on nuclear powered vessels, particularly those who tend to the nuclear reactors, and those operating in areas contaminated by military uses of radioactive materials, including nuclear weapons fallout [1].

Mammals, including humans and other animals (such as livestock), may also be exposed to ionizing radiation of human derivation from the environment. The primary source of exposure to significant amounts of such environmental radiation is from nuclear power plant accidents, such as those at Three Mile Island, Chernobyl, Tokaimura, Fukushima *etc*. Environmental exposure to ionizing radiation may also result from nuclear weapons detonations (either experimental or during wartime), discharges of actinides from nuclear waste storage and processing and reprocessing of nuclear fuel, and from naturally occurring radio-active materials such as radon gas or uranium. There is also increasing concern that the use of ordnance containing depleted uranium results in low-level radioactive contamination of combat areas [1].

As mentioned above, for most mammals the bulk of their lifetime radiation exposure derives from naturally-occurring sources. Such sources include radioactive chemical elements dispersed throughout nature, such as the slight amount of uranium that occurs naturally in granite. Small amounts of radioactive elements are found pervasively in the atmosphere, ground, and water, to lesser and greater degrees depending upon the geographical location. Other significant naturally-occurring sources derive from outer space: the sun and the cosmos. Ultraviolet radiation emitted by the sun may be particularly hazardous as relatively strong doses may be acquired accidentally throughout much of the world. Cosmic radiation, x-ray, and gamma radiation exposure is of particular risk to those living or working at high altitudes. Commercial and military flight personnel, including astronauts, are particularly susceptible to such radiation owing to the relatively long periods they spend at high-altitudes [1].

While anti-radiation suits or other protective gear may be effective at reducing radiation exposure, such gear is expensive, unwieldy, and generally not available to public. Moreover, radio-protective gear do not protect normal tissue adjacent a tumor from stray radiation exposure during radiotherapy. What is needed, therefore, is a practical way to protect subjects who are scheduled to incur, or are at risk for incurring, exposure to ionizing radiation. In the context of therapeutic irradiation, it is desirable to enhance protection of normal cells while causing tumor cells to remain vulnerable to the detrimental effects of the radiation. Furthermore, it is desirable to provide systemic protection from anticipated or inadvertent total body irradiation, such as may occur with occupational or environmental exposures, or with certain therapeutic techniques.

Pharmaceutical radio-protectants offer a cost-efficient, effective and easily available alternative to radioprotective gear. However, earlier attempts at radioprotection of normal cells with pharmaceutical compositions have not been completely successful. For example, cytokines directed at mobilizing the peripheral blood progenitor cells confer a myeloprotective effect when given prior to radiation, but do not confer systemic protection. Other chemical radio-protectants administered alone or in combination with biologic response modifiers have shown minor protective effects in mice, but application of these compounds to large mammals was less successful, and it was questioned whether chemical radioprotection was of any value.

Urgency for Radiation Protection

In today's heightened nuclear threat from terrorists and/or rogue nations as well as accidents in nuclear power-plant reactors (Fukushima Daiichi in Japan following an earthquake and tsunami in March 2011), there is an increased need to have safe and effective means to protect the nuclear-reactor workers and the population at large from the health hazards of ionizing radiation exposures. The United States Department of Energy (DOE) reports that "an unfilled dream of civil and military officials concerned with this issue is to have a globally effective pharmacologic, *i.e.,* the 'magic' radio-protective pill. This pill could be taken orally without any undue side effects prior to or after a suspected nuclear/radiological event in order to provide the individual full bodily protection against early arising acute injury and late arising pathologies[1]". Thus, a "radioprotective pill" is of urgent and vital national security interest.

The US DOE report further states: "...Currently a full range of R&D strategies are being employed in the hunt for new safe and effective radioprotectants including: a) large scale screening of newly identified chemical classes or natural products; b) reformulating or restructuring older protectants with proven efficacies to reduce unwanted toxicities; c) using nutraceuticals that are only moderately protective but that are essentially non-toxic and exceedingly well tolerated; d) using low dose combinations of potentially toxic (at high drug doses) but efficacious agents that cytoprotect through different routes in hopes of fostering radioprotective synergy; and e) accepting lower drug efficacy in lieu of non-toxicity, banking on the protection afforded by the drug can be leveraged by post-exposure therapies......Inositol hexaphosphate, IP-6 [InsP_6], and its analogs

---

[1]United States Department of Energy Report of July 13, 2005 on Inositol and Other Radioprotective Agents Workshop, Cambridge. Massachusetts.

are entering testing as drugs. One of the challenges is to cover phosphates with protecting groups, to facilitate passage of the molecule into the cell. (DOE report of July 13, 2005)." Implied is the notion that $InsP_6$ and its derivatives, including pyrophosphates, and/or inositol may currently be ineffective as radioprotectants. Nothing could be further than truth.

Inositol and $InsP_6$

Uranium

In nature uranium is found in granite and other mineral deposits being rather widely distributed (radon gas detected in some homes are produced by decay of uranium). It is weakly radioactive; most of it (approximately 99.3%) is ^{238}U, the rest being ^{235}U and minute amounts of ^{234}U. It decays by emitting α-particles; half-life of ^{238}U being 4.4 billion years and that of ^{235}U is 704 million years, the latter being the only naturally occurring fissile isotope. Isotopes of uranium and mixed plutonium-uranium oxide are used as fuel in nuclear power plants. ^{238}U has a small probability for spontaneous fission as opposed to ^{235}U and to a lesser degree ^{233}U which in sufficient critical concentration evoke a sustained nuclear chain reaction producing heat in nuclear power generators and fissile materials for nuclear weapons. Uranium also has other industrial applications.

After contamination by uranium through inhalation, ingestion or wound contamination, it is transported *via* blood to various organs where it can stay for a very long time causing toxicity. The current therapeutic strategy for treating uranium toxicity is to reduce its deposition in the organs by using chelating substances so that its excretion through the urine can be enhanced; intravenous infusion of 1.4% sodium bicarbonate solution has been the treatment of choice. However, its efficiency is rather limited; hence the need for a better chelating agent, particularly since there are no drugs immediately available for therapy of uranium contaminated persons [2].

In their quest for a more efficient strategy to combat uranium poisoning, Cebrian and colleagues thus investigated a number of chelating agents including $InsP_6$ and the currently used sodium bicarbonate [2]. Their results show that *in vitro* ability of $InsP_6$ to chelate uranium (as uranyl nitrite) was 2.0, 2.6 and 16 times higher than that for ethane-1-hydroxy-1,1-bisphosphonate (EHBP), citric acid and diethylenetriamine pentaacetic acid (DTPA), respectively. The assay is based on the ability of a potential chelating agent to displace uranium from a complex formed with a ligand. Thus, $InsP_6$ offers a very useful choice in the treatment of populations at risk of uranium poisoning from accidental or deliberate nuclear explosions.

Ultraviolet (UV) Radiation

In 2001, Chen and coworkers reported that InsP_6 strongly blocked UVB-induced activator protein-1 (AP-1) and nuclear factor-κB (NF-κB) transcriptional activities in a dose-dependent manner [3]. InsP_6 suppressed UVB-induced AP-1 and NF-κB DNA binding activities and inhibited UVB-induced phosphorylation of extracellular signal-regulated protein kinases (Erks) and c-Jun NH2-terminal kinases (JNKs). InsP_6 also blocked UVB-induced phosphorylation of I κB-α, which is known to result in the inhibition of NF-κB transcriptional activity [3]. Since AP-1 and NF-κB are important nuclear transcription factors that are related to tumor promotion, these data strongly suggest that InsP_6 prevents UVB-induced carcinogenesis by inhibiting AP-1 and NF- κB transcription activities [3].

Both UVA (320 to 400 nm) and UVB (280 to 320 nm) radiation reach the Earth from the sun; UVB radiation is known to damage skin keratinocyte DNA as well as suppress the immune system and induce chronic skin damage, resulting in cutaneous malignancy and non-melanoma skin cancer. While long-term exposure to UVB radiation results in skin photo-aging and skin cancer, the acute or short-term effects include erythema; these occur at physiologic doses of UVB intensity at 18 to 30 mJ/cm^2, corresponding to 1 to 2 min of sunbathing at sea level [1]. InsP_6 and inositol protect human cells against radiation damage *in vitro* in various ways. For example while irradiation caused increased cell death (loss of cell attachment, apoptosis and necrosis) InsP_6 and inositol treatment prevented cell death, *in vitro* [4]. *In vivo* animal experiments in laboratory demonstrated that InsP_6 significantly reduced the incidence and multiplicity of UV-induced skin tumors in mice [5].

In Vitro Experiments with Human Keratinocytes

Human keratinocyte HaCaT cells were exposed to UVB 30 mJ/cm2 (2 minutes 10 seconds) *in vitro*; non-exposed cells served as control. The cells were then treated with InsP_6, Inositol, InsP_6 + Inositol, and untreated (control) and placed in the incubator for 18 hours. The cells that remained attached to the wells were live (protected from UVB damage); the number of attached cells was measured by dissolving them in acetic acid and measuring the optical absorbance of the solution (Fig. **9.1**).

InsP_6 treatment resulted in a significant ($p < 0.05$) increase in both G_1 and G_2M phases and a significant ($p < 0.05$) decrease in the S phase of cells that after being exposed to UVB as compared untreated controls. The effect of InsP_6 on apoptosis of HaCaT cells was examined by treating with 0.5 mM InsP_6 18 h after exposure to 30 mJ/cm^2 UVB irradiation. Cells exposed to UVB radiation without InsP_6 treatment showed 33.7% viability and 32.8% apoptosis. In contrast, cell viability increased to 62.2% and apoptosis decreased to 9.9% in InsP_6 treated cells. UVB-induced caspase 3 activity was significantly ($p < 0.01$) lower in cells treated with InsP_6 [4].

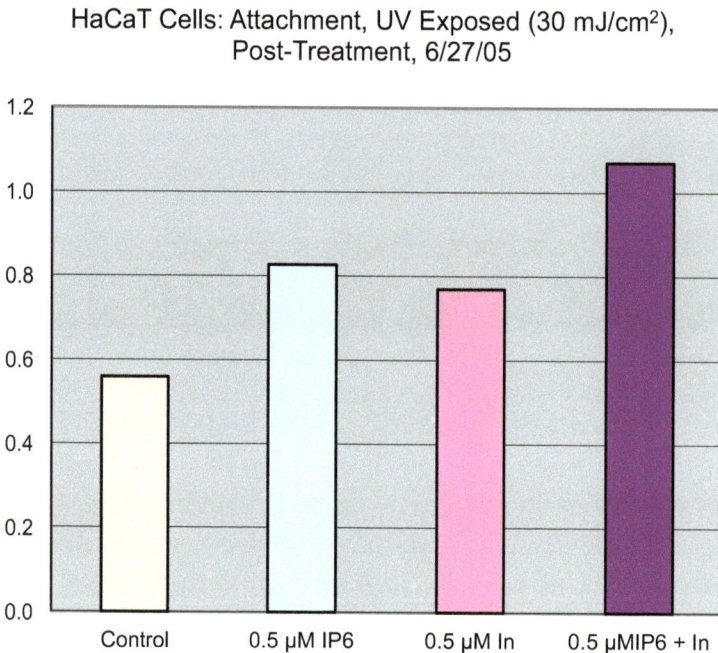

HaCaT Cells: Attachment, UV Exposed (30 mJ/cm²), Post-Treatment, 6/27/05

Fig. (9.1). Following UVB exposure, as signs of cellular injury and death, the control untreated cells show less attachment to the plate as opposed to those treated with Na-IP$_6$, Inositol (In) and IP$_6$ + Inositol. Please note that the cells treated with 1: 1 molar ratio of IP$_6$ and Inositol showed the best attachment, hence best protected. Ordinate: optical density at 595 nm.

In Vivo Experiments in SKH1 Hairless mice

The radioprotective effect of InsP_6 was tested in the *in vivo* experiments with SKH1 hairless mice either by administering in drinking water (2% Na- InsP_6) or by topic application (4% K-InsP_6) as skin cream [5].

In the *in vivo* experiment, mice were irradiated 3 times a week, initially with 1.5 kJ/m^2 dose and escalating weekly by 1.5 kJ/m^2 to a final dose of 7.5 kJ/m^2; each session lasted approximately 10 minute for 23 weeks. Animals were fed with AIN-76A diet that did not contain $InsP_6$. About 100 mg of 4% K-$InsP_6$ was applied on the dorsal surface topically in skin cream [5].

At week 32, the cumulative tumor incidence was 57% in the mice treated with topical $InsP_6$ compared with 71% in the vehicle group ($P < 0.05$) and cumulative tumor multiplicity was 86 in $InsP_6$ mice but 129 in vehicle mice [4]. $InsP_6$ in drinking water significantly decreased tumor incidence by 5-fold and tumor multiplicity by 4-fold [5].

These data obtained from both *in vitro* and *in vivo* experiments demonstrate that $InsP_6$ is protective against UVB-induced radiation damage in both acute and long-term phases.

CONCLUDING REMARKS

$InsP_6$ has shown to be a much more efficient chelator of uranium than the currently available therapy for uranium poisoning as in nuclear disasters. $InsP_6$ + inositol are also protective against UVB induced acute and chronic effects including skin cancer. As regards UV protection, the methods of administration could be either in the form of skin cream and/or orally. Combined with the data presented in the previous chapter (Chapter 8) on human clinical trial for cancer wherein $InsP_6$ + inositol reduced radiation-induced side-effects in patients, data presented here strongly support that $InsP_6$ + inositol could help protect us from various other forms of radiation damage such as radiation therapy, cosmic radiation, accidental or induced nuclear blasts *etc.*

REFERENCES

[1]   Shamsuddin AM, Vucenik I. Prevention of Nuclear, Solar and Other Radiation-Induced Tissue Damage. US Patent Application No. 11/453,843 (2006)
[2]   Cebrian D, Tapia A, Real A, *et al.* Inositol hexaphosphate: a potential chelating agent for uranium. Radiation Prot. Dosimetry 2007; 127: 477-9.
[3]   Chen N, Ma WY, Dong Z. Inositol hexaphosphate inhibits ultraviolet B-induced signal transduction. Mol Carcinog 2001; 31: 139-44.
[4]   Williams KA, Kolappaswamy K, Detolla LJ *et al.* Protective effect of inositol hexaphosphate against UVB damage in HaCaT cells and skin carcinogenesis in SKH1 hairless mice. Comp Med 2011; 61: 39-44.
[5]   Kolappaswamy K, Williams KA, Benazzi C *et al.* Effect of inositol hexaphosphate on the development of UVB-induced skin tumors in SKH1 hair-less mice. Comp Med 2009; 59: 147-52.

CHAPTER 10

Disorders of Metabolism

Abstract: InsP_6 and inositol independently may help in diabetes; InsP_6 by stimulating insulin secretion by the pancreatic β cells. Both inositol and InsP_6 can prevent oxidative damages to the cells and tissues inflicted through advanced glycosylated end products by acting as antioxidants. InsP_6 and inositol could replenish the low intracellular inositol since they do not need insulin to gain entry into the cell. They also have positive effect on the abnormalities of fat metabolism such as hypercholesterolemia and fatty liver.

Keywords: Advanced glycosylated end-products, AGE, cirrhosis, embryopathy, endothelial dysfunction, fatty liver, gestational diabetes, glycosylated hemoglobin, hypercholesterolemia, xenobiotics.

DIABETES MELLITUS

Diabetes, a Global Epidemic

Diabetes Mellitus (DM) is a chronic condition that afflicted an estimated 227 million or more people worldwide (more than 3% of the world population) in 2010. It is as a result of either insufficient secretion of insulin from the pancreas or when the body cannot effectively use the insulin it produces, resulting in increased blood sugar level - hyperglycemia.

The World Health Organization (WHO) estimates that the number of diabetics (> 180 million) is likely to more than double by the year 2030. WHO also estimated that in 2005, approximately 1.1 million people died from diabetes worldwide. Almost 80% of diabetes-related deaths occur in low and middle-income countries. There are 20.8 million adults and children in the United States, or 7% of the population, who have diabetes; unfortunately over one-third of whom (6.2 million people) are unaware that they have the disease.

"Almost half of diabetes deaths occur in people under the age of 70 years; 55% of deaths due to diabetes are in women". WHO projects that deaths due to diabetes will increase by more than 50% in the next 10 years unless means are adopted to control the disease. Worse, deaths due to diabetes are estimated to increase by over 80% in upper-middle income countries between 2006 and 2015. Thus there is an urgent need to prevent and control this global scourge.

A.K.M. Shamsuddin and Guang-Yu Yang

"There are two basic forms of diabetes: Type 1: people with this type of diabetes produce very little or no insulin. Type 2: people with this type of diabetes cannot use insulin effectively. Most people with diabetes have type 2. A third type of diabetes, gestational diabetes mellitus (GDM), develops during some cases of pregnancy but usually disappears after pregnancy".

Aside from the acute complications of diabetes, the late complications of diabetes mellitus affect multiple organ systems and are the chief causes of morbidity and mortality. The risks of the late complications vary markedly amongst individuals but are generally dependent on the duration of the disease - the longer the disease, the higher are the risks. These include: diabetic retinopathy, neuropathy, kidney disease, heart disease, stroke and peripheral vascular disease.

"Diabetic retinopathy is an important cause of blindness; after 15 years of diabetes, approximately 2% of people become blind, and about 10% develop severe visual impairment. Diabetic neuropathy affects up to 50% of people with diabetes". Combined with reduced blood flow (diabetic peripheral vascular disease), neuropathy in the feet increases the chance of foot ulcers and eventual limb amputation. And it increases the risk of heart disease and stroke; about 50% of people with diabetes die of cardiovascular disease - heart disease and stroke. In addition, people with diabetes suffer from cataract, weakened immune response - frequent bouts of infections; diabetic mothers have higher risks of miscarriages and birth-defects in off-springs (diabetic embryopathy). As if these are not grim enough, the overall risk of dying among people with diabetes is at least double the risk of those without diabetes.

What causes the complications? The inability to utilize glucose by the cells results in excessive glucose in blood and diminished glucose available for energy metabolism by the cells dependent on this fuel. On one hand, excessive glucose binds with the protein in the blood (glycosylated hemoglobin HbA_{1c} as an example, used to monitor the disease) and tissues. The latter produces irreversible formation of advanced glycosylated end products (AGE) which accumulate over the life-time of the vessel wall. AGE cross-link peptides of some proteins such as collagen, induce lipid oxidation, inactivate nitric oxide, bind nucleic acids *etc*. These are considered to be fundamental to the vascular, renal and other complications.

On the other hand, cellular metabolism is also affected in diabetes mellitus. The vast majority of cells dependent on insulin for taking up glucose suffer from a deficiency of intracellular inositol level, which in turn affects the signal transduction pathways within the cell. And those cells that are not dependent on insulin suffer from intracellular hyperglycemia leading to increased intracellular sorbitol which is directly toxic to the cells resulting in cell injury; increased sorbitol is also linked to decreased intracellular *myo*-inositol which is a precursor for phosphoinositides. There is also an increase in aldose reductase activity, platelet hyperaggregability and lipid peroxidation. These mechanisms are considered to be at play in diabetic neuropathy, retinopathy, renal vascular disease *etc.* Studies suggest that oxidative damage is also responsible for some of the complications such as cataract and embryopathy.

Inositol and its Phosphates, and Diabetes Mellitus

Insulin Secretion

Insulin is stored in very distinctive appearing (by electron microscopy) granules within the pancreatic β cells. Glucose is the signal for β cells to secrete insulin. Under normal physiological condition, as the blood glucose level rises, it elicits increased glucose uptake *via* the glucose transporters. This glucose influx raises the ATP:ADP ratio leading to the closure of ATP-sensitive K^+ channels (K_{ATP}) responsible for the resting membrane potential of β cells; closure of K_{ATP} results in membrane depolarization with Ca^{2+} influx. During the secretion of insulin by the pancreatic β cells, the hormone is extruded from the cell by a process called exocytosis. It is thought that glucose promotes insulin secretion through the coupling of changes in the ATP:ADP ratio to the influx of extracellular Ca^{2+} which causes insulin release from the granules [1].

In the vast majority of diabetics (Type 2 diabetes), loss of this glucose-stimulated insulin exocytosis from the pancreatic β cells is an early pathogenetic event. As in most other mammalian cells, InsP_6 is the dominant inositol phosphate in the insulin secreting pancreatic β cells. The concentration of InsP_6 within the pancreatic β cells transiently increases from ~50 μM to 60 μM upon glucose stimulation correlating with an influx of Ca^{2+} [2]. This small but significant increase in InsP_6 concentration (10-15%) can increase the current through voltage-gated L-type Ca^{2+} channel, due in part through differential inhibition of serine/threonine protein phosphatase activity in a concentration-dependent manner [2]. (Interestingly, none of the oral hypoglycemic sulfonylureas tested by Lehtihet *et al.,* affected protein phosphatase-1 or -2A activity at clinically relevant concentrations in these cells [3]).

Insofar as other inositol polyphosphates are concerned, that intracellular level of the ubiquitous second messenger and the most famous of the inositol phosphates, inositol 1,4,5-triphosphate ($InsP_3$) within the β-cells increases following glucose stimulation is well established; and so is the synchronized dynamic changes in intracellular Ca^{2+} ($[Ca^{2+}]_i$), electrical activity, insulin secretion *etc.* Oscillations in $InsP_3$ accompanies $[Ca^{2+}]_i$ oscillation; who follows whom is a matter that is yet to be resolved and so is their exact significance in insulin secretion [1]. The other second messenger inositol phosphate, albeit less involved is the inositol 3,4,5,6-tetrakisphosphate ($InsP_4$). Since the rise in intracellular level of $InsP_4$ is not quite as rapid, its role in insulin secretion and exocytosis under physiological condition is not considered to be in the first phase; rather it may be important in pathological hyperglycemia [1].

Efanov *et al.,* have shown that at 20 μM concentration both $InsP_6$ and $InsP_5$ can stimulate exocytosis in permeabilized insulin secreting HIT-T15 cells with more or less equal potency, though under normal physiological condition, only $InsP_6$ reaches such high intracellular concentration in the pancreatic β cells [4]. In 2007, a relatively high concentration (6 μM) of $InsP_7$ (diphosphoinositol pentakis phosphate or 5 *PP*- $InsP_5$) was reported in the β cells, which also plays a role in exocytosis of insulin as exogenously applied $InsP_7$ dose-dependently enhanced exocytosis at physiological concentrations [5].

Animal and Clinical Data

Rat Model

In vivo, Dilworth *et al.,* demonstrated that $InsP_6$ supplementation in diet caused a lowering of blood glucose level in experimental rats by commercially available $InsP_6$ as well as that extracted from sweet potato [6]. Evaluation of the levels of enzymes involved in hepatic glucose metabolism showed no significant change in the activity of 6-phosphogluconate dehydrogenase amongst the various experimental groups. $InsP_6$ supplementation showed no significant decrease in the activity of pyruvate kinase compared to the group fed formulated diets. There was a significant increase in the activity of glucose-6-phosphate dehydrogenase in the groups fed $InsP_6$ extract from sweet potato compared to the other groups. The activities of malic enzyme and ATP-citrate lyase in this study were not significantly altered among the groups [6]. Why was there a difference between the commercially available $InsP_6$ and that extracted from sweet potato insofar as the activity of glucose-6-phosphate dehydrogenase is concerned? There should not be difference in the chemical structure of the $InsP_6$ irrespective of its source; commercial $InsP_6$ is from rice or corn. The only explanation therefore is that some other yet to be identified component(s) could be responsible for this difference.

Streptozotocin is a specific toxin for pancreatic β cells and is used to induce hyperglycemia in animal models. $InsP_6$ supplement was given to streptozotocin-induced diabetic rats for 30 days. Compared to other groups, there was a downward trend in intestinal amylase activity in the group fed $InsP_6$ supplement. The spike in random blood glucose was the lowest in the $InsP_6$ supplemented group. There was also reduced serum triglyceride level, but increased total cholesterol and HDL (high density lipoprotein) cholesterol levels in the group fed $InsP_6$ supplement. Serum alkaline phosphatase and alanine amino transferase activities were significantly ($p < 0.05$) increased by $InsP_6$ supplementation. Systemic IL-1 β level was significantly ($p < 0.05$) elevated in the diabetic control and supplement treated groups. The liver lipogenic enzyme activities were not significantly altered among the groups [7]. The observed increase in total cholesterol may be due to the combined effect of streptozotocin-induced diabetes and $InsP_6$ supplementation, as this was the opposite of what has repeatedly reported by others (*vide infra* "Disorders of Fat Metabolism").

Mouse Model

C57BL/6N mice are widely used for the investigation of the pathophysiology of impaired glucose tolerance and Type 2 diabetes to look for new therapeutics. $InsP_6$ treated mice showed a marked decrease in blood glucose level, higher glucokinase and glucose-6-phosphatase activity, higher hepatic glycogen and considerably lower bodyweight than control mice [8]. A significantly higher glucokinase activity and lower phosphoenolpyruvate carboxykinase activity, glucose-6-phosphatase activity and hepatic glycogen concentration were observed in $InsP_6$ treated mice. These findings demonstrate that both rice bran and $InsP_6$ could reduce the risk of high fat diet-induced hyperglycemia *via* regulation of hepatic glucose-regulating enzyme activities [8].

Human

Epidemiological studies have consistently showed a relationship between diet and diabetes, besides other ailments. However, studies on the specific dietary component in protecting people from diabetes are inadequate, and only indirect. Notwithstanding this limitation, data suggest that $InsP_6$-rich diet have beneficial actions. Panlasigui & Thompson compared the blood glucose lowering effect of milled-rice *versus* brown rice (which retains its bran and therefore high levels of $InsP_6$ and inositol) in 10 healthy and 9 Type 2 diabetic volunteers. Their data show that the glycemic area and glycemic index (measures impact of food on blood sugar level) in healthy volunteers were 19.8% and 12.1% lower ($p < 0.05$) in brown rice than milled rice; while in diabetics, the respective values were 35.2% and 35.6% lower. The investigators attribute this to the higher amounts of $InsP_6$ in brown rice compared to milled rice, besides other substances. As an aside, the difference in some physicochemical properties of the rice samples such as minimum cooking time and degree of gelatinization may also be contributory [9].

These investigators had previously shown that endogenous and added $InsP_6$ (as well as calcium) affect the *in vitro* rate of starch digestion and *in vivo* blood glucose response to navy bean flour, prepared as unleavened bread. Unleavened navy bean flour was used since the yeast in the leavening process contains phytase destroying the $InsP_6$. Removal of $InsP_6$ from and addition of calcium to navy bean flour increased the starch digestion *in vitro* and raised the glycemic response *in vivo* while re-addition of $InsP_6$ to $InsP_6$-depleted (a.k.a. dephytinized) flour produced the opposite effect [10]. Similar studies on healthy volunteers have shown a negative correlation of glycemic index with the $InsP_6$ of the food. Addition of Na-$InsP_6$ to unleavened bread flattened the glycemic index in feeding trials [11].

$InsP_6$ has been found to be useful in treatment of diabetes from a different angle: Owing to the inconvenience of daily injection of insulin, in their quest for finding alternative methods Lee *et al.*, used $InsP_6$ as a cross-linking agent for encapsulation of insulin in a chitosan matrix for oral delivery [12]. $InsP_6$-chitosan capsules were compared with tripolyphosphate (TPP)-chitosan capsules for stable oral delivery of insulin. During a 2 h incubation in simulated gastric fluid, $InsP_6$-chitosan capsules showed better stability than TPP-chitosan capsules. $InsP_6$-chitosan capsules released less than 60% of their encapsulated insulin after 24 h incubation in simulated gastrointestinal fluids; in contrast, TPP-chitosan capsules released virtually the entire insulin content within 12 h. When studied *in vivo* using diabetic mice, $InsP_6$-chitosan capsules significantly decreased blood glucose levels while TPP-chitosan capsules caused a lesser reduction. The relative pharmacological bioactivity of $InsP_6$-chitosan capsules was 6.4% while that of TPP-chitosan capsules was 1.1% [12]. It would be interesting to know what, if any, the contribution of $InsP_6$ was in lowering the blood glucose level.

Inositol

As mentioned before, in diabetes, the low intracellular level of *myo*-inositol is responsible for the pathogenesis of the various complications. Thus, studies have been done to address this issue.

Ruf *et al.,* hypothesized that D-*myo*-inositol 1,2,6-trisphosphate (PP-56), by supplying *myo*-inositol to tissues and acting as an antioxidant, counteract some of the manifestations of diabetes [13]. Thus, they investigated the effects of PP-56 on platelet aggregation, fatty acids, and polyols in uncontrolled streptozotocin-induced diabetes in rats. A decrease in the platelet hyper-aggregation and sorbitol/*myo*-inositol ratio in platelets was observed in the rats treated by PP-56 for 7-8 weeks. There was also an associated decrease in incidence of cataract by 26 to 44%, which was statistically significant. PP-56 appeared to modulate fatty acid desaturases and aldose reductase in platelets and delay the development of cataract in this model of diabetes by a few weeks [13].

Endothelial dysfunction (ED) is an early feature of cardiovascular complications of diabetes and is attributed to excessive endothelial mitochondrial superoxide (ROS) production. Nasciemento *et al.,* [14] tested whether *myo*-inositol and dibutyryl D-*chiro*-inositol (db-DCI), would prevent or reverse ED in diabetic rats and rabbits. Orally administered inositols reduced hyperglycemia and hypertriglyceridemia and prevented ED in rat aortic rings and mesenteric beds. Inositols added *in vitro* to five diabetic tissues reversed ED. Relaxation by Ach, NO, and electrical field stimulation was potentiated by inositols *in vitro* in rabbit penile corpus cavernosa. DCI and db-DCI decreased elevated ROS in endothelial cells in high glucose and db-DCI reduced PKC activation, hexosamine pathway activity, and advanced glycation end products to basal levels. Inositols added to superoxide generated in an *in vitro* xanthine/xanthine oxidase system eliminated ROS in a dose-dependent manner, db-DCI being the most effective. The authors propose that inositols may exert antioxidant function by mechanism similar to InsP_6 as both require the presence of Fe^{3+}. However, as inert molecules as inositols are, the exact mechanism(s) how they would chelate iron to work as antioxidant remains enigmatic from a structure-function point of view, except that it may work through the xanthine/xanthine oxidase system (Chapter 16). Be that as it may, inositols also enhanced the action of nitrous oxide [14].

Additional Complications of DM

Congenital malformations involving the heart and the nervous system occur more often in children born to diabetic mothers. The incidence of an abnormality in the embryo (embryopathy) in diabetic mothers is 4 to 5 times higher than average. To investigate the mechanism(s) and possible interventional strategies, gestational diabetes was induced in Sprague-Dawley rats with streptozotocin. Of the diabetic groups, one received the usual diet, whereas the others received, 0.08, 0.16, and 0.5 mg/day supplemental *myo*-inositol orally. The incidence of neural tube defects was significantly reduced from 20.4% to 9.5% ($p < 0.01$) in the rats given *myo*-inositol supplementation (0.08 mg/day). Interestingly the investigators found that the most effective dosage of *myo*-inositol was 0.08 mg/day; increasing the dose of *myo*-inositol beyond that level did not significantly reduce the rate of neural tube defects. [15-17]. Reece and coworkers also tested the effect of oral supplemental cocktail of vitamin E (α-tocopherol, 400 mg/day), safflower oil (arachidonic acid, 1 ml/day), and *myo*-inositol (0.08 mg/day), on the outcome incidence of embryopathy in streptozotocin induced diabetes in Sprague-Dawley rats. As expected, the diabetic groups had a significantly higher ($p < 0.05$) mean blood glucose level than controls, and the insulin-treated group had glucose levels that were comparable to those of controls [17]. The un-supplemented diabetic group had a neural tube defect rate of 23.7% as opposed to 4.04% in controls and 3.55% in insulin-treated diabetics, difference being statistically significant at $p < 0.05$. The rate of neural tube defects was significantly reduced to the background level in animals receiving half-strength cocktail or stronger doses. However, diabetic animals receiving only quarter-strength cocktail did not demonstrate a significant reduction in the malformation rate [17]. Serum *myo*-inositol levels were not significantly different among the groups. However, serum levels of vitamin E were significantly higher in diabetics receiving half-strength cocktail. Superoxide dismutase activity was also significantly increased in diabetic animals receiving supplementation as opposed to animals not receiving it; there was a significant correlation ($r = 0.66$, $p < 0.05$) between the increases in vitamin E and superoxide dismutase activity [17].

On one hand, while their studies suggests that deficiency of inositol and other essential nutrients may play a very important role in diabetic embryopathy, supplementation with the deficient components and antioxidants could ameliorate the problem on the other [17]. They propose that excessive oxidation by oxygen free radicals and alterations in the levels of arachidonic acid, prostaglandins, and *myo*-inositol in the embryo to be the pathogenetic mechanism for diabetic embryopathy [18]. The investigators feel that yolk sac has an integral role in diabetic embryopathy, and arachidonic acid, *myo*-inositol, and other antioxidants offer significant promise for serving as a pharmacologic prophylaxis against diabetic embryopathy [15].

DISORDER OF FAT METABOLISM - FATTY LIVER

Fatty liver or fatty liver disease is a condition where large vacuoles of triglyceride accumulate within the liver cells. It has multiple etiological factors including alcoholism, obesity, malnutrition *etc*. Defects in fatty acid metabolism and peripheral resistance to insulin are mostly involved in the pathogenesis of fatty liver. While it is a reversible condition, if the causative factors persist, it can lead to ballooning of the hepatocytes, hepatocellular necrosis, inflammation and progression to fibrosis - cirrhosis with its own complication such as portal hypertension resulting in bleeding esophageal varicose veins.

Xenobiotics can induce hepatic lipogenic enzymes such as glucose-6-phosphate dehydrogenase and others in rats, resulting in fatty liver. Katayama and colleagues had demonstrated that equimolar amounts of $InsP_6$ and *myo*-inositol given as diet supplement reduced hepatic lipids and the lipogenic enzyme activities in rats fed 1,1,1-trichloro-2,2-*bis*(*p*-chlorophenyl) ethane (DDT) or a high-sucrose diet [19-21]. Supplemental *myo*-inositol elevated the hepatic-free *myo*-inositol level and recovered the reduced phosphatidylinositol (PI)/phosphatidylcholine (PC) ratio induced by DDT intake [22]. They also showed conclusively that dietary $InsP_6$ suppressed the elevations in serum cholesterol and phospholipids levels induced by DDT; however, *myo*-inositol had no significant influence on such elevations. The consistent and reproducibly suppressed increases in serum cholesterol reported by Katayama and colleagues supports earlier observation by Jariwalla *et al.*, who demonstrated a reduction in the serum cholesterol and triglyceride levels in animals given $InsP_6$ [23], but is in sharp contrast to that by Omuruyi *et al.*, mentioned earlier on streptozotocin-induced diabetes [7], wherein an increase in total cholesterol perhaps due to the combined effect of streptozotocin-induced diabetes and $InsP_6$ was observed.

Interestingly, and contrary to the dogma that $InsP_6$ is an "anti-nutrient", these investigators reported that dietary $InsP_6$, but not *myo*-inositol, caused significant body weight gain with or without DDT intake. Feeding $InsP_6$ with or without DDT caused a 31% or 15% increase in final body weights [22]. Food intake was lower in the DDT diet groups than in the normal groups. $InsP_6$ supplementation slightly increased the food intake in rats fed with or without DDT; however, the trends did not reach statistical significance. Food efficiency was also enhanced by $InsP_6$ especially in the rats given DDT. In contrast, *myo*-inositol supplementation had no effect on the body weight gain, food intake, or food efficiency. Additionally, supplemental $InsP_6$ and *myo*-inositol significantly increased hepatic-free *myo*-inositol regardless of DDT intake and prevented fatty liver in rats fed DDT.

While the mechanisms for these observed actions are unknown, these studies have much clinical significance in prevention and management of human fatty liver disease.

CONCLUDING REMARKS

To summarize, $InsP_6$ and inositol independently may help in diabetes by i) stimulating insulin secretion by the pancreatic β cells ($InsP_6$) and by ii) reducing the oxidative damage to the cells and tissues inflicted through AGE; both $InsP_6$ and inositol can prevent that as antioxidants. iii) Both $InsP_6$ and inositol could replenish the low intracellular inositol levels as neither of them requires insulin to gain entry into the cell; once inside the cells, $InsP_6$ can be dephosphorylated and enter into the intracellular inositol-inositol phosphate pool. That D-*myo*-inositol 1,2,6-trisphosphate (PP-56), by supplying *myo*-inositol to tissues and acting as an antioxidant have counteracted some of the manifestations of diabetes is fascinating. The positive effect of $InsP_6$ and inositol on the abnormalities of fat metabolism is also noteworthy. These exciting data beg for further studies in the field.

REFERENCES

[1]   Barker CJ, Berggren P-O. New horizons in cellular regulations by inositol polyphosphates: insight from the pancreatic β cells. Pharmacol Rev 2013; 65: 641-69.
[2]   Larsson O, Barker CJ, Sjöholm Å *et al.* Inhibition of phosphatases and increased Ca^{2+} channel activity by inositol hexakisphosphate. Science1997; 278: 471-4.
[3]   Lehtihet M, Honkanen RE, Sjöholm Å. Inositol hexakisphosphate and sulfonylureas regulate β-cell protein phosphatases. Biochem Biophys Res Comm 2004; 316: 893-7.
[4]   Efanov AM., Zaitsev SV, Berggren PO. Inositol hexakisphosphate stimulates non-Ca^{2+}-mediated and primes Ca^{2+}-mediated exocytosis of insulin by activation of protein kinase C. Proc National Acad. Sci/USA 1997; 94: 4435-9.
[5]   Illies C, Gromada J, Fiume R. Requirement of inositol pyrophosphate for full exocytotic capacity in pancreatic beta cells. Science 2007; 318: 1299-302.
[6]   Dilworth LL, Omoruyi FO, Simon OR *et al.* The effect of phytic acid on levels of blood glucose and some enzymes of carbohydrate and lipid metabolism. West Indian Med J 2005; 54: 102-6.

[7]   Omuruyi FO, Budiaman A, Eng Y, *et al*. The potential benefits and adverse effects of phytic acid supplement in streptozotocin-induced diabetic rats. Adv Pharmacol Sci 2013; 2013: 172494. doi: 10.1155/2013/172494. Epub 2013 Dec 22

[8]   Kim SM, Rico CW, Lee SC *et al*. Modulatory effect of rice bran and phytic acid on glucose metabolism in high fat-fed C57BL/6N mice. J Clin Biochem Nutr 2010; 47: 12-7.

[9]   Panlasigui LN, Thompson LU, Blood glucose lowering effect of brown rice in normal and diabetic subjects. Int J Food Sci Nutr 2006; 57: 151-8.

[10]   Thompson LU, Button CL, Jenkins CJ, Phytic acid and calcium affect the *in vitro* rate of starch digestion and blood glucose response in humans, Am J Clin Nutr 1987; 46: 467-73.

[11]   Yoon JH, Thompson LU. The effect of phytic acid on *in vitro* rate of starch digestibility and blood glucose response. Am J Clin Nutr 1983; 38: 835-42.

[12]   Lee H, Jeong C, Ghafoor K, *et al*. Oral delivery of insulin using chitosan capsules cross-linked with phytic acid. Biomed Mater Eng. 2011; 21(1): 25-36. doi: 10.3233/BME-2011-0654.

[13]   Ruf JC, Ciavatti M, Gustafsson T, *et al*. Effect of D-*myo*-inositol on platelet function and composition and on cataract development in streptozotocin-induced diabetic rats. Biochem Med Metab Biol. 1992; 48: 46-55.

[14]   Nascimento NR, Lessa LM, Kerntopf MR, *et al*. Inositols prevent and reverse endothelial dysfunction in diabetic rat and rabbit vasculature metabolically and by scavenging superoxide. Proceedings of the National Academy of Sciences USA 2006; 103: 218-23.

[15]   Reece EA, Eriksson UJ. The pathogenesis of diabetes-associated congenital malformations. Obstetrics & Gynecology Clinics of North America 1996; 23: 29-43.

[16]   Reece EA, Khandelwal M. Wu YK *et al*. Dietary intake of myo-inositol and neural tube defects in offspring of diabetic rats. Am J Obstet Gynecol 1997; 176: 536-9.

[17]   Reece EA, Wu Y-K. Prevention of diabetic embryopathy in offspring of diabetic rats with use of a cocktail of deficient substrates and an antioxidant. American J Obstetrics & Gynecology 1997; 176, 790-8.

[18]   Zhao Z, Reece EA. Experimental mechanisms of diabetic embryopathy and strategies for developing therapeutic interventions. J Soc Gynecol Invest 2005; 12: 549-57.

[19]   T. Katayama Effects of dietary myo-inositol or phytic acid on hepatic concentration of lipids and hepatic activities of lipogenic enzymes in rats fed on corn starch or sucrose. Nutr Res 1997; 17: 721-8.

[20]   Y. Okazaki, T. Katayama. Effects of dietary carbohydrate and myo-inositol on metabolic changes in rats fed 1,1,1-trichloro-2,2-bis (p-chlorophenyl) ethane (DDT). J Nutr Biochem 2003; 14: 81-9.

[21]   Okazaki Y, Kayashima T, Katayama T. Effect of dietary phytic acid on hepatic activities of lipogenic and drug-metabolizing enzymes in rats fed 1,1,1-trichloro-2,2-bis (p-chlorophenyl) ethane (DDT) Nutr Res 2003; 23: 1089-96.

[22]   Okazaki Y, Katayama T. Dietary inositol hexaphosphate, but not *myo*-inositol, clearly improves hypercholesterolemia in rats fed casein-type amino acid mixtures and 1,1,1-trichloro-2,2-bis(p-chlorophenyl) ethane. Nutr Res 2008; 28: 714-21. doi: 10.1016/j.nutres.2008.07.003

[23]   Jariwalla RJ, Sabin R, Lawson S, *et al*. Lowering of serum cholesterol and triglycerides and modulations by dietary phytate. J Appl Nutr 1990; 42: 18-28.

CHAPTER 11

Atherosclerosis and Cardiovascular Diseases

Abstract: Atherosclerotic cardiovascular disease results in serious life threatening conditions such as heart attack, stroke *etc*. Hypercholesterolemia is a risk factor; studies show that InsP_6 can lower serum cholesterol level. Oxidative damage seems to play an important role in the pathogenesis of atherosclerosis as well as in heart attack. Studies show that InsP_6 by virtue of its antioxidant function can prevent or ameliorate some of the deleterious sequel.

Keywords: Atheroma, cholesterol, HDL, heart attack, hypercholesterolemia, LDL, myocardial infarction, reperfusion injury, ROS, stroke, thrombosis.

ATHEROSCLEROSIS

Atherosclerotic cardiovascular disease is a major health problem globally, more so in the affluent societies. Atherosclerosis *i.e.,* hardening of arteries is also believed to have been prevalent since the ancient Egypt. It is characterized by *atheroma* or fibrofatty plaques in the innermost layer (intima) of the arteries. These atheromas protrude into the lumen thereby narrowing the luminal space, and weaken the media. The so diseased artery can be calcified, a fact noted as early as the sixteenth century as "degeneration of arteries into bone".

While atherosclerosis is extremely common in Europe and North America, it is much less prevalent in Africa, Asia and, South and Central America. Atherosclerosis is the cause of ischemic heart disease (atherosclerotic cardiovascular disease ASCVD) which causes morbidity and mortality; the United States has one of the highest rates. Japan on the other hand has one-fifth the mortality rate from heart attack than that of USA. However, the rate is increasing in Japan and Japanese who migrate to USA and adopt the life-style and dietary habit customs of their adopted home become more susceptible. This is true not just for atherosclerotic cardiovascular diseases, but also for cancers, especially that of the colon and rectum. This observation points to the role of environmental risk factors aside from the genetic predispositions, to pathogenesis of atherosclerosis. Diet, physical activity, cigarette smoking, diabetes, high serum cholesterol level and high blood pressure, are clearly very important in predisposing to ASCVD.

The first reported investigation into the pathogenesis of atherosclerosis made its *debut* with the publication by Ignatowski who reported a rabbit model of atherosclerosis in 1908 [1]. Subsequent studies led to the documentation of a strong relationship between elevated serum cholesterol level and atherosclerosis in various animal models [2]; and cholesterol finds its way into the wall of the arteries causing elevated plaques - atheroma.

Atheroma Formation

The atheromatous plaque consists of a focally raised plaque within the intima - the innermost of the three layers of the artery (the others are media - *tunica media* and the outer layer - *tunica adventitia*) of the arteries. It consists of a core of lipid mostly composed of cholesterol or cholesterol esters covered with a fibrous cap. Almost invariably, atheromas become calcified as the condition progresses; this is one of the commonest complications of atherosclerosis rendering the artery like pipe-stem with eggshell brittleness making the affected artery susceptible to more complications. Thus affected artery may undergo ulceration or even rupture, liberating the debris from the atheroma to the blood which may travel to distant sites (emboli) and cause damage therein. There could also be hemorrhage within the plaque acutely compromising the lumen of the vessel. And then, the thus damaged plaque could be the seat of thrombosis with partial or total obstruction of the lumen resulting in infarction of the organ *e.g.,* heart attack.

Insofar as the genesis of atherosclerosis is concerned, historically there have been two major schools of thoughts, each emphasizing either cellular proliferation in the intima or thrombosis in the arterial wall. The contemporary thought is that atherosclerosis is a chronic inflammatory response of the arterial wall secondary to an injury to the endothelial cells lining the inner surface (lumen) of the arteries. While endothelial injury has produced atheroma experimentally, early human lesions of atheroma begin at sites that show no evidence of injury, at least morphologically. That may point to other forms of injury or disruption of normalcy such as increased endothelial permeability, increased adhesion of leukocytes to the endothelium *etc*.

The adhesion of leukocytes to the endothelium is dependent on binding of the complementary adhesion molecules on the leukocytes and the endothelial cells; certain chemicals influence this binding ability by modulating the surface expression of these adhesion molecules or by modifying their avidity. There are several groups of adhesion molecules: the selectins, immunoglobulins, integrins, *etc*. Selectins are composed of three different types, depending on their location in the cell types: E-selectin (confined to endothelial cells), L-selectin in leukocytes hence "L" and P-selectin present in platelets, hence "P" (but also present in endothelium). Selectins bind through their lectin domain to sialylated forms of carbohydrate such as sialylated Lewis X.

$InsP_6$ and Atherosclerosis

Hypercholesterolemia

Hypercholesterolemia (high levels of blood cholesterol) has been considered to be a high risk factor in the pathogenesis of atherosclerotic cardiovascular disease. Cholesterol is manufactured primarily in the liver. Because cholesterol and other fats do not dissolve in water, they cannot travel through the blood plasma; however, low-density lipoprotein (LDL) formed in the liver transport cholesterol and other fats through the bloodstream. Cholesterol is returned to the liver from other body cells by another lipoprotein, high-density lipoprotein (HDL). From there, cholesterol is secreted into the bile, either unchanged or after conversion to bile acids. While cholesterol is essential for the formation of cell membranes and the manufacturing of some hormones, if blood cholesterol levels are elevated, large amounts of LDL (so-called "bad cholesterol") cholesterol deposit in the arterial walls in the atheromas.

As discussed briefly in the previous chapter (Chapter 10), Jariwalla *et al.*, demonstrated that rats fed a cholesterol-rich diet had a mean serum cholesterol level of 126 $mg.dL^{-1}$ compared to 110 $mg.dL^{-1}$ in control. There was a 19% and 32% reduction in the serum cholesterol level respectively in the control chow group of animals given $InsP_6$ and high cholesterol + $InsP_6$ group, the mean value in both groups being about 87 $mg.dL^{-1}$ [3, 4]. Similar reduction was also observed in the serum triglyceride levels.

In another model of induced hypercholesterolemia in rats by 1,1,1-trichloro-2,2-bis(*p*-chlorophenyl) ethane (DDT), Okazaki and Katayama investigated the potential hypocholesterolemic action of *myo*-inositol and $InsP_6$ [5]. Following up on a previous study showing that $InsP_6$ and myo-inositol prevent fatty liver in rats fed a casein-based diet containing DDT, they examined the comparative effects of dietary equimolar amounts of sodium $InsP_6$ and *myo*-inositol (0.2%) on the development of DDT-induced fatty liver and hypercholesterolemia in rats fed 20% casein-type amino acid mixtures. Dietary supplementation with $InsP_6$ suppressed the elevations in serum concentrations of cholesterol and phospholipids; *myo*-inositol however, had no significant effect on the elevated lipid levels. As an aside, the investigators noted that supplemental InsP_6 and *myo*-inositol significantly increased hepatic-free *myo*-inositol regardless of DDT intake and prevented fatty liver in rats fed DDT [5].

But, how does InsP_6 lower the serum cholesterol is not known. It is possible that InsP_6 acts *via* mechanisms similar to *myo*-inositol through depression of hepatic lipogenesis.

Role of Neutrophils

InsP_6 has been shown to inhibit the binding of L- and P-selectin to sialyl Lewis X [6]. Shimazawa *et al.,* investigated the role of leukocytes in the pathogenesis of atherosclerosis whether neutrophil accumulation participate in the development of intimal hyperplasia after endothelial injury in mice [7]. They also tested whether InsP_6, which inhibits the binding of L- and P-selectin to sialyl Lewis X, could inhibit the development of intimal hyperplasia. Endothelial injury was induced in the femoral artery of 70 mice *via* the photochemical reaction between systemically injected Rose Bengal and transillumination with green light (wavelength: 540 nm). An increase in the number of leukocytes adhering to the injury site was observed by scanning electron microscopy 3 days after injury. The mice were treated with antibody to produce neutropenia (reduced number of neutrophils) that resulted in a significant decrease in the intimal area in the injured vessel compared to the control group. Treatment with InsP_6 produced a significant decrease in the intimal area ($p < 0.05$ for 100 µM, $p < 0.01$ for 300 µM InsP_6); administration of *myo*-inositol had little or no effect. Their data suggest that a) neutrophil accumulation on the injured vessels may contribute to initiation and development of intimal hyperplasia and that b) either neutropenia or InsP_6 inhibit that process [7]. InsP_6 is considered to do this by blocking L- and P- selectin binding to selectin ligands to inhibit leukocyte accumulation in inflammation [7]. InsP_6 directly inhibited the migration and proliferation of cultured human coronary artery smooth muscles by platelet-derived growth actor (PDGF) PDGF-BB at ≥100 µM. Interestingly, InsP_6 but not *myo*-inositol inhibited neutrophil accumulation in the peritoneal cavity of mice after thioglycollate injection [7].

Role of Platelets

Platelets become activated at the site of vascular injury. While on one hand, platelet activation at the site of damage of a blood vessel to form clot is essential for arresting the bleeding, excessive platelet activation on the other hand can result in the unwarranted formation of arterial thrombi, the consequence of which is precipitating acute myocardial infarction, or stroke. The three main events in thrombus formation are that i) the platelets must adhere to the site of injury (platelet adhesion), ii) the platelets have to aggregate with each other and finally iii) the platelets have to be activated - main trigger being damaged endothelium. Human platelets contain matrix metalloproteinase 2 (MMP-2) and release it upon activation; active MMP-2 amplifies the platelet aggregation response and plays a critical role in thrombus formation. Therefore, adhesion of platelets to the inner layer of the artery, their aggregation and finally activation are the earliest events in the formation of thrombus and atherosclerosis. Thus, not surprisingly a lot of effort to reduce atherosclerotic cardiovascular diseases is directed towards blocking or slowing down some of these steps. The antiplatelet therapy (aspirin is the most familiar one) aims to decrease the activity of circulating platelets and block the secretion of platelet granules containing the platelet-derived thrombotic factors. And the most clinically relevant test for monitoring the state of platelet health insofar as thrombogenic potential is platelet aggregation assay.

Shamsuddin and colleagues looked at the ability of $InsP_6$ to inhibit human platelet aggregation from 10 healthy volunteers. The platelets were activated by known activators - adenosine diphosphate (ADP), collagen and thrombin, in the presence or absence of $InsP_6$. $InsP_6$ significantly ($p < 0.0001$) inhibited platelet aggregation in a dose-dependent manner [8] (Fig. **11.1**).

$InsP_6$ also significantly reduced the release of ATP from the platelet granules ($p = 0.00247$ for ADP, $p = 0.0074$ for collagen and $p = 0.0069$ for thrombin). Thus, it is quite plausible that $InsP_6$ may help prevent atherosclerotic cardiovascular diseases through this mechanism.

Calcification of Atherosclerotic Plaques

As mentioned earlier, one of the complications of atherosclerosis is calcification of the plaques which renders the arteries susceptible to thrombosis, rupture *etc.* Patients with a high degree of calcification of coronary arteries are at increased risk for these coronary artery events. Calcification of the arteries and heart valves is also age-related, with increased age there is increased calcification, the latter contributing to heart failure as well.

Grases *et al.,* evaluated the effect of $InsP_6$ on cardiovascular calcification in rats during aging. Ten weeks old rats were fed with either a balanced diet containing

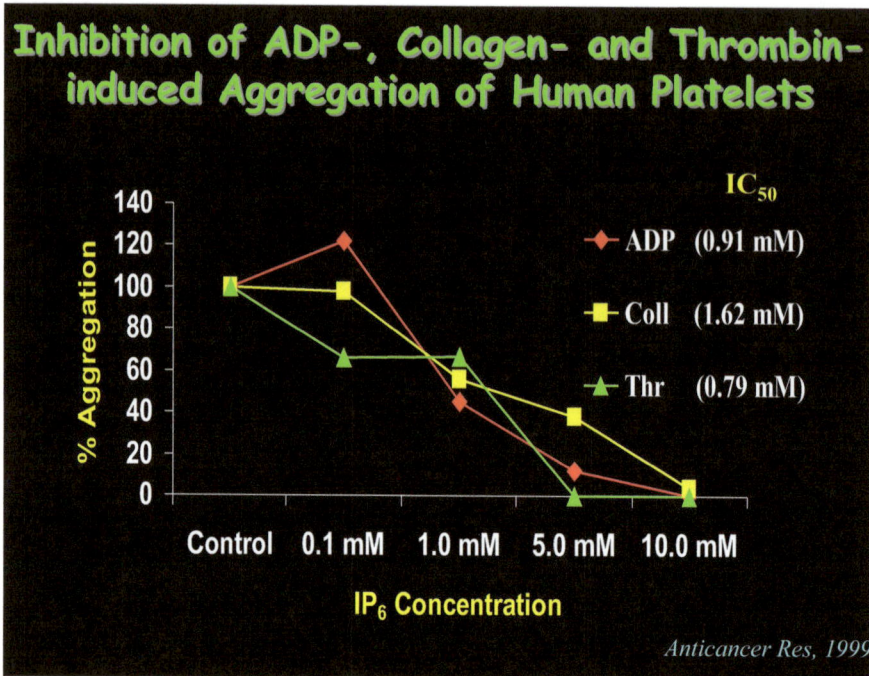

Fig. (11.1). Increased concentration of $InsP_6$ increasingly inhibits abnormal platelet aggregation.

$InsP_6$, or a purified diet that had no detectable $InsP_6$ or purified diet + $InsP_6$ and purified diet with inositol. At 76 weeks of age, all of the rats were sacrificed, and the aortas, hearts, kidneys, livers and femurs were removed for chemical analysis. The most significant result was the difference in the calcium content of the aorta. $InsP_6$-treated rats had significantly lower levels of calcium in the aorta compared to rats fed diet without $InsP_6$, thus demonstrating that $InsP_6$ treatment significantly reduced age-related calcification of the aorta [9, 10].

Grases and colleagues further extended this investigation to other tissues besides the aorta. They induced abnormal calcification in rats by giving them intramuscular injection of high doses of Vitamin D and oral nicotine [11]. After 60 hours of calcinosis treatment, all of the rats not treated with $InsP_6$ died. A highly significant increase in the calcium content in the aorta and heart muscle was observed in the control $InsP_6$ non-treated rats (21 ± 1 mg calcium/g dry aorta tissue, 10 ± 1 mg calcium/g dry heart tissue) when compared with controls (1.3 ± 0.1 mg calcium/g dry aorta tissue, 0.023 ± 0.004 mg calcium/g heart dry tissue). Compared to the $InsP_6$ *non*-treated rats, the $InsP_6$-treated animals had much lower calcium content not just in the aorta (0.9 ± 0.2 mg calcium/g dry aorta, about same as the healthy control) but also in the heart (0.30 ± 0.03 mg calcium/g dry heart tissue) [11].

Heart Attack

Free Radicals and Antioxidant Property

Occlusion of the artery feeding a tissue or organ causes ischemic injury as in heart attack (myocardial infarction) from coronary artery blockage, or stroke following carotid artery obstruction. The cells in the affected area are deprived of oxygen; as a result, they are irreversibly injured and die. The infarcted tissue however does not show a uniform picture - areas of dead cells interspersed with still viable ones can be seen microscopically. Patients with acute myocardial infarction are at a risk of heart failure, irregular heart beat (arrthymia) and death. Thus, early restoration of the blood supply that has been cut-off from the heart is not only desirable but also essential for salvaging the heart. Paradoxically, this restoration or reperfusion can exacerbate myocardial damage - a process called reperfusion injury; depending on the duration and severity of the ischemic event, a variable number of cells die *after* the blood flow resumes. One of the mechanisms responsible for this reperfusion injury is the reactive oxygen species (ROS). Furthermore, oxygen-derived free radicals reduce myocardial contractility and ventricular function, rendering the heart prone to failure.

Pre-treatment with some antioxidants, such as pyrrolidine dithiocarbamate (PDTC) or N-acetyl-cysteine, as well as some vitamins with recognized antioxidant properties, namely ascorbic acid (vitamin C), all-trans Retinoic Acid (atRA) and α-tocopherol (vitamin E) can suppress oxidative stress-induced tissue factor expression in human coronary artery endothelial cells [12]. Since the chelation of Fe^{3+} by iron-chealtors such as deferoxamine mesylate reduced myocardial re-perfusion injury, Rao and investigated whether $InsP_6$ by virtue of its antioxidant property would also protect the heart from reperfusion injury *via* the Fenton Reaction as described in Chapter 16 [13].

$InsP_6$ was injected intravenously to rats in three different dose levels 30 minutes before the hearts were removed from the body. The minimum dose was 15 mg/kg body weight, maximum dose of 150 mg/kg and an intermediate dose of 75 mg/kg. The control group received saline solution only. Global ischemia was induced in isolated hearts for 30 minutes, followed by another 30 minutes of reperfusion. Not unexpectedly reperfusion was associated with increased creatine kinase release (release of creatine kinase is a sensitive indicator of myocardial infarction) and reduced coronary artery flow. As evidenced by reduced left ventricular developed pressure and the first derivative of left ventricular pressure, the ventricular function also recovered poorly. $InsP_6$ did not alter left ventricular contractility before ischemia. However after 30 minutes of reperfusion following the ischemic insult, only the hearts from animals in the intermediate- and high-dose $InsP_6$-treatment group (equivalent to 5.25 and 10.5 g $InsP_6$ for a 70 kg human), but not the low dose-group showed better recovery of left ventricular developed pressure and the first derivative of left ventricular pressure compared to the control untreated group; the results were statistically significant at $p < 0.05$. Insofar as coronary artery flow was concerned, $InsP_6$ did not increase the coronary artery flow significantly ($p > 0.05$). However, during the re-perfusion period, hearts from $InsP_6$-treated rats showed a higher recovery of coronary artery flow as compared to the untreated saline controls [13].

The investigators examined free radical formation and quenching by producing •OH by xanthine oxidase on hypoxanthine in the presence of $FeCl_3$. A specific •OH scavenger dimethylsulfoxide decreased •OH signal. The formation of •OH in the hearts was quantitated by measuring •OH salicylate product 2,3- and 2,5-dihydroxybenzoic acid with high pressure liquid chromatography coupled with an electrochemical detector. The formation of •OH was significantly ($p < 0.05$) reduced in the hearts of animals receiving the intermediate and high dose of $InsP_6$ [13].

InsP₇ and Inhibition of Kinase

Akt is a serine/threonine specific protein kinase a.k.a. Protein Kinase B (PKB) which is responsible for a multitude of cellular functions including cell survival. Genetically modified mesenchymal stem cells over-expressing the Akt were markedly resistant to hypoxic injury. Thus, increased activation of Akt in mesenchymal stem cells may results in enhanced cardio-protection after transplantation. 5-Diphosphoinositol pentakisphosphate ($InsP_7$), formed by a family of inositol hexaphosphate kinases (IP6Ks) is a physiologic inhibitor of Akt. Zhang *et al.,* investigated the role of IP6Ks for improving mesenchymal stem cells' functional survival and cardiac protective effect after transplantation into infarcted mice hearts [14]. Bone marrow-derived mesenchymal stem cells were preconditioned with a purine analog TNP - (N6-(*p*-nitrobenzyl purine)) which is an inhibitor of IP6Ks resulting in significantly decreased $InsP_7$ production with concomitant increased Akt phosphorylation. Moreover, TNP at 10 $\mu mol.L^{-1}$ concentration significantly improved the viability and enhanced the paracrine effect of mesenchymal stem cells. The investigators demonstrate that mesenchymal stem cell therapy with IP6Ks inhibition significantly decreased fibrosis and preserved heart function in mice *in vivo*. This study demonstrates that inhibition of IP6Ks promotes mesenchymal stem cells engraftment and paracrine effect in infarcted hearts at least in part by down-regulating $InsP_7$ production and enhancing Akt activation, which might contribute to the preservation of myocardial function after MI. From a mechanistic point of view, the preceding begs the question: what happened to the level of $InsP_6$ as a result of inhibiting IP6Ks, as it is supposed to go up? Therefore, could the observed effect be the result of increased $InsP_6$.

CONCLUDING REMARKS

Experimentally at least two groups have independently demonstrated that $InsP_6$ lowers serum cholesterol and other lipids. In addition, $InsP_6$ by inhibiting platelet aggregation and atheroma formation can prevent the inception of atherosclerotic cardiovascular diseases. Following a catastrophic event of myocardial infarction, $InsP_6$ may also limit the extent of damage in the affected heart muscle through its antioxidant function. All these things considered, $InsP_6$ offers to be an important agent in the prevention and management of atherosclerotic cardiovascular diseases. Inositol pyrophosphate $InsP_7$ may also play a role in cardioprotection; this would be an exciting area for further research.

REFERENCES

[1]   Ignatowski AC. Influence of animal food on the organism of rabbits. S Peterb Izviest Imp Voyenno-Med Akad. 1908; 16: 154-173
[2]   Kapourchali FR, Surendiran G, Chen L, *et al.* Animal models of atherosclerosis. World J Clin Cases 2014; 2: 126-132. doi: 10.12998/wjcc.v2.i5.126
[3]   Jariwalla RJ, Sabin R, Lawson S, *et al.* Lowering of serum cholesterol and triglycerides and modulations by dietary phytate. J Appl Nutr 1990; 42: 18-28.
[4]   Jariwalla RJ. Inositol hexaphosphate as an anti-neoplastic and lipid lowering agent. Anticancer Res 1999; 19: 3699-702.
[5]   Okazaki Y, Katayama T. Dietary inositol hexaphosphate, but not *myo*-inositol, clearly improves hypercholesterolemia in rats fed casein-type amino acid mixtures and 1,1,1-trichloro-2,2-*bis*(*p*-chlorophenyl) ethane. Nutr Res 2008; 28: 714-21. doi: 10.1016/j.nutres.2008.07.003
[6]   Cecconi O, Nelson RM, Roberts WG, *et al.* Inositol polyanions. Non-carbohydrate inhibitors of L- and P-selectin that block inflammation. J Biol Chem 1994; 269: 15060-6.
[7]   Shimazawa M, Watanabe S, Kondo K, *et al.* Neutrophil accumulation promotes intimal hyperplasia after photochemically induced arterial injury in mice. European J Pharmacology 2005; 520: 156-63.
[8]   Vucenik I, Podczasy JJ, Shamsuddin AM. Antiplatelet activity of inositol hexaphosphate (IP$_6$). Anticancer Research 1999; 19: 3689-93.
[9]   Grases F, Sanchis P, Perelló J *et al.* Phytate reduces age-related cardiovascular calcification. Frontiers in Bioscience 2008; 13: 7115-22.

[10]   Grases F, Sanchis P, Perello J *et al.* Phytate (Myo-inositol hexakisphosphate) inhibits cardiovascular calcifications in rats. Front Biosci 2006; 11: 136-42.

[11]   Grases F, Sanchis P, Perelló J *et al.* Effect of crystallization inhibitors on vascular calcifications induced by vitamin D: a pilot study in Sprague-Dawley rats. Circ J 2007; 71: 1152-6.

[12]   De Rossa S, Cirillo P, Paglia A *et al.* Reactive oxygen species and antioxidants in the pathophysiology of cardio-vascular disease: does the actual knowledge justify a clinical approach? Curr Vasc Pharmacol 2010; 8: 259-75.

[13]   Rao PS, Liu XK, Das DK *et al.* Protection of ischemic heart from reperfusion injury by myo-inositol hexaphosphate, a natural antioxidant. Ann Thoracic Surg 1991; 52: 908-12.

[14]   Zhang Z, Liang D, Gao X *et al.* Selective inhibition of inositol hexakisphosphate kinases (IP6Ks) enhances mesenchymal stem cell engraftment and improves therapeutic efficacy for myocardial infarction. Basic Res Cardiol 2014; 109: 417. doi: 10.1007/s00395-014-0417-x. Epub 2014 May 22.

CHAPTER 12

Mineralization Disorders

Abstract: Owing to the ability to chelate calcium, InsP_6 has been studied as a preventive and therapeutic agent for abnormal calcification in various tissues such as blood vessels in atherosclerosis and skin in *calcinosis cutis*; kidney and salivary gland stones; prevention of dental caries and bone loss as in osteoporosis, *etc*. The results are very promising for wide-range practical application of InsP_6 especially against the serious debilitating problem of osteoporosis amongst the seniors, globally.

Keywords: Calcinosis cutis, caries, crystallization inhibitor, dystrophic calcification, fluoride, hydroxyapatite, kidney stone, metastatic calcification, osteoblast, osteoclast, pyrophosphate, sialolithiasis, titratable acidity, tooth decay.

INTRODUCTION

There are many roles of calcium in the body - intracellular communication (signal transduction), electrolyte balance and electrical conductivity, bone formation *etc*., functions vital to our survival. An acutely decreased serum calcium level can result in life-threatening cardiac arrhythmias or laryngeal spasm, whereas calcium deficiency over an extended period results in bone diseases. And excess calcium is not healthy either.

In the bone and teeth, calcium is present as hydroxyapatite with the molecular formula of $Ca_5(PO_4)_3(OH)$. Being a major component of bone and teeth, it imparts the rigidity of these structures. Hydroxyapatite molecules can crystalize to form clumps and when abnormally deposited in joints, tendons *etc*., result in painful inflammation.

PATHOLOGICAL CALCIFICATION

Cardiovascular Calcification

When there is abnormal deposition of calcium salts, together with a small amount of iron, magnesium and other mineral salts, in the tissue, it is called pathologic calcification. When the deposition takes place locally in diseased or dead tissues or cells, it is called *dystrophic calcification*. In disorders of calcium metabolism and hormone imbalances, calcification may take place in otherwise healthy normal tissue; this is called *metastatic calcification*. Of course if the serum

A.K.M. Shamsuddin and Guang-Yu Yang

calcium level is already high, any damage, degeneration or necrosis is likely to enhance pathological calcification.

Common examples of dystrophic calcification are those in our arteries as in advanced atherosclerosis. There are several factors that govern arterial wall calcification. Foremost is nucleation of calcium phosphate (hydroxyapatite) by dead cells and/or their membranes (dystrophic calcification). In mammalian tissues, there are several cellular proteins that regulate calcification. These proteins can stimulate or inhibit macrophage activities that destroy hydroxyapatite deposits. A deficiency in the level of calcification repressor factors, cellular defense mechanisms and/or crystallization inhibitors would be other critical components [1-4].

Félix Grases and his colleagues induced calcific atherosclerosis in rats by Vitamin D and nicotine and treated them with standard cream with or without $InsP_6$. They observed a highly significant increase in the calcium content of aorta and heart tissue in the non-$InsP_6$ treated rats (21 ± 1 mg calcium/g in aorta and 10 ± 1 mg calcium/g in heart) as compared to only 1.3 ± 0.1 mg and 0.023 ± 0.004 mg in the control group respectively. When they measured the calcium content in the $InsP_6$-treated rat aorta and heart tissue, the amounts they found were similar, if not even lower than the control: 0.9 ± 0.2 mg calcium/g in aorta and 0.30 ± 0.03 mg calcium/g in the heart. Only $InsP_6$ non-treated rats displayed important mineral deposits in aorta and heart. Thus, the findings are consistent with the action of $InsP_6$, as an inhibitor of calcification of cardiovascular system [1-3].

Since bovine pericardium is used as a source for prosthetic valves which undergoes calcification, inhibition of the process would prolong the lifespan of the implants. Thus, Grases laboratory extended their study to calcification inhibition in pericardium. Using bovine pericardium in an *in vitro* system, they tested the ability of $InsP_6$ to inhibit crystallization and compared with pyrophosphates and bisphosphonate - etidronate. The investigators show that while all the tested compounds inhibited calcification, $InsP_6$ yielded the best results [4].

Calcinosis Cutis

Calcinosis cutis is another example of pathologic calcification, of the skin and is seen in a variety of conditions such as systemic sclerosis, dermatomyositis, as complication of subcutaneous or intra-muscular injections, *etc*. Grases and coworkers tested the results of topically administered $InsP_6$ skin cream on artificially induced dystrophic calcifications in soft tissues [5]. For this experiment the sodium salt of $InsP_6$ was chosen over the calcium-magnesium salt due to the higher water solubility of the sodium salt. This study-design also allowed the evaluation of the capacity of $InsP_6$ penetration into the organism through the skin.

Rats were fed with an $InsP_6$-free or diet with $InsP_6$. Plaque formation was induced by subcutaneous injection of 0.1% $KMnO_4$ solution [5]. From 4 days before plaque induction to the end of the experiment, control rats were treated topically with a standard cream, whereas the experimental rats were treated with the same cream with 2% Na- $InsP_6$. Calcification of plaques was allowed to proceed for 10 days. Topical $InsP_6$ in the moisturizing cream resulted in a statistically significantly reduced plaque-size and weight when compared to the control group (1.6 ± 1.1 mg $InsP_6$-treated v 26.7 ± 3.0 mg control) [5]. The urinary $InsP_6$ levels of animals treated with the $InsP_6$-enriched cream were considerably (statistically significant) higher than those found in animals treated topically with the cream without $InsP_6$ (16.96 ± 4.32 mg L^{-1} $InsP_6$-treated v 0.06 ± 0.03 mg L^{-1} control) indicating the absorption of $InsP_6$ through skin, validating the usefulness of incorporating $InsP_6$ in skin cream for not just *calcinosis cutis* but also in radiation protection such as in sunscreen, moisturizing cream *etc.*, as discussed in Chapter 9.

The other more common and even more painful examples of pathological calcifications are abnormal stone formation in the kidney, gall bladder, and less commonly in the salivary glands (sialolithiasis); these are discussed next:

Kidney Stones (Urolithiasis or Renal Calculi)

Passing a kidney stone can be an excruciatingly painful event. From 1% to 5% of the population form renal calculi (stones), of which 80% to 95% of those are composed of calcium oxalate or calcium phosphate. Fortunately for many, they do not experience symptoms; other pass small sand-like sediments or stones without any complications. Unfortunately, for those who have a stone lodged in the ureter (the tube running to the bladder from the kidney) suffer severe pain, often described as 'excruciating'.

In Europe and North America, the incidence of kidney stone has been rising since the late 19th century; and in Japan since World War II. This rise in incidence correlates closely with dietary change towards a more refined diet, which typically offers less $InsP_6$. It is estimated that approximately 5% of American women and 12% of men will have a kidney stone at some time in life with an estimated cost of $2.1 billion annually, not to mention of the pain and suffering.

Hospitalization statistics of South African Blacks provides more convincing evidence as to the value of $InsP_6$ containing foods for the prevention of kidney stones. Modlin reported that from 1971 to 1979, 1 of every 510 White patients admitted to the teaching hospital of the Medical School of Cape Town, South Africa were for kidney stones; in sharp contrast, only 1 of 44,298 Black admissions was for kidney stones; an almost a 90 fold difference [6]. In Cape Town in 1970, 5.1 million Blacks and 4.5 million Whites lived in the urban areas. Moving from the rural to urban areas generally results in a more "city-like" diet for these Blacks. However, their diet is still based on daily maize or corn consumption of about 680 grams, especially in the first generation of new city dwellers. As corn may contain up to 6% $InsP_6$, it represents 40.8 grams of $InsP_6$ daily. Coupled with the *in vitro* inhibition of crystallization of calcium phosphate in metastable solution by $InsP_2$ and $InsP_3$, this seems like a logical reason to pursue the application of $InsP_6$ in prevention of kidney stones [6].

A different approach was taken by Henneman and colleagues even earlier, back in the mid 1950's. Buoyed by their success in decreasing urinary calcium excretion in sarcoidosis patients [7], they investigated the effect of $InsP_6$ in idiopathic hypercalcuria, a condition associated with increased incidence of kidney stones. Henneman and his associates published the clinical results of 35 men using Na-$InsP_6$ [8]. The patients had normal blood levels of calcium, but increased levels in their urine (idiopathic hypercalcuria). The men were given 8.8 grams of sodium $InsP_6$ orally in divided doses. The patients' urinary calcium levels returned to normal; 10 of the patients took $InsP_6$ for an extended period (on average 24 months) and only 2 of these 10 developed a kidney stone. $InsP_6$ normalized the hypercalcuria and significantly lessened kidney stone growth and recurrence [8]. The treatment of the day was a low calcium diet (avoidance of dairy) and increased fluid intake, which offered limited success. One can't help but ask the rhetorical question: why if a safe and much more effective treatment was revealed back in the 1950's, why was it not followed up and investigated further, even with the fame and prestige of Harvard Medical School, Massachusetts General Hospital and worldwide circulation of *The New England Journal of Medicine*?

Four decades later, Grases and colleagues showed a significant inhibition of calcium oxalate crystallization, which is a key event required in kidney stone formation with $InsP_6$ in an *in vitro* study. In a clinical study with 74 patients (who were calcium oxalate stone formers) he and his colleagues showed that the risk of calcium stone formation was reduced within 2 weeks after being treated with 120 mg of calcium-magnesium $InsP_6$ (Lit-Stop™) daily in contrast to the control patients [9].

Again, about a half-century after Henneman's publication, Curhan and associates also at Harvard University, published results from the Nurses' Health Study II. A total of 96,245 women who had no prior history of kidney stones filled out detailed food frequency questionnaires in 1991 and 1995. The investigators prospectively examined during an 8 year period the association between dietary factors and the risk of symptomatic kidney stones. There were two very interesting main conclusions from the study: First, "A higher intake of dietary calcium decreases the risk of kidney stone formation in younger women, but supplemental calcium is not associated with risk" [10]. This finding completely contradicts the strategy of limiting calcium intake as has often been suggested. Secondly, that $InsP_6$ may be a "new, important, and safe addition to our options for stone prevention" concluded the authors [10].

There are four main types of kidney stones: i) the vast majority (approximately 70%) are calcium-containing, either calcium oxalate (CaOx) or a mixture of calcium oxalate + calcium phosphate (CaP); ii) about 15% are the so-called triple stones (a.k.a. struvite stones); iii) uric acid stone (5% - 10%) and finally iv) the rare (1% - 2%) cysteine stones. The most important determinant is an increased urinary concentration of the stones' constituents to the point that the urine becomes supersaturated.

However, the formation of kidney stones involves both an alteration in the composition of the urine and damages in the renal tissue such as necrosis, calcification *etc*. Thus, not all people with high blood calcium or citrate or uric acid level will develop a kidney stone unless their kidneys have some damaged tissue as well. In the very first step a crystal has to form, then it has to grow and then many such growing crystals have to aggregate to each other before it becomes large enough to become a fully grown stone.

Approximately 1 to 3% of ingested $InsP_6$ is excreted from the body *via* the urine (the majority of the $InsP_6$ is used by the cells and eventually gets broken down and in fact is expired out the lungs as carbon dioxide or CO_2) [11, 12]. Experiments by Grases *et al.,* point to $InsP_6$'s inhibition of crystallization as the mechanism for kidney stone prevention [9]. While being excreted out through the kidneys, $InsP_6$ is able to bind and remove calcium atoms that make up the kidney stones. Bound together, InsP_6 and excess calcium are then both eliminated from the body *via* the urine. Since a kidney stone contains millions of atoms of calcium, such removal by InsP_6 is akin to removing grains of sand one by one from a sandbox; which clinically may take a long time. Studies by Saw *et al.,* [13] have shown that InsP_6 can remove calcium from the artificial urine. In addition to decreasing the level of ionized calcium, InsP_6 enhances the barrier to heterogeneous nucleation, inhibits its crystallization activity and inhibits *in vitro* stone growth. Thus InsP_6 provides a safe and effective treatment for kidney stone; it is simply a case of having enough InsP_6 in the urine to perform the function.

Once a kidney stone is detected, chances are that it will recur. As such, a preventive strategy is called for. As seen from the preceding, InsP_6 has the dual function of prevention as well as treatment.

Sialolithiasis (Salivary Gland Stone)

Sialolithiasis means calcific stone formation in the ducts of the salivary glands, most commonly in the submandibular gland. As expected, these stones result in pain and blockage of the salivary duct with resultant inflammation, causing more pain. Grases and colleagues chemically analyzed 21 such stones and the saliva from the patients with sialolithiasis. Salivary calcium was significantly higher in 18 of the 21 patients. More interestingly, InsP_6 concentration in the saliva of these 18 patients was significantly lower than that found in saliva of healthy individuals. This study shows that InsP_6 is present in the normal saliva, and thus taking InsP_6 supplement could raise salivary InsP_6 content and help prevent sialolithiasis [14].

Having discussed the presence of InsP_6 in saliva and its role in prevention of stone in salivary gland, it is a logical sequence discuss the other more conspicuous, and important structure - the teeth.

EROSION OF THE TEETH

Dental Caries

The commonly known 'cavity' or tooth decay is a bacterial infection that results in demineralization and destruction of the hard tissue (enamel, dentin and cementum) of the teeth. Caries-causing bacteria and fermentable carbohydrates such as sucrose are essential factors that produce caries in time. That the removal of InsP_6 from cereals and sugars during their industrial processing and refinement may be responsible for the increase in dental caries potential had been suggested as early as in the late 1950's [15]. Experimentally, $InsP_6$ adsorbs to tooth enamel reducing the rate of dissolution of the latter during the caries process *in vitro* through alteration of surface charge and free energy characteristics of the surface thereby affecting the affinity of salivary proteins and bacteria, and interfering with plaque formation [16]. Studies *in vivo* have confirmed the inhibitory effect of $InsP_6$ on dental caries in experimental rats. Kaufman and Kleinberg investigated whether there would be a difference between the ability of $InsP_1$, $InsP_3$, $InsP_4$ and $InsP_6$ in reducing enamel solubilization in acid *in vitro* [15]. $InsP_6$ yielded the best results, with $InsP_3$ and $InsP_4$ having similar effects, and $InsP_1$ being the least effective. These investigators also demonstrated that $InsP_6$ acts synergistically with fluoride, a well-known inhibitor of tooth decay [17]. And, given the data on crystallization inhibition and prevention of dystrophic calcification described above, $InsP_6$ could most likely prevent dental calculus formation as well.

Carbonated Soft Drinks

The pH of saliva should ideally be between 6.5 and 5.5. It is generally believed that the pH of 5.5 is the threshold for developing dental caries. Prolonged exposure to < pH 5.5, or frequent cycling from optimal (neutral) pH to below the threshold (pH 5.5) can result in rapid de-mineralization of the enamel. Lowered salivary pH is often a result of bacterial digestion of various sugars such as sucrose, fructose *etc.* This causes accumulation of acidic byproducts in the dental plaques. The consumption of citrus fruits and soft-drinks may be a major factor in causing dental erosion which is an irreversible process. Carbonated soft-drinks have a low pH and contain sugar and other additives thereby putting the enamel at a very high risk of dissolution and/or erosion. In the pathogenesis of dental erosion, it is the total acid level called titratable acidity, rather than the pH that is considered important [18].

The high ingestion of soft-drinks poses a major public health problem not just in the USA, but globally as it is replacing milk consumption, thus reducing the recommended dietary intake of calcium. Hara & Zero [18] therefore investigated whether adding calcium to the beverages would affect their enamel-erosive action. Ten commercially available beverages, 5 with and 5 without calcium supplementation were tested. They demonstrated a lower level of enamel demineralization with calcium supplemented beverages.

von Fraunhofer & Rogers [19] examined the relative rate of enamel dissolution in a variety of carbonated soft-drinks. Caries-free human teeth were placed in *Coca Cola, Pepsi Cola, Dr. Pepper, Mountain Dew, Ginger Ale* and *Sprite* for 14 days. The cola beverages (*Coca Cola, Pepsi Cola* and *Dr. Pepper*) showed enamel dissolution (weight loss) of about 3 mg/cm^2 whereas the non-cola drinks (*Mountain Dew, Ginger Ale* and *Sprite*) showed a 2-5 times greater dissolution. Jain *et al.,* [20] also showed that prolonged exposure to soft drinks can lead to significant enamel loss, and that the non-cola drinks are more erosive than cola drinks. Cola and non-cola drinks with sugar were more erosive than their 'diet' counterparts.

Soft-drinks are extremely popular not just in the developed countries, but thanks to effective marketing and promotion, in the developing world as well. In the United States alone, the yearly sale has been estimated to be over \$65 billion with a 30% annual growth rate. The omnipresent vending machines now also in the schools, are making the consumption even higher, particularly amongst young adults and children. Most soft-drinks are acidic in pH and their exposure to teeth may result in enamel erosion; it is therefore no surprise that our children have poor dental health, globally.

Experiments with InsP_6 have shown that addition of InsP_6 + inositol reduced the % titratable acidity (TA) of soft drinks. Subsequent studies have shown that additions of InsP_6 + inositol to a variety of beverages, including *Fresca* and *Red Bull*, reduced the % titratable acidity to close to 0. Further studies were performed on extracted human teeth, either on sections of enamel dissected off the crowns or intact teeth with the root portion of the teeth beneath the enamel/dentin junction coated with protective varnish. These enamel specimens were immersed in the various soft drinks with or without additions of 0.5 and 1.0% by weight of InsP_6. The enamel dissolution was determined as the weight loss of the enamel at different time intervals in the untreated and treated beverages [21].

As can be seen in the following Figs. (**12.1, 12.2** and **12.3**), addition of InsP_6 results in marked protection of the teeth from the erosive effect of soft-drinks.

As Grases *et al.,* have shown, even orally administered InsP_6 is protective of the salivary gland stones; and InsP_6 is excreted in the saliva to serve its function. But a more direct approach such as addition to the drinks, either at the time of manufacturing or at the time of consumption may also be effective ways to prevent dental erosion.

Titratable acidity

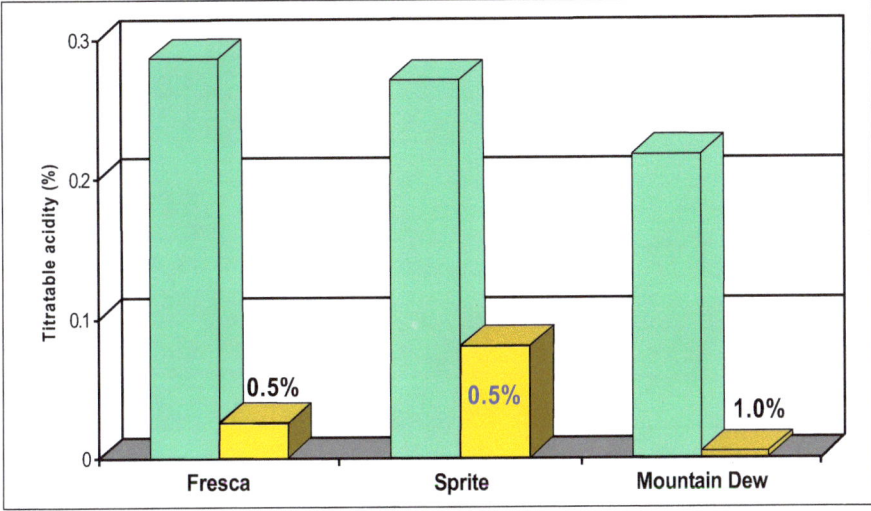

Fig. (12.1). The protective effect of InsP_6 (0.5% and 1%) as reduction in titratable acidity against *Fresca, Sprite* and *Mountain Dew* is shown here [22].

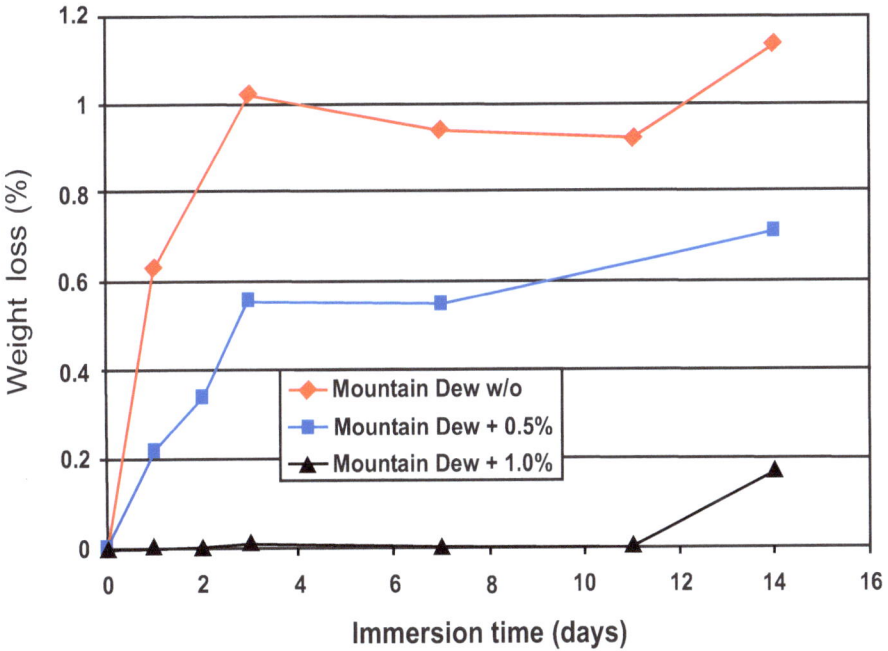

Fig. (12.2). Addition of 0.5% or 1% Na-InsP_6 results in decreased weight loss (enamel erosion) by *Mountain Dew*; higher concentration of InsP_6 confers even better protection [22].

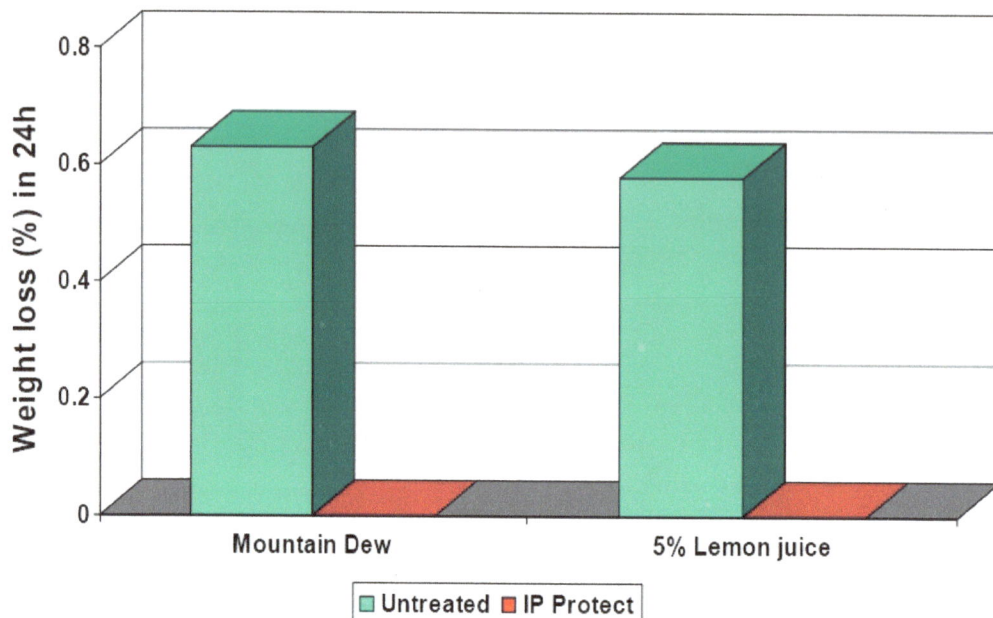

Fig. (12.3). $InsP_6$ (red almost non-existent right-hand columns) dramatically prevents the enamel erosion as measured by weight loss from Mountain Dew or 5% Lemon juice; green left hand columns are control untreated samples [22].

An additional advantage of $InsP_6$ supplement is the availability of calcium-magnesium $InsP_6$. As per Hara & Zero [18], even supplemental calcium by itself can reduce the enamel erosion. Echoing Kaufman [16], we look forward to seeing $InsP_6$ fortified mouthwashes and toothpastes to prevent tooth decay and dental calculus formation. This is not just wishful thinking; pyrophosphate which has also been shown to inhibit crystallization of hydroxyapatite *in vitro* is marketed as a component of Crest® toothpaste which inhibits dental calculus formation by 26% [15, 23]. We agree with Kaufman's assumption that $InsP_6$ could be more effective, but that needs to be tested [15].

OSTEOPOROSIS

Osteoporosis represents a group of diseases characterized by reduced bone mass per unit of bone volume resulting in weak and fragile bone with increased risks of fractures of hip, wrist and spine. "In the United States, nearly 10 million people already have osteoporosis; another 18 million have low bone mass that places them at an increased risk for developing osteoporosis". The scourge affects more women than men; 80% percent of those with osteoporosis are women.

Osteoporosis is the commonest bone disease a physician sees in his/her practice as people older than 50 years, 1 in 2 women and 1 in 8 men are predicted to have an osteoporosis-related fracture in their lifetime.

There are many risk factors for osteoporosis, most notable are old-age, estrogen deficiency, genetic predisposition, nutritional factors *etc*. "While bone may appear deceptively lifeless, it is a living tissue, for it is being continually broken down by cells called osteoclasts, and at the same time it is being reconstructed by cells called osteoblasts. It is the balance between these cells that determines whether we gain or lose bone". The parathyroid hormone parathormone (PTH) stimulates the osteoclast cells to cause bone resorption.

Osteoporosis can be either primary or secondary. Primary osteoporosis is by far the commonest, occurs mostly in elderly persons of either sex and in postmenopausal women. The exact cause(s) of primary osteoporosis are not known. Secondary osteoporosis on the other hand is caused by identifiable diseases such as hyperparathyroidism, malnutrition *etc*.

Bisphosphonates, a group of synthetic polyphosphates are the most common medications prescribed for osteoporosis treatment and they include Fosamax, Actonel, Boniva, and Reclast *etc*. While in general they are safe, as in almost all drugs, they too are not without side-effects and complications. In contrast to bisphosphonates, $InsP_6$ is a natural compound that is also a polyphosphate (6 phosphate groups). Gomes *et al.,* [24] investigated the relative abilities of various inositol phosphates ($InsP_6$, $InsP_5$, $InsP_4$, $InsP_3$, $InsP_2$ and $InsP_1$) to inhibit PTH-induced bone resorption in an organ culture system containing radii and ulnii from fetuses of rats. All of the inositol phosphates ($InsP_{1-6}$) tested inhibited PTH-stimulated bone resorption in a dose dependent manner [24]. Comparative histological and chemical analysis of the effects of $InsP_1$ (*myo*-inositol-2-monophosphate) and bisphosphonates demonstrated that while the bisphosphonates inflicted intracellular damage to the osteoclast cells, such was not seen with $InsP_1$ suggesting a different mode of action [25].

López-González *et al.,* [26] have investigated the relationship between the risk of osteoporosis and dietary $InsP_6$ consumption. Bone mineral density was determined in about 2,000 volunteers by means of dual radiological absorptiometry in the calcaneus or the lumbar column and the neck of the femur. Dietary information related to $InsP_6$ consumption was acquired by questionnaires conducted on two different occasions. The investigators found that the bone mineral density increased with increasing dietary $InsP_6$ consumption. Multivariate linear regression analysis indicated that body weight and low $InsP_6$ consumption were the risk factors with greatest influence on bone mineral density. Thus, low dietary $InsP_6$ consumption is considered an osteoporosis risk factor and $InsP_6$ has a protective effect against osteoporosis [26].

The investigators further proved it experimentally by using ovariectomized rats, a good model for post-menopausal osteoporosis as it causes estrogen deficiency [27]. Ovariectomized rats were fed standard chow with or without Ca-Mg $InsP_6$ for 12 weeks following which their femoral and vertebral bones were examined. Urinary deoxypyridinoline, a marker for bone resorption and serum osteocalcin (a marker for bone formation) were measured. The animals treated with $InsP_6$ had higher calcium and phosphorous contents in bone and bone mineral density than the untreated control animals. Not unexpectedly, the bone resorption marker deoxypyridinoline was higher in the urine of animals that did not receive $InsP_6$.

This study also asked the question whether Ca-Mg $InsP_6$ would protect against prednisone-induced bone loss. Even at very high doses of prednisone, the animals with $InsP_6$ in their food had very little bone loss as compared to the animals that did not receive $InsP_6$. Calcium or magnesium alone doesn't provide such protection. Based on the remarkable bone protective affects found to date, clinical trials are now underway to see whether Ca-Mg $InsP_6$ will protect people to the same extent as it does animals [27].

Experiments in Shamsuddin laboratory with sodium $InsP_6$ found stimulation of the bone building osteoblast cells with concomitant inhibition of the bone resorbing osteoclast cells *in vitro* [21]. It is important to note that the affect was achieved with sodium $InsP_6$ and not Ca-Mg $InsP_6$ and therefore the changes were not due to calcium or magnesium; rather this bone-protective quality was due to $InsP_6$ itself. Even more intriguing is the fact that emerging data show a strong interaction between the osteoblasts and osteoclasts on one hand and their effect *via* osteocalcin on the pancreatic β cells hence insulin secretion and bodyweight *etc.*, on the other hand [28]. Further research will unravel the additional mechanisms of $InsP_6$'s actions.

Interestingly, Addison & *McKee* [29] reported that $InsP_6$ inhibits mineralization of MC3T3-E1 osteoblast cultures by binding to growing crystals, increases gene expression of the mineralization inhibitor osteopontin. They also show that $InsP_6$ does not impair the ability of osteoblasts to synthesize a collagenous matrix, express alkaline phosphatase or differentiate to produce specific bone matrix proteins. These results are in contradiction to the other *in vitro* and *in vivo* data as well as clinical studies showing a positive effect of InsP_6 on osteoporosis prevention and bone mineral density [30].

MECHANISMS OF ACTION

Insofar as the prevention of calcification in the blood vessels as well as in the kidney, salivary glands *etc.*, is concerned, chelation of Ca^{2+} by InsP_6 appears to be the mechanism. InsP_6 is a potent inhibitor of calcium oxalate and calcium phosphate crystallization for kidney stones in experiments on artificial urine. InsP_6 decreases ionized calcium, enhances the barrier to heterogeneous nucleation, inhibits crystallization activity and inhibits *in vitro* stone growth; given the structure of InsP_6 that is expected. However, as Saw *et al.,* have demonstrated, inhibition of crystallization by InsP_6 does not depend on decreasing the effective concentration of ionized calcium. Likewise, inhibition of *in vitro* stone growth does not depend on inhibiting crystallization of the suspended crystals [13].

The mechanism(s) how InsP_6 may prevent osteoporosis is however enigmatic, for it stimulates and causes increased differentiation of the bone forming osteoblasts on one hand, and suppresses the activity of the bone destroying osteoclast cells [21, 30]. Inhibition of the conversion of InsP_6 to inositol pyrophosphates reversed the effect of InsP_6 on alkaline phosphatase mRNA levels in human umbilical cord mesenchymal cells UC-MSCs. An alteration in the basal InsP_7/InsP_8 levels may affect osteoblast differentiation [31]. Clearly, further work is needed to understand the molecular mechanism(s) of these puzzling functions.

CONCLUDING REMARKS

Notwithstanding the discrepancy of result from a single *in vitro* study, the bulk of the other *in vitro* as well as *in vivo* studies and, clinical and epidemiological data strongly support a beneficial role of InsP_6 in osteoporosis with potential as a therapeutic agent for the disease. Undisputable however are the benefits in various mineralization disorders such as kidney stone, pathological calcifications, tooth decay *etc.* To prevent tooth decay, one strategy may be to either simply add InsP_6 to the drinks at the time of bottling, and/or it could also be provided in small paper packages as for salt & pepper found in coffee shops, restaurant tables *etc.*

CONFLICT OF INTEREST

Professor Shamsuddin is the inventor of several patents related to InsP_6 and hexacitrated InsP_6.

REFERENCES

[1]   Grases F, Sanchis P, Perelló J, *et al.* Phytate reduces age-related cardiovascular calcification. Front Biosci 2008; 13: 7115-22.

[2]   Grases F, Sanchis P, Perello J, *et al.* Phytate (Myo-inositol hexakisphosphate) inhibits cardiovascular calcifications in rats. Front Biosci 2006; 11: 136-42.

[3]   Grases, F., Sanchis, P., Perelló, J *et al.*: Effect of crystallization inhibitors on vascular calcifications induced by vitamin D: a pilot study in Sprague-Dawley rats. Circ J 2007; 71: 1152-6.

[4]   Grases F, Sanchis P, Costa-Bauzá A, *et al.* Phytate inhibits bovine pericardium calcification *in vitro.* Cardiovasc Pathol. 2008; 17: 139-45. doi: 10.1016/j.carpath.2007.08.005. Epub 2007 Oct 24.

[5]   Grases F, Perelló J, Isern B, *et al.* Study of *myo*-inositol hexaphosphate based cream to prevent dystrophic calcinosis cutis. Br J Dermatol 2005; 152: 1022-5.

[6]   Modlin M: Urinary phosphorylated inositols and renal stone. Lancet 1980; 2(8204): 1113-4.

[7]   Henneman PH, Dempsey EF, Carroll EL *et al.* J Clin Invest 1956; 35: 1229-42

[8]   Henneman PH, Benedict PH, Forbes AP *et al.* N Engl J Med 1958; 17: 802-7.

[9]   Grases F, Costa-Bauzá A. Phytate (IP6) is a powerful agent for preventing calcifications in biological fluids: usefulness in renal lithiasis treatment. Anticancer Res. 1999; 19: 3717-22.

[10]   Curhan GC, Willett WC, Knight E L, *et al.* Dietary factors and the risk of kidney stones in younger women. Nurse's Health Study II. Arch Inter Med 2004; 164: 885-91.

[11]   Sakamoto K, Vucenik I, Shamsuddin AM. [^{3}H]-phytic acid (inositol hexaphosphate) is absorbed and distributed to various tissues in rats. Journal of Nutrition 1993; 123: 713-20.

[12]   Nahapetian A, Young VR. Metabolism of ^{14}C-phytate in rats: effect of low and high dietary calcium intakes. J Nutr 1980; 110: 1458-72.

[13]   Saw NK, Chow K, Rao PN, *et al.* Effects of inositol hexaphosphate (phytate) on calcium binding, calcium oxalate crystallization and *in vitro* stone growth. J Urology 2007; 177: 2366-70.

[14]   Grases F, Santiago C, Simonet BM *et al.* Sialolithiasis: mechanism of calculi formation and etiologic factors. Clinica Chimica Acta 2003; 334: 131-36.

[15]   Kaufman HW. Interaction of inositol phosphates with mineralized tissues. In: Graf E Ed. Phytic Acid chemistry & applications. Minneapolis, Pilatus Press 1986; pp. 303-320.

[16]   Magrill DS. The effect of pH and of orthophosphate on the adsorption of phytate by hydroxyapatite during prolonged exposure. Arch Oral Biol. 1973; 18: 1269-73.

[17]   Kuzmiak-Jones H, Reynolds GR, Bowen WH. The effects of a combination of fluoride and phytate on dental caries in rats. J Dent Res 1977; 56 (Special Issue B), AADR Abstract 336.

[18]   Hara AT, Zero DT. Analysis of the erosive potential of calcium-containing acidic beverages. Euro J Oral Sci. 2008; 116: 60-65.

[19]   von Fraunhofer JA, Rogers MM. Dissolution of dental enamel in soft drinks. Gen Dentistry 2004; 52: 308-12.

[20]   Jain P, Nihill P, Sobkowski J, *et al.* Commercial soft drinks: pH and *in vitro* dissolution of enamel. Gen Dentistry 55: 150-154, 2007.

[21]   Shamsuddin AM, von Fraunhofer JA. Prevention of Tooth Decay and Other Bone Degeneration. US 20070212449 (2007).

[23]   Mallatt ME, Beiswanger BB, Stookey GK, *et al.* Influence of soluble pyrophosphate on calculus formation in adults. J Dent Res 1985; 64: 1159-62.

[24]   Gomes BC, Kaufman HW, Bloom JR *et al.* Inhibitory effect of inositol phosphates on parathyroid hormone-induced bone resorption in organ culture. J Dent Res 1984; 66: 890-93.

[25]   Gomes BC, Kaufman HW, Archard HO *et al.* Histologic study of the inhibition of bone resorption in organ culture by myo-inositol-2-monophosphate. J Oral Pathol 1986; 15: 54-8.

[26] López-González AA, Grases F, Roca P, *et al*. Phytate (*myo*-inositol hexaphosphate) and risk factors for osteoporosis, J Med Food 2008; 11: 747-52.

[27] Grases F, Sanchis P, Prieto RM, *et al*. Effect of tetracalcium dimagnesium phytate on bone characteristics in ovariectomized rats. J Med Food 2010; 13: 1301-6. doi: 10.1089/jmf.2009.0152

[28] Karsenty G, Ferron M. The contribution of bone to whole-organism physiology. Nature 2012; 481: 314-20.

[29] Addison WN, McKee MD. Inositol hexaphosphate inhibits mineralization of MC3T3-E1 osteoblast cultures. Bone 2012; 46: 1100-7. doi: 10.1016/j.bone.2010.01.367. Epub 2010 Jan 14.

[30] Arriero Mdel M, Ramis JM, Perelló J, *et al*. Inositol hexaphosphate inhibits osteoclastogenesis on RAW 264.7 and human primary osteoclasts. PLoS One. 2012; 7(8): e43187. doi: 10.1371/journal.pone.0043187. Epub 2012 Aug 14.

[31] Arriero Mdel M, Ramis JM, Perelló J, *et al*. Differential response of MC3T3-E1 and human mesenchymal stem cells to inositol hexaphosphate. Cell Physiol Biochem 2012; 30: 974-86. doi: 10.1159/000341474. Epub

CHAPTER 13

The Roles of Inositol and its Phosphates in Neuropsychiatric Disorders

Abstract: There are experimental data showing beneficial actions of inositol and inositol hexaphosphate (InsP_6) in Alzheimer's disease and Parkinson's disease. The different stereoisomers of inositol, *myo-, epi-, scyllo-* and *chiro*-inositol have shown efficacy in Alzheimer's disease. The antioxidant and cation chelating properties of InsP_6 are considered to be some of the mechanisms; those of *myo*-inositol are not clear. Insofar as psychiatric diseases, lithium salts have been in use since the 1950s for bipolar disorder. While on one hand there is reduction of *myo*-inositol by lithium, *myo*-inositol on the other hand has been found useful in psychiatric disorders, and at least in one study of bipolar disorder. Thus, further studies of *myo*-inositol, and InsP_6 in the treatment of bipolar disorder are warranted. A clinical trial of InsP_6 in bipolar disorder is underway.

Keywords: Alzheimer's disease, BACE 1, bipolar disorder, CGI-S, HAM-D, Levodopa, lithium, MADRS, OCD, panic disorder, Parkinson's disease, Y-BOCS.

INTRODUCTION

In this chapter, we will discuss the current use and further potential therapeutic applications of *myo*-inositol and InsP_6 in Alzheimer's disease, Parkinson's disease and several psychiatric conditions.

Alzheimer's Disease (AD)

In 1907, the German psychiatrist and neuropathologist Dr. Alois Alzheimer described a dementia in the elderly, eponymously known as Alzheimer's disease (AD), the commonest type of dementia in our seniors today. It is not a new disease by any means. The ancient Greek and Roman philosophers had noted increasing dementia along with advancing age. Pathological hallmark of the disease is loss of nerve cells and their communicating junctions with each other (synapses) in the brain resulting in its shrinkage or atrophy; the characteristic lesions are neurofibrillary tangles and amyloid plaques. "The tangles are aggregates of the microtubule-associated protein *tau*, which has become hyper-phosphorylated and accumulated within the cells". Although many older individuals develop some plaques and tangles as a consequence of aging, the brains of AD patients have a greater number of them in specific areas.

A.K.M. Shamsuddin and Guang-Yu Yang

AD can be considered as a disease of protein misfolding since the amyloid plaques are composed of 4-kDa misfolded β-amyloid (Aβ) protein of 39-43 residue peptides. "Mutations in codon 717 of the amyloid precursor protein gene (presenilin 1 and presenilin 2), result in an increased production of residues 1-42 (Aβ42) over Aβ1-40 (Aβ40)". Aβ42 is more amyloidogenic and is found in extracellular deposits throughout the central nervous system resulting in selective loss of neuronal subpopulations including cholinergic fibers, proliferation of reactive astrocytes and microglia. These lead to progressive mild cognitive impairment; and eventually to AD. Insofar as to our understanding of how the disease starts, this is where the consensus ends. It is not the purpose of the eBook to delve into the controversies of AD pathogenesis; thus we restrict to the information relevant to inositol and its phosphates. However, we will discuss the rationales for therapeutic approaches as they deem germane to the topic.

InsP_6

Based on the observation that the pineal gland hormone melatonin level decreases with advancing age, but more so in AD patients; Grases *et al.,* had put forward a tantalizing hypothesis [1]. This somewhat neglected melatonin has multiple important roles aside from being an antioxidant and a neuroprotective agent; even more to the point is that melatonin inhibits the formation Aβ. But, why does the pineal gland stop secreting this crucial neuroprotectant? That the pineal gland, the source of melatonin becomes calcified, correlating with decreased secretion of melatonin with age is well known. Calcification may take place if there is a lack of inhibitors of calcium salt crystallization, of which InsP_6 is one. Grases and colleagues thus proposed that InsP_6 deficiency may be a risk factor for AD [1].

Interestingly, soon after, Anekonda *et al.,* proposed treatment of AD with InsP_6. "Aβ is believed to interfere with neuronal activity because of its stimulatory effect on the production of free radicals, resulting in oxidative stress and neuronal cell death". Owing to the antioxidant property of InsP_6, it has the potential in preventing Aβ-induced pathology - they hypothesized. Anekonda *et al.,* also rationalized that InsP_6 could favorably affect AD pathology by mimicking caloric restriction, promoting autophagy, and modulating clathrin-coated endocytosis of APP and its cleavage products. Thus they set out to investigate in both *in vitro* and *in vivo* models of AD [2].

Human neuroblastoma MC65 cells were used for *in vitro* and Tg2576 mouse model for *in vivo* studies. The investigators showed that 100 µM Na-InsP_6

protected MC65 cells against the harmful effect of Aβ and caused a modest reduction in the level of Aβ and plaques in the Tg2576 mice treated with InsP_6. In MC65 cells, 48-72-hour treatment with Na-InsP_6 provided complete protection against amyloid precursor protein-C-terminal fragment-induced cytotoxicity by attenuating levels of increased intracellular calcium ($[Ca^{2+}]_i$), H_2O_2, •O_2 and β-amyloid oligomers; the expression of the autophagy protein beclin-1 was also moderately up-regulated [2].

Initially, wild type mice were treated with 2% Na-InsP_6 in drinking water for 70 days. Treatment of mice with InsP_6 did not affect the ceruloplasmin (copper-carrying protein in blood) activity, brain copper and iron levels and brain superoxide dismutase and ATP levels. On the other hand, there was a significant increase in brain levels of cytochrome oxidase and a decrease in lipid peroxidation with InsP_6 administration [2]. Brain levels of copper, iron, and zinc were unaffected even after 6 months of treatment of Tg2576 and wild type mice with 2% InsP_6. The effects of InsP_6 were modest on the expression of APP trafficking-associated protein AP180, autophagy-associated proteins (beclin-1, LC3B), sirtuin 1, the ratio of phosphorylated AMP-activated protein kinase (PAMPK) to AMPK, soluble Aβ1-40, and insoluble Aβ1-42 [2]. These results suggest that InsP_6 "may provide a viable treatment option for AD", concluded the authors [2].

Comparison of the incidence of AD in Japanese in Japan and those Japanese who had immigrated to the United States show that the incidence of AD in the latter population increases to the same level as the mainstream US population [3]. Since rice is the staple food in Japan as opposed to wheat-based food in the Western countries, Abe & Taniguchi hypothesized that rice grain may contain some component that might prevent Aβ accumulation in brain [4]. Since Aβ is excised from the amyloid-β precursor protein through sequential cleavage by β-secretase 1 (BACE 1) and γ secretase, BACE 1 inhibitors could therefore prevent Aβ accumulation in the brain. The investigators show that rice extracts and digest inhibit BACE1 activity *in vitro* and that the rice digest inhibited Aβ production in cultured human neuroblastoma SH-SY5Y cells. By ion exchange chromatography, they identified InsP_6 as one of the components inhibiting BACE1 activity *in vitro* and Aβ production in cultured human neuroblastoma cells in a dose-dependent manner [4]. Abe & Taniguchi also tested InsP_{3-5} (D-myo-inositol 1,4,5-tris-phosphate trisodium salt - InsP_3, D-*myo*-inositol 1,4,5,6-tetrakisphosphate potassium salt - InsP_4 and D-*myo*-inositol 1,3,4,5,6-pentakisphosphate pentapotassium salt - InsP_5) for BACE 1 inhibition, but these lower inositol phosphates did not inhibit BACE 1 activity at 1 μg.mL^{-1} while InsP_6 did at the same concentration. Of note is the anatomical correspondence of the presence of InsP_6 in rice bran which is in unpolished rice; its extract was more potent than that from the polished rice [4].

Inositol and Alzheimer's Disease

Aggregation of Aβ into insoluble amyloid fibers is a nucleation-dependent event that may be modulated by the presence of molecules associated with amyloid. Binding studies of Aβ40 and Aβ42 to phosphatidylinositol (PI), and different inositol head-groups show that at pH 6.0 and in the presence of PI vesicles, both Aβ40 and Aβ42 adopted an amyloidogenic β-structure; whereas, at neutral pH only Aβ42 folded into a β-structure in the presence of PI vesicles [5]. McLaurin *et al.,* also found that at pH 7.0, *myo*-inositol was sufficient to induce β-structure in Aβ42 but had no effect on the conformation of Aβ40. *Myo*-inositol mono-, di- and triphosphates (InsP_{1-3}) were not as efficient in inducing β-structure in both peptides, suggesting that interaction of Aβ40 and Aβ42 with PI acts as a seed for fibril formation, while *myo*-inositol stabilizes a soluble Aβ42 micelle inhibiting fibril formation [5].

Noting that a) inositol is dysregulated in Down's syndrome with high *myo*-inositol in brain, and in AD, b) uptake of *myo*-inositol is increased in the fibroblast cells in Down's, and c) a large amount of Aβ is present in the brain of Down's syndrome patients prior to the deposition of plaques, McLaurin and colleagues further studied 4 stereoisomers (*myo*-, *epi*-, *scyllo*- and *chiro*-) of inositol since the charge distribution across the surface of the sugar ring varies, affecting the ability to inhibit Aβ42 fibrillogenesis [6]. They used PC-12 pheochromocytoma cells from rat adrenal medulla that in response to nerve growth factors stop growing and terminally differentiate, making them a good model to study nerve cells; and primary human neuronal culture. Inositol interacts with Aβ; the resultant Aβ-inositol complex is non-toxic to both nerve growth factor-differentiated PC-12 cells and primary human neuronal cultures. The investigators believe that the reduction of toxicity is the result of Aβ-inositol interaction. Since inositol stereo-isomers are naturally occurring molecules that readily cross the blood-brain barrier, they also believe that inositols may represent a viable treatment for AD [6].

They further expanded their investigation to TgCRND8 mice, a model for AD; and demonstrated that 1-4-di-*O*-methyl-*scyllo*-inositol attenuated spatial memory impairment and significantly decreased cerebral amyloid pathology [7]; the exact logic for choosing the *scyllo*- and other isomers over the *myo*- however remains unclear.

Parkinson's Disease

Parkinson's disease (PD) is a debilitating neurological disease that affects an estimated 1% of the population of over 50 years of age in USA. The characteristic microscopic finding is selective degeneration of dopaminergic cells in special areas of the brain called *substantia nigra* and *locus ceruleus*. Usual residents of these areas are cells with dark pigment (hence the name *substantia nigra*) who become degenerated, dead and replaced by gliosis, a special form of brain fibrosis with resultant irreversible motor dysfunction. Since the basic pathology is loss of dopamine producing cells, the treatment option is dopamine replacement therapy, which only gives symptomatic relief as the cells are irreversibly lost. Although compounds such as Levodopa (L-DOPA) improve the striatal dopamine content they fail to prevent the progression of the degenerative process. And their long-term usage is associated with progressive decrease in drug response, motor fluctuations, impairment of voluntary movements resulting in fragmented or jerky motion (dyskinesia), and drug-induced toxicity.

How and why do these cells undergo degeneration and die? It is believed that oxidative damage may be a causative factor in the pathogenesis of PD. "Oxidative stress in the neuron of *substantia nigra* of patients with Parkinson's disease increases the iron content, superoxide dismutase activity and lipid peroxidation of neuronal cells, but decreases the levels of the reduced form of glutathione (GSH)". Thus, reactive oxygen species (ROS), especially •OH radicals are strongly implicated in the pathogenesis of Parkinson's disease.

Experimentally, 1-methyl-4-phenyl 1,2,3,6-tetra-hydropyridine (MPTP) produces a Parkinson's syndrome after its conversion to a dopamine-selective neurotoxin 1-methyl-4-phenylpyridinium ion (MPP^+) through the formation of •OH. Since $InsP_6$ is a potent antioxidant, T Obata investigated if $InsP_6$ would have any potential benefit in PD in an *in vivo* rat model. Sodium salicylate in Ringer's solution (0.5 nmol/µl/min) was infused through a micro-dialysis probe to detect the generation of •OH as reflected by the non-enzymatic formation of 2,3-dihydroxybenzoic acid (DHBA) in the rat brain (striatum). $InsP_6$ at 100 µM concentration did not significantly decrease the levels of MPP^+-induced •OH formation trapped as 2,3-DHBA. Infusion of Fe^{2+} enhanced MPP^+-induced •OH generation through Fenton reaction; and $InsP_6$ significantly suppressed the Fe^{2+}-enhanced •OH formation after MPP^+ treatment [8].

MPP^+-induced cell death of immortalized rat mesencephalic/dopaminergic cells was also investigated by Xu *et al.,* [9]. Compared to MPP^+ treatment, $InsP_6$ increased cell viability by 19% and decreased cell death by 22%. MPP^+-induced increase in caspase-3 activity and DNA fragmentation was decreased by 55% and 52%, respectively with $InsP_6$; along with a dose-dependent increase in cell survival by 18% - 42% in iron-excess conditions. Likewise, $InsP_6$ treatment also caused a 45% reduction in DNA fragmentation. Similar protection was also observed with the differentiated cells against 6-hydroxy-dopamine-induced apoptosis in both normal and iron-excess conditions in immortalized rat mesencephalic dopaminergic neuronal cell line (1RB3AN27 also known as N27), demonstrating a significant neuroprotective effect of $InsP_6$ in a cell culture model of PD [10].

PSYCHIATRIC DISORDERS

Depression

Brain has one of the highest levels of $InsP_6$, and it has been considered to play a role in neuronal signaling. In the late 1980's and early 1990's $InsP_6$ was found to stimulate Ca^{2+} uptake in cultured cerebellar neurons and anterior pituitary cells [11, 12]. $InsP_6$ increased both intracellular free Ca^{2+} and prolactin secretion in perifused pituitary cells. Subsequently, $InsP_6$-binding site - the $InsP_6$ receptor was isolated. Radiolabelled $InsP_6$ ($[^3H]$- $InsP_6$) bound to specific and saturable recognition sites in membranes prepared from cerebral hemispheres, anterior pituitaries and cultured cerebellar neurons in all subcellular fractions including the mitochondria [12, 13]. On the other hand, the level of *myo*-inositol in brain cells is reduced in a variety of neuropsychiatric conditions. The level of inositol in cerebrospinal fluid (that bathes the brain and the spinal cord) has been found to be reduced in patients with depression. Experimentally, pilocarpine causes a limbic seizure syndrome in lithium (Li) treated rats; these Li-pilocarpine seizures are reversible by intracerebroventricular administration of inositol to rats. Despite the apparently poor transfer of *myo*-inositol across the blood-brain barrier, large doses of intra-peritoneal inositol can also reverse Li-pilocarpine seizures, with an elevation of brain inositol level following intra-peritoneal administration [14]. Following an open-label, add-on trial of inositol in depression suggested a beneficial effect, in a subsequent 1-month, parallel-groups, double-blind, placebo-controlled study of 28 patients, inositol was effective as sole therapy for depression ($p = 0.043$). Inositol was also effective for panic disorder in a double-blind, random-assignment, placebo-controlled crossover study of 21 patients, with 4 weeks in each phase ($p = 0.02$); the effect was comparable to that of imipramine [15].

A double-blind controlled trial of 12 g/day of *myo*-inositol in 28 depressed patients for four weeks showed a significant overall benefit for inositol compared to placebo as early as at week 4 on the Hamilton Depression Scale - the standard measure to assess the effectiveness of an antidepressant substance [16]. No changes were noted in hematologic parameters or, kidney or liver functions. Since many antidepressants are effective in panic disorder, twenty-one patients with panic disorder with or without agoraphobia (abnormal fear of being helpless in an embarrassing situation leading to avoidance of open or public places) completed a double-blind, placebo-controlled, four week, random-assignment crossover treatment trial of 12 g of *myo*-inositol 12 g per day. The frequency and severity of panic attacks and severity of agoraphobia declined significantly with inositol compared to placebo, with minimal side-effects [16].

The conventional treatments for patients with anxiety disorders in the United States include antidepressants, which are not without their own side-effects such as sleep disturbances, sexual dysfunction, gastrointestinal problems, weight gain *etc.* In that regard natural remedies may be more appealing and inositol offers a great potential.

Panic Disorder

Panic disorder is an anxiety disorder wherein people have sudden and repeated attacks of fear that last for several minutes (panic attacks); sometimes symptoms may last longer. These panic attacks are characterized by a fear of disaster or of losing control even when there is no real danger. A person may also have a strong physical reaction during a panic attack; it may feel like having a heart attack. Panic attacks can occur at any time, and many people with panic disorder worry about and dread the possibility of having another attack.

Benjamin and colleagues examined the effectiveness of inositol *versus* placebo in panic disorder in a double-blind, crossover trial involving 21 patients. The patients were assigned to either 12g/day of inositol or placebo for 4 weeks each. All of the patients then crossed over to the alternate treatment arm for additional 4 weeks. The baseline number of panic attacks per week for all subjects was 9.7 ± 15, which decreased to 3.7 ± 4 in subjects taking inositol (significant at $p = 0.04$) as opposed to 6.3 ± 9 in the placebo arm. The mean scores of the severity of panic attacks also decreased significantly from 72 ± 140 at baseline for all subjects to 31 ± 50 during the placebo arm, and to 11 ± 11 during inositol treatment; the difference being quite significant at $p = 0.007$ [17].

In another study by the same group of investigators, 20 patients with panic disorder with or without agoraphobia were given either inositol (up to 18/day) or fluvoxamine (up to 150 mg/day) for 4 weeks in a double-blind, controlled, random-order crossover study. Compared to the conventional drug fluvoxamine that decreased weekly panic attacks from 5.8 ± 4 at baseline to 3.4 ± 4 at endpoint, the frequency of panic attacks in the patients in inositol group decreased from 7.2 ± 4 to 3.1 ± 3. Statistical analysis showed that inositol was marginally more effective than fluvoxamine from baseline to endpoint for treating the number of panic attacks ($P = 0.049$). The two treatments were considered equally effective at reducing Hamilton Anxiety scores, agoraphobia scores, and Clinical Global Impression of Severity (CGI-S) scores. Not unexpectedly, side effects were minimal with inositol treatment, as opposed to frequent nausea and tiredness associated fluvoxamine. All things considered, inositol was more effective than fluvoxamine in reducing the number of panic attacks with less adverse effects [18].

Obsessive-Compulsive Disorder (OCD)

OCD is an anxiety disorder characterized by unusual apprehension, fear or worry (obsession) and repetitive behavior to address those unfounded fears. People with OCD feel the need to check things repeatedly (compulsion), or have certain thoughts, or perform routines and rituals over and over again. The thoughts and rituals associated with OCD cause distress and get in the way of daily life. People with OCD cannot control these obsessions and compulsions; the rituals end up controlling them. For example, if people are obsessed with germs or dirt, they may develop a compulsion to wash their hands over and over again - a common sign of OCD. If they develop an obsession with intruders, they may lock and relock their doors many times before retiring to bed.

The effectiveness of inositol in OCD was studied in 13 patients in a double-blind controlled cross-over trial of 18g/day of inositol, or placebo for 6 weeks each. The patients had significantly lower scores on the Y-BOCS (Yale-Brown obsessive compulsive scale) test of severity of OCD when taking inositol than when taking placebo ($p = 0.009$) [19].

Carey *et al.,* studied the effect of inositol treatment on brain function in OCD through single-photon emission computed tomography [20]. Fourteen OCD subjects underwent single-photon emission computed tomography before and after receiving inositol for 12 weeks. Following treatment with inositol, there was deactivation in OCD responders relative to non-responders in several areas of the brain with significant reductions in Y-BOCS and CGI-S (clinical global impression scale - commonly used measures of symptom severity, treatment-response and the efficacy of treatments in studies of patients with mental disorders) scores. These results not only show the efficacy of inositol in OCD, but also suggest that inositol may exert its clinical effects through alternate neuronal circuitry to the selective serotonin reuptake inhibitors (SSRIs) - the anti-depressants [20].

Since serotonin re-uptake inhibitors benefit OCD patients and inositol reverses desensitization of serotonin receptors, 13 patients with OCD completed a double-blind controlled crossover trial of 18 g/day inositol or placebo for six weeks each. Inositol significantly reduced the scores of OCD symptoms compared with placebo. It appears that inositol has therapeutic effects in the spectrum of illness responsive to serotonin selective re-uptake inhibitors, including depression, panic and OCD; however it was not found to be beneficial in schizophrenia or autism [16].

BIPOLAR DISORDER

Bipolar disorder, also known as manic-depressive disorder, or manic depression, is a mental illness of unknown etiology that affects approximately 3% people worldwide with similar proportion across racial and gender characteristics. The disorder is characterized by episodes of an elevated mood known as mania, usually alternating with episodes of depression. During mania an individual feels abnormally happy, energetic, or irritable, but often makes poor decisions due to unrealistic ideas or poor regard of consequences. Such alternating episodes of mania and depression can impair the individual's ability to function in ordinary life, with morbid suicidal thoughts.

Lithium has been used in the treatment of bipolar disorder since the 1950's and has proven to be effective in the reduction of suicide and suicidal behaviors.

Though lithium is the most effective mood stabilizer it is toxic at only twice the therapeutic dosage and has many undesirable side effects.

The exact mechanism of lithium's action is however not understood. It appears to affect the intracellular second messenger systems. One of the targets of lithium is inositol mono-phosphatase within the phosphatidylinositol signaling pathway. This inhibition leads to depletion of inositol with resulting reduction in the re-synthesis of phosphatidylinositol bisphosphate (PIP2), and preventing regeneration of the second messenger inositol-(1,4,5) triphosphate (InsP_3). At therapeutic doses, lithium is a potent inhibitor of various enzymes involved in inositol phosphate metabolism, including the intracellular enzymes inositol monophosphatase and inositol polyphosphatase 1-phosphatase. The former, inositol monophosphatase regenerates *myo*-inositol from inositol monophosphates, which in turn leads to the re-synthesis of phosphatidylinositol [21]. It was hypothesized that lithium treatment would regulate cellular signaling presumably induced by hyperactive neuronal signaling in the brain of bipolar patients. Despite the extensive evidence in support of inositol depletion as a viable explanation of lithium's pharmacodynamic actions, other observations have been inconsistent and often contradictory [22].

In a pilot clinical trial, Chengappa and co-workers enlisted 24 adult men and women with bipolar depression (bipolar I = 21; bipolar II = 3) and randomly assigned them to receive either 12 g/day of inositol or D-glucose as placebo for 6 weeks [23]. Among the 22 subjects who completed the trial, six (50%) of the inositol-treated subjects responded with a 50% or greater decrease in the baseline Hamilton Depression Rating Scale (HAM-D) score and a Clinical Global Improvement (CGI) scale score change of 'much' or 'very much' improved, as compared to three (30%) subjects assigned to placebo, a statistically non-significant difference [23]. On the Montgomery-Asberg Depression Rating Scale (MADRS), eight (67%) of twelve inositol-treated subjects had a 50% or greater decrease in the baseline MADRS scores compared to four (33%) of twelve subjects on placebo ($p = 0.10$) [23]. These encouraging data suggest a controlled study with an adequate sample size, and the appropriate rating scale may demonstrate efficacy for *myo*-inositol in bipolar depression.

Hypothesizing that InsP_6 may have similar function as *myo*-inositol Dr. Michael McCarthy at the San Diego Veterans Healthcare System has embarked on a clinical trial of InsP_6 on Bipolar patients. At the time of composing this eBook, patient enrolment has just begun; for follow-up information one may visit the web site: http://www.clinicaltrials.gov/ct2/show/NCT02081287.

CONCLUDING REMARKS

The preceding has shown that *myo*-inositol offers therapeutic benefit in various psychiatric disorders. Along with InsP_6, it may also be of benefit in Alzheimer's disease. The potential beneficial action of InsP_6 in Parkinson's disease is also encouraging. The mechanisms of the action are not completely understood, though antioxidant and signaling pathways seems to be at work.

REFERENCES

[1]   Grases F, Costa-Bauzà A, Prieto RM. A potential role for crystallization inhibitors in treatment of Alzheimer's disease. Medical Hypotheses 2010; 74: 118-9.
[2]   Anekonda TS, Wadsworth TL, Sabin R, *et al.* Phytic acid as a potential treatment for Alzheimer's pathology: evidence from animal and *in vitro* models. J Alzheimer's Dis 2011; 23: 21-35.
[3]   Farrer L. Intercontinental epidemiology of Alzheimer disease - a global approach to bad gene hunting. JAMA 2001; 285: 796-8.
[4]   Abe TK, Taniguchi M. Identification of myo-inositol hexakisphosphate (IP6) as a β-secretase 1 inhibitory molecule in rice grain extract and digest. FEBS Open Bio 2014; 4: 162-7. doi: 10.1016/j.fob.2014.01.008. eCollection 2014
[5]   McLaurin J, Franklin T, Chakravatty a *et al.* Phosphatidylinositol and inositol involvement in amyloid beta-fibril growth and arrest. J Mol Biol 1998; 278: 183-94.
[6]   McLaurin, J., Golomb, R., Jurewicz, A., *et al.* Inositol stereoisomers stabilize an oligomeric aggregate of Alzheimer amyloid β peptide and inhibit Aβ-induced toxicity. J Biol Chem 2000; 275(24): 18495-18502.
[7]   Hawkes CA, Deng LH, Shaw JE *et al.* Small molecule beta-amyloid inhibitors that stabilize protofibrillar structures *in vitro* improve cognition and pathology in a mouse model of Alzheimer's disease. Eur J Neurosci 2010; 31: 203-13 doi: 10.1111/j.1460-9568.2009.07052.x. Epub 2010 Jan 13.
[8]   Obata T. Phytic acid suppresses 1-methyl-4-phenylpyridinium iron-induced hydroxyl radical generation in rat striatum. Brain Research 2003; 978: 241-4.
[9]   Xu Q., Kanthasamy A.G., Reddy M.B.: Neuroprotective effect of natural iron chelator phytic acid in a cell culture model Parkinson's disease. Toxicology 2008; 245: 101-8.
[10]   Xu Q, Kanthasamy AG, Reddy MB. Phytic acid protects against 6-hydroxy dopamine-induced dopaminergic neuron apoptosis in normal and iron excess conditions in a cell culture model. Parkinson's Disease. 2011: 431068, Published online 7 February 2011. doi: 10.4061/2011/431068.
[11]   Nicoletti F, Bruno V, Fiore L *et al.* Inositol hexakisphosphate (phytic acid) enhances Ca^{2+} influx and D-aspartate release in cultures cerebellar neurons. J Neurochem 1989; 53: 1026-30.
[12]   Nicoletti F, Bruno V, Cavallaro S *et al.* Specific binding sites for inositol hexakisphosphate in brain and anterior pituitary. Mol Pharmacol 1990; 37: 689-93.
[13]   Hawkins PT, Reynolds Dj, Poyner DR *et al.* Identification of a novel inositol phosphate recognition site: [^{3}H]inositol hexakisphosphate binding to brain regions and cerebellar membranes. Biochem Biophys Res Commun 1990; 167: 819-27.
[14]   Agam G, Shapiro Y, Bersudsky Y, *et al.* High-dose peripheral inositol raises brain inositol levels and reverses behavioral effects of inositol depletion by lithium. Pharmacol Biochem Behavior 1994; 49: 341-3.
[15]   Benjamin J, Agam G, Levine J *et al.* Inositol treatment in psychiatry. Psychopharmacol Bull. 1995; 31: 167-75.

[16]   Levine J. Controlled trials of inositol in psychiatry. European Neuropsychopharmacol 1997; 7: 147-55.

[17]   Benjamin J, Levine J, Fux M, *et al*. Double-blind, placebo-controlled, cross-over trial of inositol treatment for panic disorder. Am J Psychiatry 1995; 152: 1084-6.

[18]   Palatnik A, Frolov K, Fux M, *et al*. Double-blind, controlled, crossover trial of inositol *versus* fluvoxamine for the treatment of panic disorder. J Clin Psychopharm 2001; 21: 335-9.

[19]   Fux, M., Benjamin, J., Belmaker, R. H.: Inositol *versus* placebo augmentation of serotonin reuptake inhibitors in the treatment of obsessive-compulsive disorder: a double-blind cross-over study. Int J Neuropsychopharmacol *1999*; 2: 193-5.

[20]   Carey, P. D., Warwick J., Harvey, B. H., Stein, D. J., and Seedat, S.: Single photon emission computed tomography (SPECT) of anxiety disorders before and after treatment with inositol. Metab Brain Dis 2004; 19: 125-34.

[21]   Berridge, MJ, Downes CP, Hanley, MR. Neural and developmental actions of lithium: a unifying hypothesis. Cell 1989; 59: 411-9.

[22]   Marmol F. Lithium: bipolar disorder and neurodegenerative diseases. Possible cellular mechanisms of the therapeutic effects of lithium. Prog Neuropsychopharmacol Biol Psychiatry. 2008; 32: 1761-71.

[23]   Chengappa KN, Levine J, Gershon S *et al*. Inositol as an add-on treatment for bipolar depression. Bipolar Disord. 2000; 2: 47-55.

CHAPTER 14

Sickle Cell Disease and Inositol Phosphates

Abstract: Sickle cell disease (SCD) is a hereditary blood disorder due to a mutation in the hemoglobin gene, and characterized by abnormal hemoglobin S (HbS). Under hypoxic conditions, HbS polymerizes, inducing distortion of RBCs into rigid sickle-shaped cells; in particular, homozygous SCD patients (SS) may experience painful episodes and complications due to vaso-occlusion (VOC) during their lifetime. There is no cure, but symptomatic management to improve anemia and lower the complications through blood transfusions is currently the most accepted therapy for those patients. Chronic red blood cell (RBC) exchanges or transfusion are effective in preventing the recurrence of VOC. While simple blood transfusions are important therapy for SCD, the goal of efficient transfusion is to increase the oxygen delivery and to prevent the sickling. It has been demonstrated that $InsP_6$-loaded red blood cells ($InsP_6$-RBCs) reduce the risks of sickling of sickle RBCs (SS RBCs) exposed to hypoxia.

Keywords: $InsP_6$, $InsP_6$-loaded red blood cells, malaria, oxy-hemoglobin dissociation curve, sickle cell anemia.

INTRODUCTION

Sickle cell disease (SCD) or sickle cell anemia (SCA) is a hereditary blood disorder due to a mutation in the hemoglobin gene [1-5]. This disease is characterized by red blood cells (RBCs) that assume an abnormal, rigid, sickle shape [5]. The normal RBCs are disc shaped like doughnuts but without the holes; and they move easily through the blood vessels. RBCs contain the iron-rich protein hemoglobin that delivers oxygen from the lungs to the various tissues and organs throughout the body.

In contrast to normal RBCs, sickle cells contain an abnormal hemoglobin called sickle hemoglobin or hemoglobin S which causes the cells to assume a crescent or sickle shape [5]. These abnormally shaped cells are not only stiff but also sticky, resulting in blocking of the small blood vessels in the limbs and organs causing pain and tissue damage. SCD is an autosomal recessive hereditary disease; people who have the disease inherit two genes of Hemoglobin S, one from each parent. SCD is referred to as a form of homozygosity for the hemoglobin gene mutation that forms Hemoglobin S 'HbSS', or 'SS disease'. In heterozygous people, there is only one sickle gene and one normal adult hemoglobin gene referred to as 'HbAS' or 'sickle cell trait'. People with sickle cell trait live a normal life as they have few, if any symptoms of the disease. Insofar as inheriting the disease, if one parent has sickle-cell anemia (SS) and the other sickle-cell trait, the child will have a 50% chance of having sickle-cell disease and a 50% chance of having sickle-cell trait [3, 6]. When both parents have sickle-cell trait, a child has a 25% chance of sickle-cell disease. There is a 25% chance that the child will not carry any sickle cell alleles, and a 50% chance that it will have the heterozygous condition [3, 6].

There is a highly interesting paralleled distribution of SCD and malaria [7]. SCD patients are more common among people of tropical and sub-tropical, and sub-Saharan ancestry where malaria is common [8]. In general, carrying a single sickle-cell allele (sickle cell trait) confers a selective advantage [9]. Specifically, humans with one of the two alleles of sickle-cell disease show less severe symptoms when infected with malaria [7].

SCD may lead to various acute and chronic complications, several of them have a high mortality rate, particularly the painful events due to vascular occlusion (VOC) [1, 5]. Besides many other factors, hypoxia is the most important triggering factor for VOC. Inositol hexaphosphate ($InsP_6$) reduces the oxygen-hemoglobin affinity leading to the oxygen release in the blood stream and in the tissues [10]. It has been demonstrated that $InsP_6$-loaded red blood cells ($InsP_6$-RBCs) reduce the risks of sickling of sickle RBCs (SS RBCs) exposed to hypoxia, indicating that $InsP_6$-RBCs are potentially useful in sickle cell anemia [11]. In this chapter we discuss SCD from epidemiology, genetics, pathogenesis and clinical management to potential usefulness of $InsP_6$-loaded RBCs in transfusion for sickle cell anemia patients.

EPIDEMIOLOGY

There are millions of people suffering from SCD throughout the world. SCD is particularly common among people whose ancestors came from sub-Saharan Africa, Spanish-speaking regions in the Western Hemisphere (South America, the Caribbean, and Central America), Saudi Arabia; India; and the Mediterranean countries such as Turkey, Greece, and Italy [12].

In the world, the highest frequency of SCD is identified in tropical regions, including sub-Saharan Africa, India and the Middle-East [12]. Population migration from these high to low prevalence countries in Europe and the United States has dramatically increased in recent decades; in some European countries SCD is a more familiar genetic condition than hemophilia and cystic fibrosis [13].

In high-income countries that provide neonatal diagnosis and care for the newborn patients, most survive well into adult life; whereas, the vast majority of affected children born in low-income countries still die undiagnosed. If the average survival rate is achieved in only half of the African newborns, over six million Africans will be living with SCD; clearly, care for these disorders must become part of primary-care wherever they are common [12]. The exact number of people living with SCD in the United States is unknown. Based on the data from the Center for Disease Control (CDC) and the National Institutes of Health (NIH), it has been estimated that a) there are 90,000 to 100,000 Americans living with SCD, and 2.5 million Americans are heterozygous carriers for the sickle cell trait (SCT); b) SCD occurs among about 1 out of every 500 Black or African-American births; c) SCD occurs among about 1 out of every 36,000 Hispanic-American births; and d) SCT occurs among about 1 in 12 Blacks or African Americans, and 1 of 100 Hispanic-Americans. Routine neonatal screening now identifies most infants with SCD born in the United States [14]; universal neonatal screening for SCD is provided in 44 states along with the District of Columbia, Puerto Rico and the Virgin Islands [14].

GENETICS AND PATHOGENESIS

Normally, humans have Hemoglobin A, which consists of two alpha and two beta chains, Hemoglobin A2, which consists of two alpha and two delta chains and Hemoglobin F, consisting of two alpha and two gamma chains in their bodies. Of these, Hemoglobin A makes up around 96-97% of the normal hemoglobin in humans.

Sickle-cell gene mutation probably occurred spontaneously in different geographic areas. The allele responsible for sickle-cell anemia is a gene on the short arm of chromosome 11 for coding the β-globin chain of hemoglobin [15]. In people carrying heterozygous sickle cell gene for hemoglobin S (HbS, carriers of sickling hemoglobin), the polymerization problems are minor, because the normal allele is able to produce over 50% of the normal hemoglobin [16]. In people carrying homozygous sickle cell genes for HbSS, the presence of long-chain polymers of HbSS distort the shape of the red blood cell from a smooth doughnut-like shape to ragged and fragile; and susceptible to breaking within capillaries [16].

The sickle cell gene is a known mutation of a single nucleotide (single-nucleotide polymorphism - SNP) (A to T) of the β-globin gene [16]. A point mutation in the β-globin chain of hemoglobin causes the hydrophilic amino acid glutamic acid to be replaced with the hydrophobic amino acid valine at the sixth position. Genetically, this single nucleotide mutation leads from a CTC to CAC codon on the coding strand, which is transcribed from the template strand into a GUG codon. The association of two wild-type α-globin subunits with two mutant β-globin subunits forms hemoglobin S, referred to as HbS, as opposed to the normal adult HbA. Under the conditions of normal oxygen concentration, this mutation apparently does not affect the secondary, tertiary, or quaternary structure of hemoglobin. However, under conditions of low oxygen concentration, the deoxy form of hemoglobin exposes a hydrophobic patch on the protein between the E and F helices. The hydrophobic residues of the valine at position 6 of the β chain in sickle cell hemoglobin (HbS) are able to associate with the hydrophobic patch, promoting the non-covalent polymerization (aggregation), which distorts red blood cells into a sickle shape and decreases their elasticity.

The central pathogenesis event in SCD is the loss of red blood cell elasticity [17]. Elasticity is the normal nature of red blood cells that allows them to deform as they pass through fine capillaries. In SCD, low-oxygen tension promotes red blood cell sickling and decreases the cell's elasticity. These sickling cells fail to return to normal shape even when normal oxygen tension is restored. As a consequence, these rigid sickle red blood cells are unable to deform as they pass through narrow capillaries, resulting in vessel occlusion and ischemia. Another key pathogenesis event is hemolysis - the destruction of the red blood cells due to their misshape or sickling. In general sickle cells only survive 10-20 days, as compared to healthy red blood cells that live 90-120 days.

ANIMAL MODELS OF HUMAN SCD

Berkeley sickle cell mice (BERK mice) are a useful animal model for human sickle cell disease [18]. BERK mice have the targeted deletions of murine α and β globins ($\alpha^{-/-}$, $\beta^{-/-}$) with a transgene containing human α, β^s, $^A\gamma$, $^G\gamma$, and β globins [19]. Thus, these mice express human sickle hemoglobin almost exclusively, or highly mimic human SCD. Another useful model of human SCD is the transgenic SAD mouse model, which expresses modified sickle hemoglobin, Hb SAD, displays *in vivo* hemoglobin polymerization and erythrocyte sickling [20]. Both the models would be useful not only for determining the pathogenesis but also for evaluating the potential therapeutic approaches.

CLINICAL SIGNS, SYMPTOMS AND MANAGEMENT OF SCD

The signs and symptoms of SCD vary clinically. Some people have mild symptoms. Others may present with very severe symptoms and often are hospitalized for treatment. In general SCD is present at birth, but many infants do not show any signs until after 4 months of age. The most common signs and symptoms are anemia and episodes of pain. Sickle cells are fragile and break apart easily and die; that leads to a chronic shortage of red blood cells in the body, clinically known as anemia. Periodic episodes of pain, called vaso-occlusive crisis, is the result of obstruction of capillaries by sickle-shaped red blood cells causing restriction of blood flow to an organ. This in turn results in ischemia, pain, necrosis and often organ damage. There is considerable inter-individual variation of the frequency, severity, and duration of these crises.

Swollen hands and feet, called 'hand-foot syndrome', often are the first signs of SCD in babies [21]. This syndrome is also caused by sickle-shaped red blood cells blocking blood flow out of their hands and feet. The most critical crises are splenic sequestration crisis and acute chest syndrome (ACS) [22, 23]. Splenic sequestration crisis presents as acute, painful enlargements of the spleen [22]. It is the result of intra-splenic trapping of red cells causing a precipitous fall in hemoglobin levels with the potential for hypovolemic shock [22]. Obviously this sequestration crisis is a clinical emergency event. These crises are transient; they continue for 3-4 hours and up to a day. If not treated, patients may die within 1-2 hours due to circulatory failure. ACS accounts for about 25% of death in patients with SCD [23]. ACS is characterized by new pulmonary infiltrate with pulmonary symptoms such as tachypnea and dyspnea [22, 23]. The majority of cases present with vaso-occlusive crises are followed by ACS [24]. The management for these crises is supportive, sometimes with blood transfusion.

Anemia is the most common clinical presentation. Aplastic crises are acute worsening of the patient's baseline anemia producing pallor, tachycardia, and fatigue. Parvovirus B19 is a crucial trigger factor for this crisis. Parvovirus B19 directly affects erythropoiesis *via* invading the erythrocyte-precursors, multiplying in them and destroying them [25]. Even in normal individuals, parvovirus infection almost completely prevents red blood cell production for 2-3 days (this is of little consequence for normal individuals). Under this infectious situation, sickle-cell patients who have already the shortened red cell life will suffer from an abrupt, life-threatening crisis. Reticulocyte counts drop dramatically during the disease (causing reticulocytopenia), and the rapid turnover of red cells leads to a significant lowering of hemoglobin. This crisis takes 4 days to one week to disappear. While most patients can be managed supportively, some may need blood transfusion [25].

Various complications occur in SCD patients. The common complications include infection, stroke, cholelithiasis (gallstones) and cholecystitis, avascular necrosis, eye complications (particularly retinopathy), pulmonary hypertension, sickle cell nephropathy, *etc.*

TRANSFUSION THERAPY AND InsP$_6$-LOADED RED BLOOD CELLS

Blood transfusions are often used in the management of sickle cell disease in acute crises and severe anemia, and to prevent complications by decreasing the number of red blood cells that can sickle by adding normal red blood cells [26]. In children, prophylactic chronic RBC transfusion therapy has been shown to be successful to some extent in reducing the risk of first stroke or silent stroke when transcranial Doppler (TCD) ultrasonography shows abnormally increased cerebral blood flow velocities. In those patients who have sustained a prior stroke event it also reduces the risk of recurrent stroke and additional silent strokes [27, 28].

Since hypoxia is a major cause of painful vaso-occlusive crisis in sickle cell disease, simple transfusion and red blood cell exchange are commonly used as preventive therapies *via* diluting hemoglobin HbS-containing RBCs with normal RBCs (AA-RBCs) to prevent sickling. Obviously there are two aims for transfusion: 1) improving oxygen delivery and 2) decreasing the sickling of patient's red blood cells.

Inositol hexaphosphate (InsP_6) as an allosteric Hb effector would be useful to improve the effectiveness of transfusion for SCD patients. Kumpati *et al.,* in 1982 studied the effect of InsP_6-loaded red blood cells (InsP_6-RBCs) on oxygen delivery [10]. Through incorporating InsP_6 into RBCs by a liposomal transport system to prepare low oxygen affinity red cells or InsP_6-RBCs and determining oxygen equilibrium curves and percentage sickling as a function of pO$_2$, the *in vitro* mixture of InsP_6-RBC and SCD patient red blood cells (1:1 mixture) results in higher venous pO$_2$ values and less sickling throughout the range of oxygen delivery [10]. The potential mechanism of action of InsP_6-RBCs transfusion they proposed are: 1) direct oxygen transfer from InsP_6-RBCs to sickle red blood cells helping to prevent sickling and 2) high oxygen supply to tissues that decrease the rate of oxygen delivered from sickle red blood cells, as seen in Fig. (**14.1**).

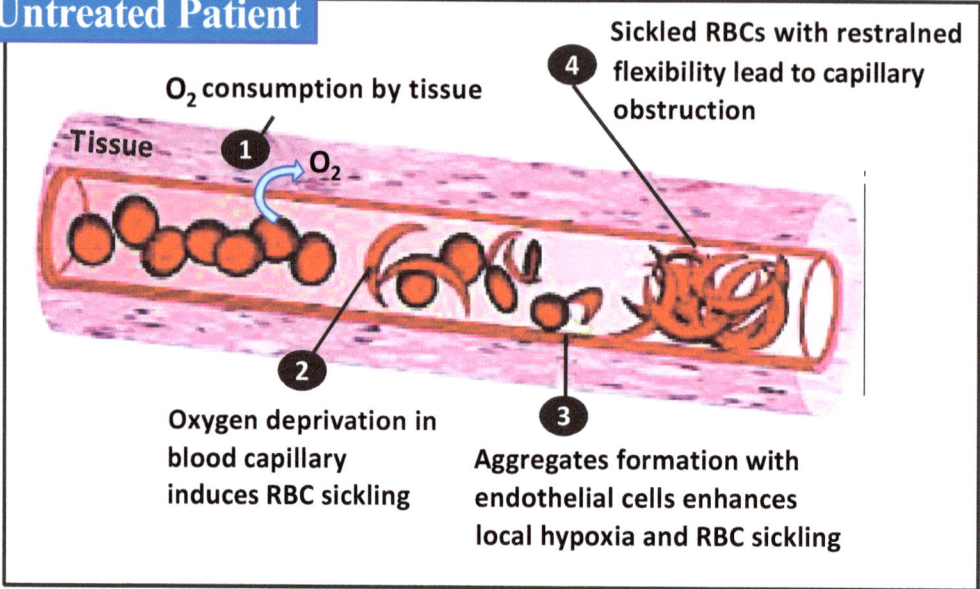

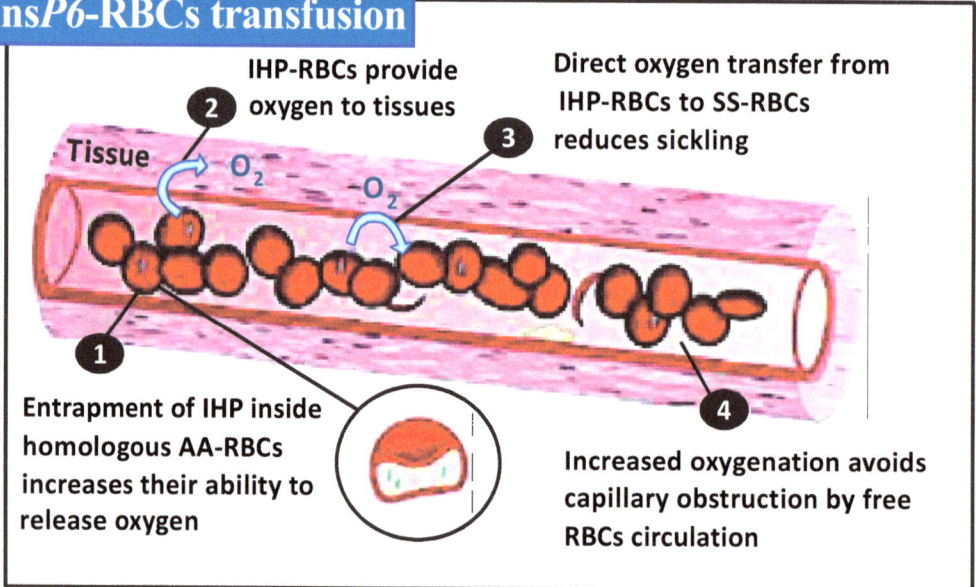

Fig. (14.1). The potential mechanism of InsP6-RBCs (also called IHP-RBCs) on prevention of vaso-occlusive crisis in SCD [4]. Reprinted with permission from "Transfusion Volume 50, Issue 10, pages 2176-2184, October 2010".

Teisseire *et al.*, further investigated the long-term physiological effects of $InsP_6$-loaded RBCs on enhancing O_2 release in a pig model [29]. $InsP_6$-loaded RBCs are achieved through a continuous lysing and resealing procedure resulting in incorporation of $InsP_6$ in these cells. $InsP_6$-loaded RBCs lead to significant rightward shifts of the HbO_2 dissociation curves with *in vitro* P_{50} (partial pressure of O_2 at 50% Hb saturation), values increasing from 32.2 ± 1.8 torr for control erythrocytes to 86 ± 60 torr (pH 7.40; pCO_2 40 torr at 37 °C; 1 torr = 1.333 x 10W Pa), as seen in Fig. (**14.2**) (curves 2-6). The life span of $InsP_6$-loaded RBCs is equal to that of control erythrocytes or lysed/resealed erythrocyte without $InsP_6$. The long-term physiological effects of the $InsP_6$-loaded RBCs show an increase of O_2 release and reduction of cardiac output. The reduced O_2 affinity of the $InsP_6$-loaded RBCs is effective 20 days after transfusion.

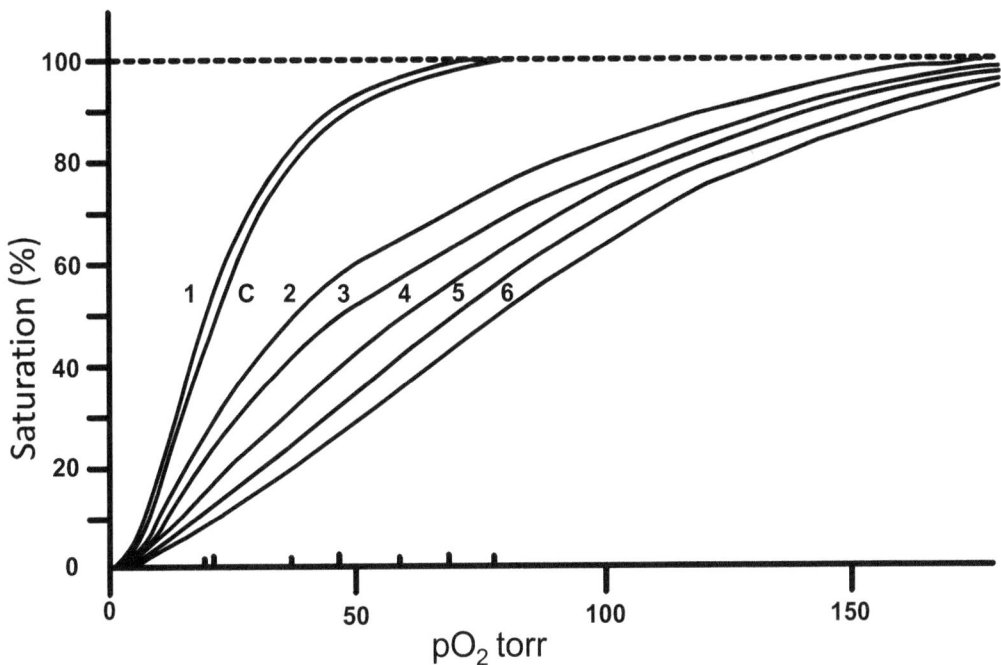

Fig. (14.2). Hb-O_2 dissociation curves and the P_{50} changes induced by $InsP_6$-loaded RBCs. A high value of the regression coefficient (r = 0.99) is obtained from these four points of $InsP_6$-loaded RBCs (Curves 2 -6). C, Control; 1, lysed and resealed erythrocyte without $InsP_6$; and 2 - 6, $InsP_6$-loaded RBCs. Adapted from ref [29].

Since Hemoglobin S polymerization is markedly dependent on intracellular hemoglobin concentration, further *in vitro* study indicated that sickle cells modified by an osmotic pulse in the presence of $InsP_6$ decreased intracellular hemoglobin S concentration and decreased *in vitro* sickling without prolonging *in vivo* survival [30, 31].

Bourgeaux *et al.,* further investigated the *in vitro* effectiveness of InsP_6-RBCs to prevent *in vitro* sickling. InsP_6-RBCs were prepared using the method of poration (reversible hypotonic lysis) with AN69 hollow fibers. Compared to the stored RBCs, InsP_6-RBCs treatment was seven times more effective on sickling reduction and sickling was inhibited in a dose-dependent manner [4]. Hemorheologic properties of InsP_6-RBCs have been studied under normoxia and/or after hypoxic challenges and results further indicate that InsP_6-RBC decrease RBC aggregation and diminish the adverse effects of hypoxia on RBC deformability and blood viscosity [11].

The milestone study for determining the effectiveness of InsP_6-RBCs treatment *in vivo* has been performed in BERK mice and SAD mice, models highly mimicking human SCD. Transfusion of InsP_6-RBCs to BERK mice resulted in a significant improvement of survival rate and brain development, prevention of severe anemia and a greatly lowered risk of vascular occlusion. Transfusion of InsP_6-RBCs in SAD mice subjected to acute hypoxic stress led to significantly decreased mRNA levels of molecules involved in intravascular disorders in the lungs.

In summary, all of these results indicate that InsP_6-RBCs transfusion prevents *in vitro* sickling, increases the oxygen delivery and reduces SCD disorders in sickle transgenic mice. InsP_6-RBCs could be useful in SCD patients to improve conventional transfusion therapy in terms of transfused volume, frequency, and efficacy.

Germane to the delivery of InsP_6 to the red blood cells is the issue whether or not InsP_6 can be transported across the red blood cell membrane; the current dogma is that it cannot be. However, as discussed in Chapter 5 (Pharmacokinetics), it is evident that InsP_6 is transported across the cell membrane. Thus it might be feasible to deliver therapeutic dosage of InsP_6 by conventional means without going through delivery *via* liposomes, poration, *etc*.

Mechanism

The underlying mechanism of sickling is the gelation of the concentrated HbS when deoxygenated. A small change in polymer stability evokes profound alteration in its solubility. The red blood cell metabolite 2,3 diphosphoglycerate

(DPG) regulates the oxygen affinity of HbA by binding preferentially to the deoxy (T) conformer of the allosteric equilibrium, R↔T. Since, only the T state is incorporated into the deoxy-HbS polymer, the presence of DPG is thought to facilitate polymerization under hypoxic condition existing in microcirculation; the results have been conflicting. However, $InsP_6$ as a heterotropic allosteric effector of the oxygen affinity of Hb, drastically diminishes the solubility of deoxy-HbS [32]. $InsP_6$ is almost three times more effective in decreasing the solubility than DPG, and ten times more effective in lowering the oxygen affinity than DPG; thereby inducing greater constraints in the tertiary structure of the β chain of Hb [32].

CONCLUDING REMARKS

$InsP_6$-loaded red blood cells show a promising effect on preventing sickling and delivering high oxygen supply to tissues in both *in vitro* and *in vivo* models. Given the plethora of data, clinical trials are overdue.

REFERENCES

[1]   Platt OS, Brambilla DJ, Rosse WF *et al*. Mortality in sickle cell disease. Life expectancy and risk factors for early death. N Engl J Med 1994; 330(23): p. 1639-44.
[2]   Pauling L, Itano HA, Singer SJ *et al*. Sickle cell anemia a molecular disease. Science 1949; 110: 543-8.
[3]   Stuart MJ, Nagel RL. Sickle-cell disease. Lancet 2004; 364: 1343-60.
[4]   Bourgeaux V, Hequet O, Campion Y *et al*. Inositol hexaphosphate-loaded red blood cells prevent *in vitro* sickling. Transfusion 2010; 50: 2176-84.
[5]   Malowany JI, Butany J. Pathology of sickle cell disease. Semin Diagn Pathol 2012; 29: 49-55.
[6]   Lesi FE, Bassey EE. Family study in sickle cell disease in Nigeria. J Biosoc Sci 1972; 4: 307-13.
[7]   Wellems TE, Hayton K., Fairhurst RM. The impact of malaria parasitism: from corpuscles to communities. J Clin Invest 2009; 119: 2496-505.
[8]   Kwiatkowski DP. How malaria has affected the human genome and what human genetics can teach us about malaria. Am J Hum Genet 2005; 77: 171-92.
[9]   Poncon N, Toty C, L'Ambert G *et al*. Biology and dynamics of potential malaria vectors in Southern France. Malar J 2007; 6: 18.
[10] Kumpati J, Franco RS, Weiner M *et al*. Sickling as a function of oxygen delivery: effect of simulated transfusions of stored, fresh and inositol-hexaphosphate-loaded (low affinity) red cells. Blood Cells 1982; 8: 263-72.
[11] Lamarre Y, Bourgeaux V, Pichon A, *et al*. Effect of inositol hexaphosphate-loaded red blood cells (RBCs) on the rheology of sickle RBCs. Transfusion 2013; 53: 627-36.
[12] Weatherall DJ, Clegg JB. Inherited haemoglobin disorders: an increasing global health problem. Bull World Health Organ 2001; 79: 704-12.
[13] Roberts I, de Montalembert M. Sickle cell disease as a paradigm of immigration hematology: new challenges for hematologists in Europe. Haematologica 2007; 92: 865-71.
[14] American Academy of Pediatrics. Health supervision for children with sickle cell disease. Pediatrics 2002; 109: 526-35.
[15] Levings PP, Bungert J. The human beta-globin locus control region. Eur J Biochem 2002; 269: 1589-99.

[16]   Ashley-Koch A, Yang Q, Olney RS. Sickle hemoglobin (HbS) allele and sickle cell disease: a HuGE review. Am J Epidemiol 2000; 151: 839-45.

[17]   Bunn HF, Pathogenesis and treatment of sickle cell disease. N Engl J Med 1997; 337: 762-9.

[18]   Paszty C. Transgenic and gene knock-out mouse models of sickle cell anemia and the thalassemias. Curr Opin Hematol, 1997; 4: 88-93.

[19]   Paszty C, Brion CM, Manci E *et al*. Transgenic knockout mice with exclusively human sickle hemoglobin and sickle cell disease. Science 1997; 278: 876-8.

[20]   De Paepe ME, Trudel M. The transgenic SAD mouse: a model of human sickle cell glomerulopathy. Kidney Int 1994; 46: 1337-45.

[21]   Jadavji T, Prober CG. Dactylitis in a child with sickle cell trait. Can Med Assoc J, 1985; 132: 814-5.

[22]   Khatib R, Rabah R, Sarnaik SA. The spleen in the sickling disorders: an update. Pediatr Radiol 2009; 39: 17-22.

[23]   Mekontso Dessap A, Leon R, Habibi A *et al*. Pulmonary hypertension and cor pulmonale during severe acute chest syndrome in sickle cell disease. Am J Respir Crit Care Med 2008; 177: 646-53.

[24]   Paul RN, Castro OL, Aggarwal A *et al*. Acute chest syndrome: sickle cell disease. Eur J Haematol; 2011; 87: 191-207.

[25]   Slavov SN, Kashima S, Pinto AC *et al*. Human parvovirus B19: general considerations and impact on patients with sickle-cell disease and thalassemia and on blood transfusions. FEMS Immunol Med Microbiol 2011; 62: 247-62.

[26]   Drasar E, Igbineweka N, Vasavda N *et al*. Blood transfusion usage among adults with sickle cell disease - a single institution experience over ten years. Br J Haematol 2011; 152: 766-70.

[27]   Gyang E, Yeom K, Hoppe C *et al*. Effect of chronic red cell transfusion therapy on vasculopathies and silent infarcts in patients with sickle cell disease. Am J Hematol 2011; 86: 104-6.

[28]   Mirre E, Brousse V, Berteloot L *et al*. Feasibility and efficacy of chronic transfusion for stroke prevention in children with sickle cell disease. Eur J Haematol 2010; 84: 259-65.

[29]   Teisseire B, Ropars C, Villereal MC *et al*. Long-term physiological effects of enhanced O2 release by inositol hexaphosphate-loaded erythrocytes. Proc Natl Acad Sci U S A, 1987; 84: 6894-8.

[30]   Franco R, Barker-Gear R, Silberstein E *et al*. Sickle cells modified by an osmotic pulse in the presence of inositol hexaphosphate have decreased intracellular hemoglobin concentration and decreased *in vitro* sickling without prolonged *in vivo* survival. Adv Exp Med Biol 1992; 326: 325-31.

[31]   Franco RS, Barker-Gear R, Green R. Inhibition of sickling after reduction of intracellular hemoglobin concentration with an osmotic pulse: characterization of the density and hemoglobin concentration distributions. Blood Cells 1993; 19: 475-88; discussion 489-91.

[32]   Poillon Wn, Robinson MD, Kim BC. Deoxygenated sickle hemoglobin. J Biol Chem 1985; 260: 13897-900.

CHAPTER 15

Polycystic Ovary Syndrome and Therapeutic Potential of Inositol Compounds

Abstract: Polycystic ovary syndrome (PCOS) is one of the most common endocrine disorders among females. Although there is a diverse range of causes that are not entirely understood, recent studies indicate the potential role of *myo*-inositol in the pathophysiology of PCOS: i) insulin resistance in PCOS women can be attributed to a deficiency of *myo*-inositol's intracellular metabolites, D-*chiro*-inositol (DCI) and inositol-phosphoglycan (IPG), mediators of insulin action; ii) PCOS patients have a higher urinary clearance of DCI; and iii) DCI functions as an intracellular messenger in mammalian oocytes, playing a role in the follicular milieu, meiotic resumption, and oocyte maturation. Recently, *myo*-inositol has been used in PCOS patients more and more as a natural insulin sensitizer and until today, there are more than 70 clinical trials on determining the effect of *myo*-inositol and/or DCI as insulin-sensitizing compound in PCOS women.

Keywords: AGE, anovulation, BMI, *chiro*-inositol, DCI, inosituria, oligoovulation, PCOS, sorbitol, Stein-Leventhal.

INTRODUCTION

Though polycystic ovary syndrome (PCOS) was first described by Stein and Leventhal in 1935 [1], until today, there are still several aspects of this syndrome, including diagnosis and treatments that are still debated. As clear in the name, PCOS is formation of multiple cysts in ovaries commonly identified on ultrasound examination. But polycystic formation is not an absolute requirement in all definitions of this disorder. Currently, the diagnosis is based on the outcome of a consensus meeting held in Rotterdam in 2003 sponsored by the European Society for Human Reproduction and Embryology and the American Society for Reproductive Medicine. The Rotterdam definition and diagnostic criteria of PCOS is based on the clinical presentation that PCOS to be present in any two of the following criteria: a) oligoovulation and/or anovulation, b) excess androgen activity and c) polycystic ovaries [2]. In addition to the above three clinical presentations, the insulin resistance is another one of the most common symptoms; 16% to 80% of PCOS patients show insulin resistance, regardless of the body mass index (BMI) [3]. Insulin resistance is associated with high cholesterol levels, non-insulin dependent diabetes (Type II) and obesity. There is a marked inter-individual variation in symptoms and severity of the PCOS among affected women.

The etiology of PCOS is largely unknown; however, there is strong evidence indicating it to be a genetic disease. Although the exact gene affected has not yet been identified, possible PCOS genes have been proposed including *CYP11A*, the insulin gene, and a region near the insulin-receptor gene [4]. The evidence of PCOS as a genetic disease includes 1) familial clustering of PCOS cases, 2) great concordance in monozygotic compared with dizygotic twins, and 3) heritability of endocrine and metabolic features of PCOS [5]. Prevalence of PCOS is high and there are approximately 5% to 10% of women of reproductive age (approximately 12 to 45 years) producing PCOS symptoms. The World Health Organization (WHO) estimates that as of 2010 there are about 116 million women with PCOS (3.4% of women) worldwide [6]. It is one of the leading causes of female subfertility [7-10]. Recently, *myo*-inositol has been used in PCOS patients more and more as a natural insulin sensitizer. *Myo*-inositol is not only a natural carbohydrate but also is an isomerized and dephosphorylated precursor of glucose-6-phosphate. Recent studies focused on the role of *myo*-inositol in the pathophysiology of PCOS indicate that i) insulin resistance in PCOS women can be attributed to a deficiency of *myo*-inositol's intracellular metabolites, D-*chiro*-inositol (DCI) and inositol-phosphoglycan (IPG), mediators of insulin action [11-14]; ii) PCOS patients have a higher urinary clearance of DCI [15]; and iii) DCI functions as an intracellular messenger in mammalian oocytes, playing a role in the follicular milieu, meiotic resumption, and oocyte maturation [16, 17]. In this chapter, we mainly discuss the pathogenesis of PCOS and the therapeutic potential of *myo*-inositol in PCOS; the rationale for *myo*-inositol therapy may be seen by understanding the pathogenesis.

PATHOGENESIS

The common sign of PCOS is multiple ovarian cysts or polycystic formation on ultrasound examination. Pathologically these "cysts" are immature follicles or cystic follicles. Due to the disturbed ovarian function, the development of follicles (arising from primordial follicles) has arrested at an early antral stage, and these immature follicles are commonly located along the ovarian periphery, appearing as tiny cysts. In general, functional ovarian overproduction of androgens is responsible for the multiple cyst formation in the ovaries. Intra-ovarian androgen excess promotes the growth of multiple small follicles, yet hampers selection of a dominant follicle, subsequently leading to these small follicles being arrested in development, and resulting in "polycystic" formation of the ovaries [4, 18, 19].

Overproduction of androgens, oligoovulation and/or anovulation, and insulin resistance are key pathogenic events in PCOS. Since PCOS is a heterogenic endocrine disorder, there are multiple pathophysiologic mechanisms involved that include i) an alteration in gonadotropin-releasing hormone secretion leading to increase luteinizing hormone (LH) secretion, ii) an alteration in insulin secretion and insulin action leading to hyperinsulinemia and insulin resistance, and iii) a defect in androgen synthesis leading to increased ovarian androgen production. Increased LH relative to follicle-stimulating hormone (FSH) is a characteristic hallmark of the laboratory abnormality identified in PCOS. It has been considered that increased LH secretion occurs as a result of increased frequency of hypothalamic gonadotropin-releasing hormone (GnRH) pulses. Elevated LH plays a crucial role in the pathogenesis of PCOS by increasing androgen production and secretion by ovarian theca cells, which are responsible for all the clinical features of PCOS including hirsutism (excessive growth of hair), anovulation, and polycystic ovaries [20, 21].

A characteristic metabolic disturbance in PCOS is insulin resistance. Both obese and non-obese PCOS women have a higher incidence of insulin resistance and hyperinsulinemia than age-matched controls [22]; however, obese PCOS women have significantly decreased insulin sensitivity compared with non-obese PCOS women [22-24]. In many cases, a complex positive feedback loop of insulin resistance and hyperandrogenism seems to be at play; treatment with either insulin-sensitizing agents or anti-androgens improves both hyperandrogenism and insulin resistance [22-24]. Insulin acts synergistically with LH to enhance androgen production in the ovarian theca cells. Insulin also decreases hepatic synthesis and secretion of sex hormone-binding globulin; the latter binds testosterone in the circulation, thereby increasing the amount of biologically available free testosterone. Typically, women with PCOS and hyperinsulinemia have elevated levels of free testosterone in the blood; however, the total testosterone concentration may be only modestly elevated, or at the upper range of normal. The insulin-resistant hyperinsulinemia induced by an obesogenic diet has been shown to cause hyperandrogenic anovulation by signaling through the theca cell insulin receptor in transgenic mice [25]. Over-activation of the insulin/IGF-I system augments LH stimulation of ovarian androgen production by interfering with the normal down-regulation of the theca cell response to LH [25].

Chronic inflammation may play some role in PCOS, and the evidence includes i) correlating inflammatory mediators with anovulation and, ii) relation between PCOS and increased level of oxidative stress [26-29]. PCOS has also been associated with a specific FMR1 sub-genotype. Women with heterozygous-normal/low FMR1 have polycystic-like symptoms of excessive follicle-activity and hyperactive ovarian function [30].

Pathogenic Role of *myo*-Inositol and D-*chiro*-Inositol in Diabetes and PCOS

Myo-inositol's intracellular concentration is dependent on several factors that include extracellular *myo*-inositol uptake, *de novo* biosynthesis, regeneration (phosphoinositide cycle) and, efflux and degradation [31]. Alteration of one or several of these processes can lead to inositol's intracellular abnormalities. Functionally, *myo*-inositol and D-*chiro*-inositol may be involved in maintaining glucose homeostasis; abnormalities in their metabolism is associated with insulin-resistance and long-term diabetic microvascular complications in both diabetic subjects and in women with PCOS [32].

In diabetic animal models and humans, a concomitant depletion of intracellular *myo*-inositol and accumulation of intracellular sorbitol is commonly observed. Excretion of excessive amounts of *myo*-inositol and decreased amounts of D-*chiro*-inositol in urine (called inosituria) are also commonly observed in addition to tissue-specific *myo*-inositol depletion [33]. Inhibition of cellular *myo*-inositol uptake, altered *myo*-inositol biosynthesis, enhanced *myo*-inositol efflux due to intracellular accumulation of sorbitol, and increased *myo*-inositol degradation are putative mechanisms of *myo*-inositol intracellular depletion [33]. *Myo*-inositol depletion under hyperglycemic conditions in insulin insensitive tissues appears contributory to the development of diabetic microvascular complications, together with increased advanced glycation end products (AGEs) formation, activation of protein kinase C (PKC), increased hexosamine and sorbitol pathways [32, 33].

Elevated concentrations of *myo*-inositol in human follicular fluid play a role in follicular maturity and provide a marker of good-quality oocytes [17]. Unlike other tissues such as muscle and liver, ovaries never become insulin resistant [34-36]. This phenomenon is probably due to the presence of hyperinsulinemia in PCOS patients that enhances *myo*-inositol to D-*chiro*-inositol epimerization in the ovary, and results in an increased D-*chiro*-inositol/*myo*-inositol ratio (*i.e.*, overproduction of D-*chiro*-inositol), sequentially leading to a *myo*-inositol deficiency in the ovary [32, 34-36]. This *myo*-inositol depletion could eventually be responsible for the poor oocyte quality observed in these patients [37].

EFFECTS OF *MYO*-INOSITOL AND D-*CHIRO*-INOSITOL IN WOMEN WITH P.C.O.S: RANDOMIZED CONTROLLED CLINICAL TRIALS

Myo-Inositol

Although the diagnostic criteria established in Rotterdam are widely accepted, insulin resistance, a crucial condition related to PCOS is not considered in this criterion. Insulin resistance is very commonly seen in PCOS women regardless of the body mass index (BMI). Hyperinsulinemia due to insulin resistance is identified in approximately 80% of PCOS women with central obesity and in 30%-40% of lean PCOS women [38, 39]. Obesity significantly worsens insulin resistance, and is a crucial factor in the pathogenesis of anovulation and hyperandrogenism [40]. The impairment in insulin pathway, particularly a defect in the inositol phosphates signaling, may play an important role in controlling glucose metabolism. In PCOS women, a defect in tissue availability or altered metabolism of *myo*-inositol and D-*chiro*-inositol may contribute to insulin resistance [12]. *Myo*-inositol and D-*chiro*-inositol have been used as insulin sensitizer drugs in the treatment of PCOS.

Until today, there are more than 70 clinical trials on determining the effect of *myo*-inositol and/or D-*chiro*-inositol as insulin-sensitizing compound in PCOS women [11, 41-45]. All of the subjects analyzed in these clinical trials are PCOS patients. Most of the trials are double-blind randomized controlled trials of *myo*-inositol *versus* folic acid; and only one trial is a randomized controlled trial of *myo*-inositol *versus* metformin (an oral hypoglycemic agent). The ranges of *myo*-inositol dose are 2 - 4 grams per day, D-*chiro*-inositol dose is 600 mg per day; that of folic acid is 200 - 400 μg/day and metformin 1500 mg/day.

The outcomes of these randomized controlled trials as summarized in Table **15.1** indicate that *myo*-inositol significantly improved insulin resistance, ovarian function, metabolic and hormonal parameters [11, 41-45].

D-*Chiro*-Inositol

As stated before, hyperinsulinemia in PCOS patients results in an overproduction of D-*chiro*-inositol, resulting in *myo*-inositol deficiency in the ovary [32, 34-36]

Table 15.1.   **Outcome of the randomized controlled clinical trials of *myo*-inositol in PCOS patients [11, 41-45].**

Trial	Dose	Subjects	Outcome		
			Insulin resistance*	*Endocrine**	*Ovarian function**
Genazzani	2g/d	20	Reducing insulin level, increasing the index of insulin sensitivity glucose/insulin ratio, and lowering both insulin response and the area under the curve (AUC) after oral glucose load.	Reducing plasma luteinizing hormone (LH) and prolactin (PRL) levels, LH/follicle-stimulating hormone (FSH) ratio.	Inducing menstrual cycle and restoring the normal menstrual cycles, and reducing Ovarian volumes
Costantino	2g/d	42	Lowering both insulin response and the area under the curve (AUC) after oral glucose load, decreased plasma triglyceride and cholesterol, and increased the composite whole body insulin sensitive index	Reducing testosterone, dehydroepiandrosterone	Restoring ovulation
Papaleo	2g/d	60	Not tested	Reducing both the total recombinant FSH and the number of stimulation days. Reducing estradiol (E2) levels after human chorionic gonadotropin administration	Reducing number of immature oocytes and degenerated oocytes
Gerli	2g/d	92	Decreased BMI, lowering circulating leptin concentration, and increased e high-density lipoprotein (HDL)	Increase E2 levels	Improving ovulation
Raffone	4g/d	120	Note tested	Not tested	Improving the restored spontaneous ovulation activity and pregnancy rate

* Only parameters with statistical significance were summarized in the table.

which could be responsible for the poor oocyte quality in PCOS patients. It would therefore seem paradoxical to treat PCOS patients with D-*chiro*-inositol. Be that as it may, there have nevertheless been clinical trials of D-*chiro*-inositol in PCOS patients, albeit a few, with small numbers of patients. One trial included 20 lean women (body mass index, 20.0 to 24.4 kg/m 2) with PCOS, and 600 mg of D-*chiro*-inositol or placebo orally administered once daily for 6 to 8 weeks [46]. The results show that the plasma insulin curve after oral administration of glucose decreased significantly from 8,343 ± 1,149 mU/mL per min to 5,335 ± 1,792 mU/mL per min in comparison to no significant change in the placebo group ($P = 0.03$ for difference between groups); the serum free testosterone concentration decreased by 73% from 0.83 ± 0.11 ng/dL to 0.22 ± 0.03 ng/dL, a significant change in comparison to essentially no change in the placebo group ($P = 0.01$); significant improvement of ovulation in the D-*chiro*-inositol group (60%) in comparison with the placebo group (20%, $P = 0.17$), and significant improvement in systolic and diastolic blood pressures, as well as decreased plasma triglyceride concentrations [46]. Another trial aimed to determine the efficacy of the integrative treatment with D-*chiro*-inositol (500 mg/day for 12 weeks) on hormonal parameters and insulin sensitivity and showed a significant improvement of several endocrine parameters including LH, LH/FSH ratio, androstenedione and insulin, insulin response to oral glucose tolerance test, gonadotropin-releasing hormone (GnRH)-induced LH response and a significant decrease of BMI [47]. A recent trial enrolled 50 PCOS patients and randomly divided them into two groups: 25 were treated with 4 g of *myo*-inositol/day plus 400 µg of folic acid/day orally for six months, 25 with 1 g of D-*chiro*-inositol/day plus 400 µg of folic acid/die orally for six months. They analyzed serum LH, LH/FSH ratio, total and free testosterone, dehydroepiandrosterone sulfate (DHEA-S), Δ-4-androstenedione, SHBG, prolactin, glucose/immunoreactive insulin (IRI) ratio, homeostatic model assessment (HOMA) index, and the resumption of regular menstrual cycles and concluded that both the isoforms of inositol are effective in improving ovarian function and metabolism in patients with PCOS, although *myo*-inositol showed the most marked effect on the metabolic profile, whereas D-*chiro*-inositol better reduced hyperandrogenism [48].

However, in term of ovulation, so far only *myo*-inositol has been shown to be present in the follicular fluid; and in a direct comparison between *myo*-inositol and D-*chiro*-inositol, only *myo*-inositol is able to improve oocyte and embryo quality [49]. This conclusion is also evidenced by a clinical study focusing on determining the role D-*chiro*-inositol at ovarian level. Not surprisingly, this trial showed that PCOS patients receiving 300, 600, 1200, 2400 mg of D-*chiro*-inositol exhibited progressively worsening oocyte quality and ovarian response [50].

Inositol Safety: Clinical Evidences

No serious side-effects have been reported for inositol, even with a therapeutic dosage that equals about 18 times the average dietary intake. The main outcome was that only the highest dose of *myo*-inositol (12 g/day) induced mild gastrointestinal side effects such as nausea, flatus and diarrhea. The severity of side effects did not increase with the dosage [51]. The United States Food and Drug Administration (FDA) have accorded GRAS (generally recognized as safe) status to inositol [52].

CONCLUDING REMARKS

Increasingly, *myo*-inositol is being used in PCOS patients as a natural insulin sensitizer. More than 70 clinical trials to date showed promising therapeutic effect. Further mechanistic studies and larger clinical trials are needed to understand and validate these encouraging results.

REFERENCES

[1]   Stein IF, Leventhal ML. Amenorrhea associated with bilateral polycystic ovaries. Am J Obstet Gynecol 1935; 29: 181-91.
[2]   Revised 2003 consensus on diagnostic criteria and long-term health risks related to polycystic ovary syndrome (PCOS). Hum Reprod 2004; 19: 41-7.
[3]   Dunaif A. Insulin resistance and the polycystic ovary syndrome: mechanism and implications for pathogenesis. Endocr Rev 1997; 18: 774-800.
[4]   Legro RS, Strauss JF. Molecular progress in infertility: polycystic ovary syndrome. Fertil Steril 2002; 78: 569-76.
[5]   Kollmann M, Martins WP, Raine-Fenning N. Terms and thresholds for the ultrasound evaluation of the ovaries in women with hyperandrogenic anovulation. Hum Reprod Update 2014; 20: 463-4.
[6]   Vos T, Flaxman AD, Naghavi M *et al.* Years lived with disability (YLDs) for 1160 sequelae of 289 diseases and injuries 1990-2010: a systematic analysis for the Global Burden of Disease Study 2010. Lancet 2012; 380: 2163-96.
[7]   Azziz R, Woods KS, Reyna R *et al.* The prevalence and features of the polycystic ovary syndrome in an unselected population. J Clin Endocrinol Metab 2004; 89: 2745-9.
[8]   Diamanti-Kandarakis E, Kandarakis H, Legro RS. The role of genes and environment in the etiology of PCOS. Endocrine 2006; 30: 19-26.
[9]   Nafiye Y, Sevtap K, Muammer D *et al.* The effect of serum and intrafollicular insulin resistance parameters and homocysteine levels of nonobese, nonhyperandrogenemic polycystic ovary syndrome patients on *in vitro* fertilization outcome. Fertil Steril 2010; 93: 1864-9.
[10]   Carmina E, Koyama T, Chang L *et al.*, Does ethnicity influence the prevalence of adrenal hyperandrogenism and insulin resistance in polycystic ovary syndrome? Am J Obstet Gynecol 1992; 167: 1807-12.
[11]   Genazzani AD, Lanzoni C, Ricchieri F *et al.* Myo-inositol administration positively affects hyperinsulinemia and hormonal parameters in overweight patients with polycystic ovary syndrome. Gynecol Endocrinol 2008; 24: 139-44.
[12]   Baillargeon JP, Nestler JE, Ostlund RE *et al.* Greek hyperinsulinemic women, with or without polycystic ovary syndrome, display altered inositols metabolism. Hum Reprod 2008; 23: 1439-46.
[13]   Nestler JE; Jakubowicz DJ; Reamer P *et al.*, Ovulatory and metabolic effects of D-chiro-inositol in the polycystic ovary syndrome. N Engl J Med 1999; 340: 1314-20.
[14]   Romero G, Larner J. Insulin mediators and the mechanism of insulin action. Adv Pharmacol 1993; 24: 21-50.
[15]   Baillargeon JP, Diamanti-Kandarakis E, Ostlund RE Jr *et al.*, Altered D-chiro-inositol urinary clearance in women with polycystic ovary syndrome. Diabetes Care 2006. 29(2): p. 300-5.

[16]   Aouameur R, Da Cal S, Bissonnette P *et al.* SMIT2 mediates all myo-inositol uptake in apical membranes of rat small intestine. Am J Physiol Gastrointest Liver Physiol 2007 293: G1300-7.

[17]   Chiu TT, Rogers MS, Law EL *et al.*, Follicular fluid and serum concentrations of myo-inositol in patients undergoing IVF: relationship with oocyte quality. Hum Reprod 2002. 17(6): p. 1591-6.

[18]   Yildiz BO, Yarali H, Oguz H *et al.* Glucose intolerance, insulin resistance, and hyperandrogenemia in first degree relatives of women with polycystic ovary syndrome. J Clin Endocrinol Metab 2003; 88: 2031-6.

[19]   Nestler JE, Jakubowicz DJ, de Vargas AF *et al.*, Insulin stimulates testosterone biosynthesis by human thecal cells from women with polycystic ovary syndrome by activating its own receptor and using inositolglycan mediators as the signal transduction system. J Clin Endocrinol Metab 1998; 83: 2001-5.

[20]   Carmina E, Lobo RA, Polycystic ovary syndrome (PCOS): arguably the most common endocrinopathy is associated with significant morbidity in women. J Clin Endocrinol Metab 1999; 84: 1897-9.

[21]   Haakova L, Cibula D, Rezabek K *et al.* Pregnancy outcome in women with PCOS and in controls matched by age and weight. Hum Reprod 2003. 18(7): p. 1438-41.

[22]   Dunaif A, Segal KR, Futterweit W *et al.* Profound peripheral insulin resistance, independent of obesity, in polycystic ovary syndrome. Diabetes 1989; 38: 1165-74.

[23]   Legro RS, Kunselman AR, Dodson WC *et al.* Prevalence and predictors of risk for type 2 diabetes mellitus and impaired glucose tolerance in polycystic ovary syndrome: a prospective, controlled study in 254 affected women. J Clin Endocrinol Metab 1999; 84: 165-9.

[24]   Hoeger K. Obesity and weight loss in polycystic ovary syndrome. Obstet Gynecol Clin North Am 2001; 28: 85-97, vi-vii.

[25]   Wu S, Divall S, Nwaopara A *et al.* Obesity-induced infertility and hyperandrogenism are corrected by deletion of the insulin receptor in the ovarian theca cell. Diabetes 2014; 63: 1270-82.

[26]   Sathyapalan T, Atkin SL. Mediators of inflammation in polycystic ovary syndrome in relation to adiposity. Mediators Inflamm 2010; 2010: p. 758656.

[27]   Fukuoka M, Yasuda K, Fujiwara H *et al.*, Interactions between interferon gamma, tumour necrosis factor alpha, and interleukin-1 in modulating progesterone and oestradiol production by human luteinized granulosa cells in culture. Hum Reprod 1992; 7: 1361-4.

[28]   Murri M, Luque-Ramírez M, Insenser M *et al.* Circulating markers of oxidative stress and polycystic ovary syndrome (PCOS): a systematic review and meta-analysis. Hum Reprod Update 2013; 19: 268-88.

[29]   Gonzalez F, Rote NS, Minium J *et al.*, Reactive oxygen species-induced oxidative stress in the development of insulin resistance and hyperandrogenism in polycystic ovary syndrome. J Clin Endocrinol Metab 2006; 91: 336-40.

[30]   Gleicher N, Weghofer A, Lee IH *et al.* FMR1 genotype with autoimmunity-associated polycystic ovary-like phenotype and decreased pregnancy chance. PLoS One, 2010; 5: p. e15303.

[31]   Clements RS Jr., Darnell B. Myo-inositol content of common foods: development of a high-myo-inositol diet. Am J Clin Nutr, 1980; 33: 1954-67.

[32]   Croze ML, Soulage CO. Potential role and therapeutic interests of myo-inositol in metabolic diseases. Biochimie 2013. 95: 1811-27.

[33]   Larner J. D-chiro-inositol in insulin action and insulin resistance-old-fashioned biochemistry still at work. IUBMB Life 2001. 51: 139-48.

[34]   Harwood K, Vuguin P, DiMartino-Nardi J. Current approaches to the diagnosis and treatment of polycystic ovarian syndrome in youth. Horm Res 2007; 68: 209-17.

[35]   Matalliotakis I, Kourtis A, Koukoura O *et al.*, Polycystic ovary syndrome: etiology and pathogenesis. Arch Gynecol Obstet, 2006. 274(4): p. 187-97.

[36]   Rice S, Christoforidis N; Gadd C *et al.*, Impaired insulin-dependent glucose metabolism in granulosa-lutein cells from anovulatory women with polycystic ovaries. Hum Reprod 2005; 20: 373-81.

[37]   Chattopadhayay R, Ganesh A, Samanta J *et al.*, Effect of follicular fluid oxidative stress on meiotic spindle formation in infertile women with polycystic ovarian syndrome. Gynecol Obstet Invest 2010. 69(3): p. 197-202.

[38]   Genazzani AD, Battaglia C, Malavasi B *et al.* Metformin administration modulates and restores luteinizing hormone spontaneous episodic secretion and ovarian function in nonobese patients with polycystic ovary syndrome. Fertil Steril, 2004. 81(1): p. 114-9.

[39]   Ciampelli M, Fulghesu AM, Cucinelli F *et al.*, Impact of insulin and body mass index on metabolic and endocrine variables in polycystic ovary syndrome. Metabolism, 1999. 48(2): p. 167-72.

[40]   Genazzani AD, Santagni S, Ricchieri F *et al.*, *Myo*-inositol modulates insulin and luteinizing hormone secretion in normal weight patients with polycystic ovary syndrome. J Obstet Gynaecol Res 2014; 40: 1353-60.

[41]   Costantino D, Minozzi G, Minozzi E *et al.*, Metabolic and hormonal effects of myo-inositol in women with polycystic ovary syndrome: a double-blind trial. Eur Rev Med Pharmacol Sci 2009; 13: 105-10.

[42]   Papaleo E, Unfer V, Baillargeon JP *et al.* *Myo*-inositol may improve oocyte quality in intracytoplasmic sperm injection cycles. A prospective, controlled, randomized trial. Fertil Steril 2009; 91: 1750-4.

[43]   Gerli S, Papleo E, Ferrari A *et al.* Randomized, double blind placebo-controlled trial: effects of myo-inositol on ovarian function and metabolic factors in women with PCOS. Eur Rev Med Pharmacol Sci 2007; 11: 347-54.

[44]   Gerli S, Mignosa M, Di Renzo GC. Effects of inositol on ovarian function and metabolic factors in women with PCOS: a randomized double blind placebo-controlled trial. Eur Rev Med Pharmacol Sci 2003; 7: 151-9.

[45]   Raffone E, Rizzo P, Benedetto V. Insulin sensitiser agents alone and in co-treatment with r-FSH for ovulation induction in PCOS women. Gynecol Endocrinol 2010; 26: 275-80.

[46]   Iuorno MJ, Jakubowicz DJ, Baillargeon JP *et al.* Effects of d-chiro-inositol in lean women with the polycystic ovary syndrome. Endocr Pract 2002; 8: 417-23.

[47]   Genazzani AD, Santagni S, Rattighieri E *et al.* Modulatory role of D-chiro-inositol (DCI) on LH and insulin secretion in obese PCOS patients. Gynecol Endocrinol 2014; 30: 438-43.

[48]   Pizzo A, Lagana AS, Barbaro L. Comparison between effects of myo-inositol and D-chiro-inositol on ovarian function and metabolic factors in women with PCOS. Gynecol Endocrinol, 2014; 30: 205-8.

[49]   Galletta M, Grasso S, Vaiarelli A *et al.*, Bye-bye chiro-inositol - myo-inositol: true progress in the treatment of polycystic ovary syndrome and ovulation induction. Eur Rev Med Pharmacol Sci 2011; 15: 1212-4.

[50]   Isabella R, Raffone E. Does ovary need D-chiro-inositol? J Ovarian Res 2012; 5: 14.

[51]   Carlomagno G, Unfer V. Inositol safety: clinical evidences. Eur Rev Med Pharmacol Sci 2011; 15: 931-6.

[52]   http://www.fda.gov/Food/IngredientsPackagingLabeling/GRAS/SCOGS/ucm084104.htm

CHAPTER 16

Antioxidant Properties & Function of Inositol and Inositol Phosphates

Abstract: Inositol and its phosphates, specifically $InsP_6$ have antioxidant properties, the latter is a very strong antioxidant indeed. This is brought about by chelating iron through the Fenton Reaction. The exact mechanism how inositol acts as an antioxidant is not clear; however it could be by phosphorylation to inositol phosphates. Inositol hexaphosphate-citrate, a new molecule has double the valence than $InsP_6$; thus it offers to be an even more powerful antioxidant. Owing to the antioxidant properties, these molecules have important biological functions beneficial to human and animal health, as well as industrial and environmental application.

Keywords: Eutrophication, Fenton Reaction, Free radical, Haber-Weiss Reaction, $InsP_6$ citrate, lipid peroxidation, oxidants, reactive oxygen species, ROS, SOD, superoxide dismutase.

INTRODUCTION

Oxygen is obviously indispensable for human life and for the existence of most other organisms. However, as paramount as its role is for our survival and well-being, once inside the human body, there needs to be a strict regulation of quantities and specific activities. The reason for this is that an excess of oxygen can cause substantial damage to our cells and their intracellular components, the harmful effect being known as 'oxidative stress.' However, oxidative stress is a normal phenomenon in the body and is kept under check by various enzyme systems participating in the *in vivo* redox homeostasis. Thus, oxidative stress can also be viewed as an imbalance between the pro-oxidants and antioxidants in the body.

But, what is stress? Stress can be defined as a process of altered biochemical homeostasis produced by psychological, physiological, or environmental stressors. "Any stimulus, no matter whether social, physiological, or physical, that is perceived by the body as challenging, threatening, or demanding can be labeled as a stressor. The presence of a stressor leads to the activation of neurohormonal regulatory mechanisms of the body, through which it maintains the homeostasis" [1, 2].

A.K.M. Shamsuddin and Guang-Yu Yang

FREE RADICALS

When molecules containing oxygen interact with each other in certain chemical reactions "free radicals" may be produced. Free radicals are substances with unpaired electrons which readily react with healthy molecules in a destructive way. They can be produced in large quantities in cells by different mechanisms, such as exposure to oxygen, radiation, environmental toxins, *etc*. In situations where each molecule of oxygen has an equal number of shared electrons, it is in a "content" state. "Discontent" may however arise if one of the electrons is pulled away from the molecule, thus leaving the other unpaired. What happens in the "discontent stage" is that the "free" or unpaired electron will endeavor to secure another oxygen molecule's electron so that it may form a new electron pairing. When an atom or a molecule has an unpaired electron and is thus in a state where it seeks to acquire electrons from other atoms or molecules, it is called a free radical. There are three major stages of free radical reactions: initiation, propagation, and termination. Once formed, irrespective of how it is initiated, the free radicals can propagate themselves indefinitely in the presence of oxygen until those radicals reach a high concentration to react with each other and produce a non-radical species.

The organism however has evolved to counteract the danger by being equipped with mechanisms to deal with these highly reactive molecules by detoxifying them through enzymatic means. These include a network of compartmentalized antioxidant enzymes and non-enzyme molecules that are usually distributed within the cytoplasm and various cell organelles. In eukaryotic organisms, several ubiquitous primary antioxidant enzymes, such as superoxide dismutase (SOD), catalase, glutathione reductase and several peroxidases *e.g.,* glutathione peroxidase catalyze a complex cascade of reactions to convert some of these free radicals such as reactive oxygen species to more stable molecules, *e.g.,* water and O_2. And that is critical for maintaining the delicate intracellular redox balance and minimizing undesirable cellular damage caused by ROS to sustain health and life.

Reactive Oxygen Species (ROS)

The most abundant free radicals in cells are the reactive oxygen species (ROS) which are all unstable metabolites of molecular oxygen (O_2) that have higher reactivity than O_2. These highly reactive molecules containing oxygen are generated during normal intracellular metabolism; and mitochondria are the major organelle responsible for ROS production. ROS play essential roles in cell differentiation, proliferation, and host defense response such as killing of bacteria within the phagocytic cells. While the ROS are generated as byproducts of normal aerobic metabolism, their level increases under stress which turns out to be a basic health hazard as various components of the cell are damaged by oxygen-derived free radicals resulting in lipid peroxidation, DNA damage, and protein oxidation.

A particular type of free radical is an energy-laden, highly reactive form of oxygen called "singlet oxygen" - written as "$\cdot O_2$". Two other names for singlet oxygen are "activated oxygen" or "reactive oxygen species". Singlet oxygen is just one of the many known free radicals. Other examples of free radicals are: hydrogen peroxide (H_2O_2) - it has two unpaired electrons and the hydroxyl radical ($\cdot OH$) - it has three unpaired electrons. Both of these reactive species are members of a group called the superoxides. Superoxides may cause damage to DNA, as well as to other cell components. The fundamental chemical reaction underlying such damage is called an oxidation reaction. The free radicals produced in such a process are called oxidants. Excessive generation of free radicals can induce lipid peroxidation and oxidative damage of other biomolecules. "Lipid peroxidation leads to the formation of a number of different aldehydes and other carbonyl compounds, which contribute to peroxidative cell damage by inhibiting DNA, RNA, and protein synthesis, blocking respiration and depleting glutathione pool". These compounds, especially 4-hydroxyalkenals, are sufficiently long-lived products to attack target molecules distant from the site of formation and to impair their structure/function. The two most toxic 4-hydroxyalkenals are 4-hydroxynonenal (HNE) and 4-hydroxyhexanal (HHE).

Aside from the environmental toxic agents (both physical and chemical) that produce oxygen free radicals, reactive oxygen and nitrogen compounds are also formed by the macrophages and polymorphonuclear leukocytes at the site of inflammation and infection. In the DNA, these chemicals can cause adduct formation that can impair the ability of DNA for correct base pairing. The DNA replication and transcription may also be blocked, nucleotides may be lost or there may even be DNA single strand breaks.

Lipid Peroxidation

Polyunsaturated fatty acids, the main component of cell membranes, are vulnerable to free radical attack because they contain multiple double bonds with extremely reactive hydrogen atoms making them susceptible to be attacked by free radicals, especially hydroxyl radicals, which will lead to the destruction of cell membrane permeability, and eventually cellular dysfunction. Oxygen-free radicals, particularly superoxide anion radical ($\cdot O_2$), hydroxyl radical ($\cdot OH$), and alkylperoxyl radical ($\cdot OOCR$), are potent initiators of lipid peroxidation; these play a very important part in the pathogenesis of a variety of diseases. Once lipid peroxidation is initiated, a propagation of chain reactions will take place until termination products are produced. As a result, end products of lipid peroxidation, such as malondialdehyde (MDA), 4-hydroxy-2-nonenol (4-HNE), and F2-isoprostanes, are accumulated in biological systems [2]. MDA is a three-carbon compound formed mostly from arachidonic acid. It is one of the end products of membrane lipid peroxidation. Since MDA levels are increased in various diseases with excess of oxygen free radicals and many relationships with free radical damage were observed, it serves as a marker for oxidative stress [2].

Uncontrolled lipid oxidation has been incriminated in the development of age-related diseases, malignancy, infective diseases *etc.*

DNA Damage

The ROS-induced DNA damage mainly includes strand break, cross-linking, base hydroxylation, and base excision. The induction of these DNA damages will result in mutagenesis and consequently transformation, especially if combined with a deficient apoptotic pathway. The predominant detectable oxidation product of DNA bases *in vivo* is 8-hydroxy-2-deoxyguanosine. Oxidation of DNA bases can cause mutations and deletions in both nuclear and mitochondrial DNA. Mitochondrial DNA is especially prone to oxidative damage due to its proximity to a primary source of ROS and its deficient repair capacity compared with nuclear DNA [2]. These oxidative modifications lead to functional changes in various types of proteins, which can have substantial physiological impact. Similarly, redox modulation of transcription factors produces an increase or decrease in their specific DNA binding activities, thus modifying the gene expression [2].

Protein Oxidation

The proteins in cells are also believed to be the main targets of free radicals. Aromatic amino acids, cysteine, and disulphide bonds are susceptible to the attack of free radicals, which will lead to protein denaturation and enzyme inactivation. Furthermore, the reactive protein derivatives generated might act as intermediates to induce propagation of oxidative damages to other cell components - chain reactions.

ANTIOXIDANT PROTECTION

Antioxidants terminate these chain reactions by removing free radical intermediates, and inhibit other oxidation reactions. Aside from the enzymatic scavenging of the free radicals by SOD, catalases, peroxidases, *etc.*, as mentioned earlier, there are other endogenous mechanisms of non-enzymatic reduction of ROS; these include GSH, NADPH, thioredoxin, vitamins E and C, and trace metals, such as selenium.

How do antioxidants stop oxidants? They do so by supplying oxidants with the missing electrons they seek (as outlined above). In this way, anti-oxidants are able to prevent oxidative chain reactions which can induce cellular damage. Aside from the antioxidants mentioned earlier, both inositol and InsP_6 are also antioxidants, InsP_6 being one of the most powerful.

But, before we discuss the antioxidant properties of inositol and InsP_6, let's look further at free radicals. Interestingly, free radicals do not always pose a threat to our health; they also have properties that can be very beneficial. For instance, free radicals can be produced by neutrophils - that serve the immune system by engulfing invaders (such as bacteria) which have managed to penetrate the body's first line of defense. Neutrophils use superoxides (and other free radicals) to kill invaders they have trapped. Within the neutrophils are the lysosomes which store these oxidants until they are needed. The safe storage of free radicals in this fashion ensures that they cannot damage cell structures; as well, the free radicals are not inadvertently rendered inactive by endogenous antioxidants - they are stored until ready to be used. The body must handle free radicals with great care. While they serve to fulfill vital protective functions within our immune system, an excessive or otherwise abnormal buildup of free radicals could be very damaging to the host.

But free radicals are not only generated within our bodies; there are numerous external sources that also produce them, *e.g.,* cigarette smoke, UV radiation, indiscriminate use of oxygen therapy, radiation from cancer therapy, certain drugs and; our everyday diet, it too, can generate free radicals during the food metabolism. By damaging our DNA and causing mutations, which in turn may promote a wide array of diseases, from cataracts to cancers, free radicals can also contribute directly to the aging process.

If a cell has sustained DNA damage, and if that damage is not repaired, the cell may be prevented from replicating or copying its own DNA. In cases where a cell does not need to divide, the consequences may be benign. However, we have many cells that must divide - if they are prevented from doing so (*i.e.*, if their DNA cannot be replicated), the consequences can be far direr: Cell death can be the result. Under normal circumstances, enzymes naturally found in our bodies are able to excise those parts of a DNA strand that have sustained oxidative damage. This repair mechanism can however become undermined when the DNA mutations caused by oxidative damage accumulate over time - as they tend to. Serious health problems - cancers and other diseases commonly associated with the aging process - can be the result.

Aside from the health benefit, antioxidants also have many industrial uses, such as preservatives in food, wine and cosmetics, and to prevent the degradation of rubber, gasoline, *etc.*, [3]. The industrial uses of $InsP_6$ have been discussed in Chapter 21.

How Does $InsP_6$ Function as Antioxidant?

To understand this, we must discuss the mitochondria - the cells' "energy factories". Within the mitochondria, a complex series of chemical reactions takes place, further processing the compounds originating from our diet. The end result of these reactions is the extraction of energy that ultimately sustains us. Respiration at the cellular level takes place within the mitochondria; it involves the transference of electrons from molecule to molecule. During this reaction, it is critically important that the mineral iron be present. Cells procure the iron they need from the blood plasma. During the respiration process, free radicals are generated as a byproduct. Whereas some of these free radicals are in fact necessary for the respiration itself, an excessive production of reactive species represents a dangerous potential. In a process called the Fenton reaction, hydrogen peroxide (H_2O_2) is formed. The H_2O_2 then reacts with the iron, sparking the production of highly reactive and damaging hydroxyl radicals ($\cdot OH$).

Fenton Reaction and Haber-Weiss Reaction

In 1894 Henry John Horstman Fenton discovered that several metals have a special property for transfer of oxygen which improves the use of H_2O_2 [4]. Ferrous iron (Fe^{2+}) is oxidized by H_2O_2 to Fe^{3+} forming a hydroxyl radical ($\cdot OH$) and hydroxyl ion (OH^-) as shown in the first step of the reaction below - also known as Haber-Weiss Reaction as it was described by Fritz Haber and his student Joseph Joshua Weiss [5]. However, Fe^{3+} is reduced back to Fe^{2+} by another molecule of H_2O_2 (step 2) forming a superoxide radical •OOH and a proton.

The Fe^{2+}-catalyzed H_2O_2 reaction is known as the Fenton Reaction (step 1).

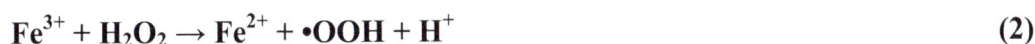

$$Fe^{2+} + H_2O_2 \rightarrow Fe^{3+} + \cdot OH + OH^- \tag{1}$$

$$Fe^{3+} + H_2O_2 \rightarrow Fe^{2+} + \cdot OOH + H^+ \tag{2}$$

Graf *et al.,* were the first to report on the antioxidant property of $InsP_6$ [6, 7]. They demonstrated that $InsP_6$ inhibits •OH generation and subsequent lipid peroxidation - peroxidation of arachidonic acid driven by ascorbic acid and iron [7]. They measured the effect of added $InsP_6$ upon iron-mediated •OH production and arachidonic acid peroxidation. A substantial amount of •OH was produced by a superoxide-generating system in the presence of iron alone. However, the generation of •OH was completely blocked by micromolar amounts of $InsP_6$. $InsP_6$ brings about this by chelating free iron (as it also binds to other cations; Fig. **16.1**).

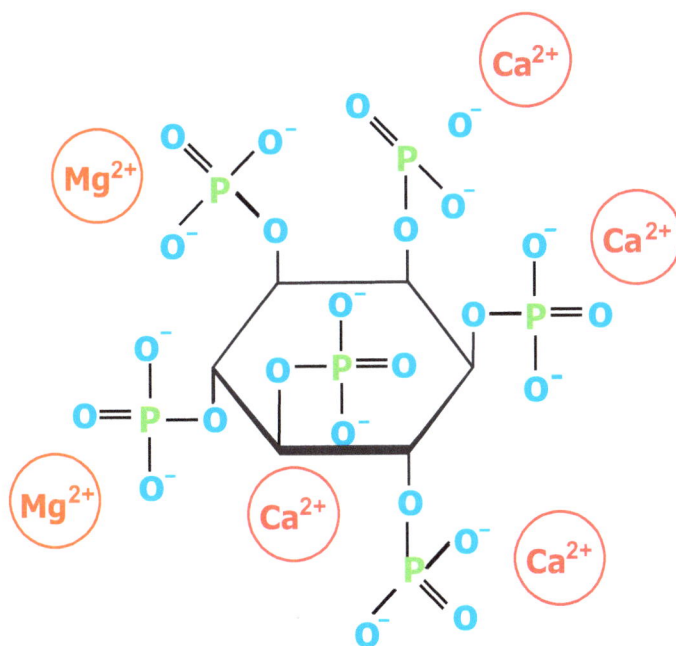

Fig. (16.1). Chemical structure of calcium magnesium inositol hexaphosphate P = Phosphorus; O = oxygen.

Graf *et al.,* concluded that since $InsP_6$ occupies all 6 coordination sites in the Fe^{3+}-$InsP_6$ complex, the chelate cannot participate in the Fenton Reaction [6], thus inhibiting •OH generation and subsequent lipid peroxidation [7]. At higher iron to $InsP_6$ ratio, iron is however likely to gain an aquo coordination site and to catalyze oxidative events as iron remains catalytically active as long as iron within the complex retains an available coordination site [7].

Slices of raw potato when submerged in tap water turns brown with time, is a process dependent on oxidative iron-mediated reaction. The addition of 50 mM $InsP_6$ retarded this brown coloration for several days [7]. Furthermore, the vegetable maintained textural firmness, and the appearance of putrid odor was delayed; similar results were obtained with apple, avocado and banana, thus opening the door for $InsP_6$ to be used as a food preservative (please see Chapter 21). The formation of hydroxyl radical generation was monitored by reaction with dimethylsulfoxide (DMSO) which yields formaldehyde that was quantitated colorimetrically.

Electron Spin Resonance Spectroscopy

Electron spin resonance (ESR) also known as electron paramagnetic resonance (EPR) is a technique to study unpaired electrons in a material. Akin to the NMR (nuclear magnetic resonance) in which the spin of the atomic nuclei are excited, it is the electron spins that are excited in EPR/ESR. An unpaired electron can move between two energy levels by either absorbing or emitting a photon of energy; it can move between the two spin states. Typically, there is a net absorption of energy which is monitored and converted into a spectrum.

Experiments of electron spin resonance spectroscopy in Shamsuddin's lab in collaboration with Mary Hinzman and Peter Gutierrez showed a ≥ 2.5 fold reduction in signal for •OH by 100 nM Na-$InsP_6$ in an aerated aqueous solution of KO_2 and 100 mM DMPO (Figs. **16.2**, **16.3**) [8].

At the site of the Fenton reaction, $InsP_6$ binds to excess iron, thus inhibiting the formation of hydroxyl radicals. $InsP_6$'s activity in this regard is highly beneficial, because it prevents an ensuing (and also harmful) reaction - lipid peroxidation. Lipid peroxidation is sustained by the lipids that are natural constituents of cell membranes where membrane lipids come under attack by free radicals.

ESR SPECTRA SHOWING REDUCTION OF HYDROXYL RADICALS BY SODIUM INOSITOL HEXAPHOSPHATE

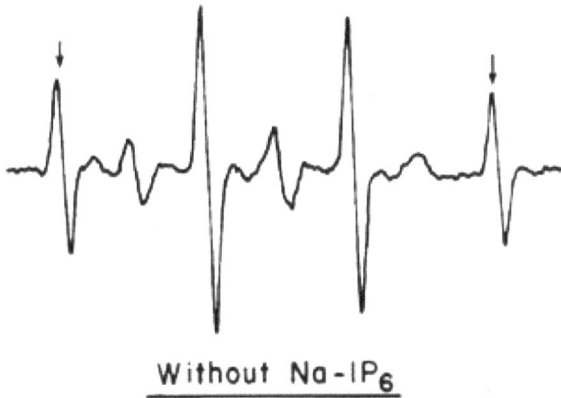

Without Na-IP$_6$

Fig. (16.2). ESR spectra of •OH-DMPO obtained from an aerated aqueous solution of KO$_2$ containing 100 mM DMPO showing evidence for the formation of short-lived hydroxyl radicals (control).

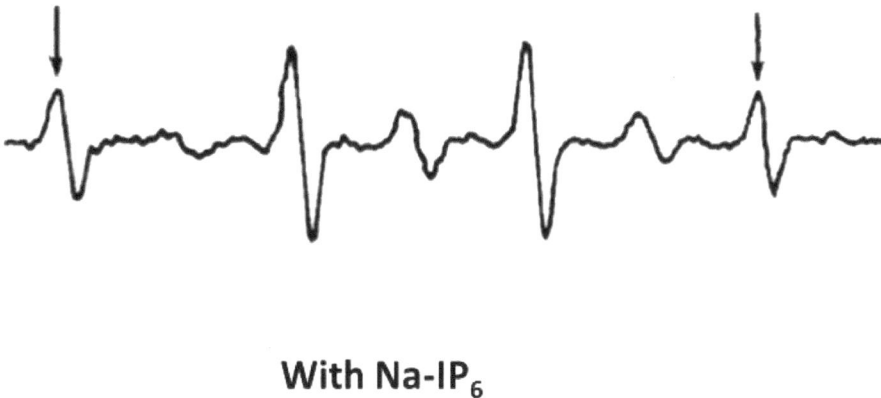

With Na-IP$_6$

Fig. (16.3). Under identical condition except for the addition of 100 nM Na-InsP_6 the intensity of •OH -DMPO was reduced by at least 2.5 fold [8].

Lipid Peroxidation

Excessive generation of free radicals can induce lipid peroxidation and oxidative damage of other biomolecules leading to the formation of a number of different saturated and unsaturated aldehydes and other carbonyl compounds, which in their turn contribute to peroxidative cell damage by affecting DNA, RNA, and protein synthesis, blocking respiration and depleting glutathione pool. Many different pathological conditions such as age-related diseases, malignancy, infectious diseases, *etc.*, have been attributed to uncontrolled lipid peroxidation.

Zajdel *et al.,* [9] utilized Caco-2 cells which are derived from a human colon adenocarcinoma and are commonly used as *in vitro* model as they spontaneously differentiate to form confluent monolayer of polarized cells structurally and functionally resembling the intestinal epithelium. Using HPLC/MS/MS they investigated whether InsP_6 is capable of inhibiting linoleic acid autoxidation and Fe^{2+}ascorbate-induced peroxidation, as well as Fe^{2+}ascorbate-induced lipid peroxidation in Caco-2 cells. The antioxidant properties of InsP_6 were tested within a range of 1-500 μM. InsP_6 at 100 μM and 500 μM concentration effectively inhibited the decay of linoleic acid, both in the absence and presence of Fe^{2+}/ascorbate. InsP_6 did not change linoleic acid hydroperoxides concentration levels after 24 hours of Fe^{2+}/ascorbate-induced peroxidation. In the absence of Fe^{2+}/ascorbate, InsP_6 at 100 μM and 500 μM concentrations significantly suppressed decomposition of linoleic acid hydroperoxides [9]. Moreover, InsP_6 at concentrations of 100 μM and 500 μM significantly decreased 4-hydroxyalkenal levels in Caco-2 cells [9]. Thus the investigators showed that InsP_6 inhibited linoleic acid oxidation and reduced the formation of 4-hydroxyalkenals, another mechanism through which it may help to prevent intestinal diseases induced by oxygen radicals and lipid peroxidation products [9].

Ethanol

Excessive reactive oxygen species (O_2.-, •OH and H_2O_2) are also formed as a result of acute and chronic exposure to ethanol causing enhanced peroxidation of lipids, proteins and DNA. These may have a pathogenetic role in ethanol-induced liver diseases. Using SK-Hep-1 cells (human hepatic carcinoma cell line), Lee *et al.,* [10] investigated the protective effect of InsP_6 besides others, against ethanol-induced cytotoxicity. Exposure of the cells to excess ethanol resulted in a significant increase in cytotoxicity, ROS production, lipid peroxidation; and a reduction in mitochondrial membrane potential and the amount of reduced glutathione. Co-treatment of cells with InsP_6 significantly inhibited oxidative ethanol metabolism-induced cytotoxicity by blocking ROS production. When the cells were treated with ethanol after pretreatment of 4-methylpyrazole, increased cytotoxicity, ROS production, antioxidant enzyme activity, and loss of mitochondrial membrane potential were observed [10]. The addition of InsP_6 to these cells caused suppression of non-oxidative ethanol metabolism-induced cytotoxicity, similar to oxidative ethanol metabolism, suggesting that InsP_6

protects against ethanol metabolism-induced oxidative damage in these cells *via* antioxidant mechanism. The use of $InsP_6$ in wine industry is further discussed in Chapter 21.

Alkaptonuria

Alkaptonuria is an extremely rare (1:250,000 - 1:1,000,000) genetic disease (autosomal recessive) with inborn error of catabolism of the amino acids phenylalanine and tyrosine leading to the accumulation of homogentisic acid and resultant tissue damage in various organs. It is the oxidation of homogentisic acid to benzoquinone acetic acid resulting in free radical production that is considered to be the pathogenetic event in the tissue damage and complicating inflammations. Homogentisic acid-induced production of melanin-based ochronotic pigment, characteristic of alkaptonuria is associated with lipid peroxidation as well. Spreafico *et al.,* [11] propose that alkaptonuria is a secondary serum amyloid (SAA)-based amyloidosis; in support they found co-localization of SAA amyloid and ochronotic pigment.

Thus the investigators set out to evaluate the anti-amyloid capacity of antioxidants including $InsP_6$ in a human chondrocytic (cartilaginous) cell model of alkaptonuria. Along with other antioxidants tested, $InsP_6$ significantly reduced the production of amyloid, and inhibited homogentisic acid-induced pro-inflammatory cytokines. Thus, it may be a novel approach to the treatment of alkaptonuria [11].

Asbestosis

Asbestosis is mostly an occupational chronic lung disease caused by inhalation of asbestos fibers, which are natural mineral products resistant to heat and corrosion. Inhaled asbestos fibers, especially the thin and straight amphiboles penetrate deep into the lungs, and induce foreign-body reaction in the alveoli with resultant inflammation and fibrosis. The disease can range from mild to severe. As a result of chronic inflammation and fibrosis of the lung parenchyma, there is shortness of breath; and increased risk of cancer of the lung and pleura - mesothelioma.

ROS are considered important in the pathogenesis of the asbestosis. Asbestos itself may augment ROS release from inflammatory cells, and iron complexed to asbestos catalyzes the formation of •OH through Fenton Reaction as shown before.

Kamp *et al.,* [12] demonstrated that asbestos damaged cultured human pulmonary epithelial-like cells (WI-26 cells) and, InsP_6 significantly diminished the asbestos-induced •OH production and ameliorated DNA damage in the cultured pulmonary alveolar epithelial cells *in vitro*. In a subsequent *in vivo* model of asbestosis, the investigators gave a single intra-tracheal instillation of amosite (a type of amphibole fibers) asbestos to Sprague-Dawley rats and treated with 500 μM InsP_6. Compared to controls, asbestos elicited a significant pulmonary inflammatory response and significantly increased the fibrosis score at 2 wk. Whereas in contrast to asbestos-alone, InsP_6-treated asbestos elicited significantly less inflammatory cell response and caused significantly less fibrosis [13].

To determine whether asbestos causes mitochondrial dysfunction in the alveolar epithelial cells, these investigators then exposed A549 cells to amosite asbestos and assessed mitochondrial membrane potential changes. Asbestos, as opposed to titanium dioxide - an inert particulate, reduced mitochondrial membrane potential after a 4 h exposure period. InsP_6 blocked asbestos-induced reductions in mito-chondrial membrane potential and attenuated apoptosis, thereby demonstrating the role of mitochondrial ROS in the pathogenesis of asbestosis and the inhibitory effect of InsP_6 [14].

There are many other examples of diseases *e.g.,* myocardial infarction that are due to oxidative damage wherein InsP_6 ameliorates them; these will be discussed in subsequent chapters as germane to the discussion.

Seed Protection

InsP_6's oxidative capacities also come into play in the plant kingdom. In here, InsP_6 serves to preserve and protect seeds, allowing them to remain viable (*i.e.,* keep their ability to grow) for extended periods of time [7]. Morris & Ellis had demonstrated that the major portion of iron in seeds is complexed with InsP_6 [15]. As any farmer knows, storage of seeds for a prolonged time results in degradative changes as painfully noted by decreased seed vigor and reduced germination. Oxygen is one of the three major factors, humidity and temperature being the other two, responsible for this degradation; moisture and oxygen act synergistically to enhance the process [16]. The protective effect of InsP_6 may be responsible for prolonged viability of some seeds for > 1,000 years, even under deleterious conditions [7].

Colon Carcinogenesis and Superoxide Dismutase (SOD)

There are two major types of SOD: CuZnSOD (SOD1), which mainly exist in cytoplasm (copper and zinc being in the active site), and the MnSOD (SOD2), located in mitochondrial matrix, with manganese in the active site. They can catalyze the reaction to decompose superoxide anion radicals into H_2O_2, which will then be converted to water and oxygen by catalase or glutathione peroxidase [17]. Amaral *et al.*, investigated the expression of SOD1 in Wister rats injected with the colon carcinogen azoxymethane (AOM) and its modulation by InsP_6 [18]. The expression of SOD1 was evaluated by using quantitative immunohistochemistry by computer image processing of the colonic crypts. While the SOD1 expression in the control group was 16.0 ± 3.7, there was higher expression in the carcinogen-group (26.7 ± 3.5) [18]. Control animals on InsP_6 alone showed virtually identical expression at 16.9 ± 3.1 and, as a result of InsP_6 treatment, SOD1 expression in InsP_6 + AOM was 20.9 ± 3.3 demonstrating the antioxidant property of InsP_6 *in vivo*.

INOSITOL AS AN ANTIOXIDANT

Inositol has also been demonstrated to have antioxidative properties. It appears that inositol may do this through the xanthine oxidase (XO) pathway.

Xanthine oxidoreductase (XOR) is a molybdoflavin enzyme that catalyzes the terminal two reactions in purine degradation: oxidation of hypoxanthine $\rightarrow$ xanthine and the subsequent oxidation of xanthine $\rightarrow$ uric acid [19]. The enzyme is transcribed as a single gene product, xanthine dehydrogenase (XDH) where substrate-derived electrons reduce NAD^+ to NADH. During inflammatory conditions, oxidation of key cysteine residues (535 and 992) and/or limited proteolysis converts XDH to xanthine oxidase (XO) [19]. In the oxidase form, affinity for oxygen is significantly enhanced resulting in univalent and divalent electron transfer to O_2 generating $\cdot O_2$ and hydrogen peroxide (H_2O_2). This capacity to reduce O_2 led to XOR being identified as a significant source of reactive species mediating ischemia/reperfusion injury [20, 21]. XO is abundant in the intestine [22]. Since dietary iron remains largely unabsorbed in the intestine, $\cdot OH$ radicals are generated through the interaction of iron with $\cdot O_2$ produced by XO in the intestine [23]. Since the role of InsP_6 in the XO pathway was not clearly defined, Muraoka & Miura [23] set to investigate the role of InsP_6 as an antioxidant in XO-induced lipid peroxidation and DNA damage. They reported that InsP_6 inhibited XO-induced superoxide-dependent lipid peroxidation [23].

Though the exact rationale for testing *myo*-inositol is not clear, the investigators also reported that it did not inhibit XO-induced superoxide-dependent lipid peroxidation [23]. However, both InsP_6 and *myo*-inositol inhibited XO-induced superoxide-dependent DNA damage. Though not entirely clear as to how, their data suggest that inositol may also act as hydroxyl radical scavenger. EDTA-Fe^{2+} efficiently catalyzes the decomposition of H_2O_2 to •OH; and deoxyribose degradation induced by XO-X with EDTA-Fe^{2+} was blocked by *myo*-inositol [23]. This could explain how *myo*-inositol acts as a free radical scavenger.

Vascular Complications of Diabetes Mellitus

A markedly increased risk of cardiovascular diseases is seen in diabetes. These include myocardial infarction, stroke, peripheral vascular disease - resulting in amputation of limbs, *etc*. All of these are causally related to the abnormal functionality of endothelial cells (cells lining the inner wall of our blood vessels). Hyperglycemia and hyperlipidemia in diabetes mellitus result in these cardiovascular complications which are in turn due to endothelial cell dysfunction; excessive endothelial mitochondrial superoxide production is considered to be the key mechanism.

Following up on Muraoka & Miura's work demonstrating the antioxidant function of *myo*-inositol [23] Nascimento and collaborators [24] too showed that *myo*-inositol exerts its antioxidant function, by inhibiting xanthine oxidase and scavenging superoxide both *in vitro* and *in vivo*, and preventing formation of ADP-iron-oxygen complexes that initiate lipid peroxidation. They tested whether a variety of inositol compounds including *myo*-inositol, D-*chiro*-inositol and dibutyryl D-*chiro*-inositol would prevent endothelial cell dysfunction in diabetic models of rats and rabbits [24]. Oral inositols reduced hyperglycemia and hypertriglyceridemia.

In diabetic animals, the mesenteric vascular tree undergoes hypertrophy preceded by attenuated endothelial dependent relaxation; ROS contribute to this enhanced basal vascular tone; and superoxide dismutase (SOD) reverses the impaired activity of nitrous oxide in renal arterioles. Decreased endothelial-dependent relaxation of aortic rings and mesenteric micro-vessels is a common assay for endothelial dysfunction. Inositols reduced ROS in endothelial cells incubated in high glucose concentration, in a dose dependent manner; and scavenged superoxide in an *in vitro* xanthine/xanthine oxidase system; dibutyryl D-*chiro*-inositol being the most effective of the three. Additionally, inositol compounds enhanced nitrous oxide action and protected nitrous oxide signaling [24]. The iron chelator desferrioxamine blocked the action of dibutyryl D-*chiro*-inositol and that added Fe^{3+} in turn reversed the effect of desferrioxamine. The mechanism of action of the dibutyryl D-*chiro*-inositol is thus similar to action of $InsP_6$ as both require the presence of Fe^{3+} for their antioxidant action. But exactly how that is brought about, especially how *myo*-inositol mediates this free radical scavenging action is intriguingly unclear.

Lee *et al.,* [25] have also demonstrated that lipid peroxidation was dramatically decreased by $InsP_6$ in a rat hepatocarcinogenesis model initiated by diethylnitrosamine (DEN) and promoted by partial hepatectomy, as by *myo*-inositol. Raj *et al.,* [26] demonstrated that addition of *myo*-inositol decreased the peroxidant effect of hydrogen peroxide in human erythrocytes and human cataract lenses. The extent of lipid peroxidation was monitored as levels of thiobarbituric reacting substances. Addition of *myo*-inositol decreased the peroxidation effect of hydrogen peroxide in a dose dependent manner suggesting an antioxidant property for inositol. One plausible mechanism is that inositol could facilitate the synthesis of inositol phosphates ($InsP_{2-6}$) which then exert antioxidant effects by chelating iron through the Fenton Reaction. Miyamoto *et al.,* [27] demonstrated that the hydrolysis products of $InsP_6$ ($InsP_{2-6}$) possess chelating ability and prevent iron-induced lipid peroxidation. Once inside cells, inositol may undergo re-phosphorylation to form inositol phosphates [28]. That these in turn, may chelate the cations involved in the generation of ROS is plausible [25].

In summary, the protective benefits exerted by $InsP_6$ and inositol as antioxidants extend to diabetes, cancer, cardiovascular diseases, cataracts, just to name a few.

$InsP_6$ CITRATE - THE SUPER ANTIOXIDANT

A new $InsP_6$ molecule has been created by addition of citrate to $InsP_6$ molecule called inositol hexaphosphate citrate ($InsP_{6c}$). While the parent $InsP_6$ molecule has 12 valences, the new one ($InsP_{6c}$) has 24, making it a better chelator and hence an even better antioxidant, perhaps the strongest available to date.

While $InsP_6$ is a strong chelator of bivalent and trivalent cations, citric acid is a very potent chelator for monovalent cations such as sodium, potassium, lithium *etc*. Following extensive experimentation and computer modeling it was discovered that citric acid could be combined with $InsP_6$ to form a new compound $InsP_{6c}$ which would retain the basic properties of both citric acid and $InsP_6$; $InsP_{6c}$

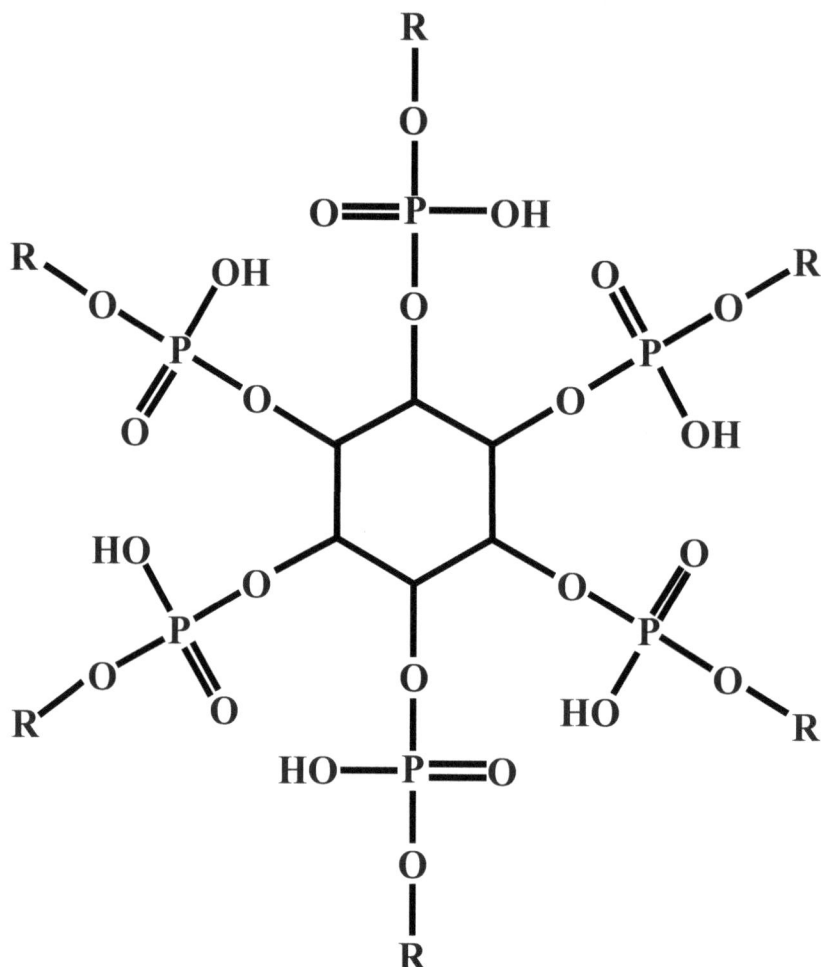

Fig. (16.4). InsP$_6$ Citrate; each R can be either H or Citrate, thus there could be IP$_6$Cit$_{1-6}$.

for instance is a strong chelator of mono-, di- and trivalent cations (Figs. **16.4**, **16.5**) [29]. InsP$_{6c}$ had the strongest chelating ability when compared to the various chelating agents, including InsP$_6$, EDTA, citric acid, lactic acid, acetic acid, glucoronic acid and polyphosphoric acid [30, 31].

Thus, as a superior antioxidant, InsP$_{6c}$ not only retains all the functional abilities of the InsP$_{6c}$ but also being more efficient as its better dissolution rate; hence it's utility as an artery plaque remover [30, 31]. Additional utility of InsP$_{6c}$ lies in environmental science.

$$COOH$$
$$HOOCCH_2 - C - CH_2COOH$$
$$O$$
$$O$$
$$O - P - OH$$
$$O$$

$$CH_2COOH$$
$$HOOC - C - O - O = P$$
$$CH_2COOH$$

$$OH$$
$$P = O - O - C - COOH$$
$$O$$

$$CH_2COOH$$
$$COOH$$
$$CH_2COOH$$

$$CH_2COOH$$
$$HOOC - C - O - O = P$$
$$CH_2COOH$$

$$OH$$
$$P = O - O - C - COOH$$
$$O$$

$$CH_2COOH$$
$$COOH$$
$$CH_2COOH$$

$$O - P - OH$$
$$O$$
$$O$$
$$HOCOOH_2 - C - CH_2COOH$$
$$COOH$$

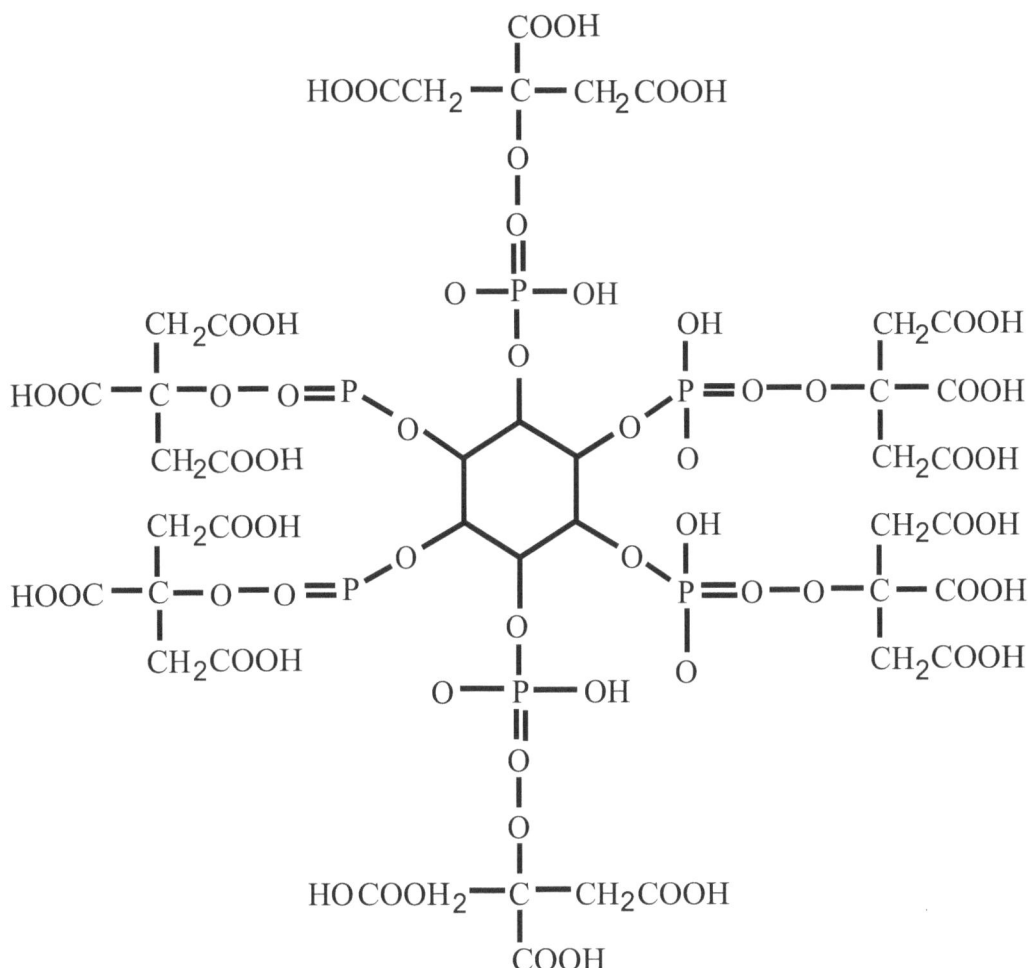

Fig. (16.5). Hexacitrated InsP_6.

Approximately 75% of the phosphates in the feed for cattle, pigs and poultry are from InsP_6 in corn seeds. Non-ruminants such as poultry and pigs do not have efficient enzyme system to utilize that phosphate, resulting in environmental problems such as water pollution and eutrophication (increased nutrient initially causing increased productivity such as crops, followed by environmental damages as poor water quality, overgrowth of algae *etc.*,) with decreased bone phosphates in the animals themselves [32]; addition of citric acid to the feed improves that "phytate-phosphorous" utilization [33]. Thus, the new molecule InsP_{6c} could be an important improvement that is likely to be not only healthy for the cattle, pigs and poultry, but also provide better environment. And preliminary data

[Shamsuddin, unpublished observation] indicate that, indeed, $InsP_6$ Citrate may be more potent in cancer inhibition than $InsP_6$.

As an aside, it is to be noted that the phosphate from $InsP_6$ in the manure can be the answer to some our environmental problems [34].

CONCLUDING REMARKS

In summary, the antioxidant property of inositol, $InsP_6$ and $InsP_{6c}$ offers numerous benefits in diverse fields, not only in health as discussed in this chapter, but also in many other areas as described in Chapter 21.

CONFLICT OF INTEREST

Professor Shamsuddin is the inventor of several patents related to $InsP_6$ and hexacitrated $InsP_6$.

REFERENCES

[1]   Dimitrios NT, Geogrios KC, Dmitrios IXH. Neurohormonal hypothesis in heart failure. Hellenic J Cardiol 2003; 44: 195-205.

[2]   Rahal A, Kumar A, Singh V, *et al.* Oxidative stress, prooxidants, and antioxidant: the interplay. Biomed Res Int 2014; 2014: 761264. Published online Jan 23, 2014 doi: 10.1155/2014/761264; Epub 2014 Jan 23.

[3]   Dabelstein W, Reglitzky A, Schütze A, *et al.* "Automotive Fuels". Ullmann's Encyclopedia of Industrial Chemistry. doi: 10.1002/14356007.a16_719.pub2.ISBN 3527306730.

[4]   Fenton HJH: Oxidation of tartaric acid in presence of iron. J. Chem. Soc.Trans.1894; 65: 899-911, doi: 10.1039/ct8946500899.

[5]   Haber F and Weiss J: "Über die Katalyse des Hydroperoxydes". *Naturwissenschaften* 1930; 20 (51): 948-950. 1932 doi: 10.1007/BF0150471.

[6]   Graf E, Mahoney JR, Bryant RG *et al.* Iron-catalyzed hydroxyl radical formation. Stringent requirement for free iron coordination site. J Biol Chem Chem.1984; 259: 3620-4.

[7]   Graf E, Empson KL, Eaton JW: Phytic acid. A natural antioxidant. J Biol Chem 1987; 262: 11647-50.

[8]   Shamsuddin AM, Sakamoto K: Antineoplastic action of inositol compounds, In: Wattenberg L, Lipkin M, Boone CW, *et al.* editors. Cancer Chemoprevention. Boca Raton, Florida: CRC Press, 1992, pp. 285-308.

[9]   Zajdel A, Wilczok A, Węglarz L, *et al.* Phytic Acid inhibits lipid peroxidation *in vitro*. Biomed Res Int. 2013: 147307. doi: 10.1155/2013/147307. Epub 2013 Oct 24.

[10]   Lee KM, Kang HS, Yun CH, *et al.* Potential *in vitro* Protective Effect of Quercetin, Catechin, Caffeic Acid and Phytic Acid against Ethanol-Induced Oxidative Stress in SK-Hep-1 Cells. Biomol Ther (Seoul). 2012 Sep; 20(5): 492-8. doi: 10.4062/biomolther.2012.20.5.492.

[11]   Spreafico A, Millucci L, Ghezzi L *et al.* Antioxidants inhibit SAA formation and pro-inflammatory cytokine release in a human cell model of alkaptonuria. Rheumatology 2013; 52: 1667-73. doi: 10.1093/rheumatology/ket185. Epub 2013 May 23.

[12]   Kamp DW, Israbian VA, Preusen SE, *et al.* Asbestos causes DNA strand breaks in cultured pulmonary epithelial cells: role of iron-catalyzed free radicals. Am J Physiol (Lung Cellular and Molecular Physiology) 1995; 268: L471-L480.

[13]   Kamp DW, Israbian VA, Yeldandi AV, *et al.* Phytic acid, an iron chelator, attenuates pulmonary inflammation and fibrosis in rats after intratracheal instillation of asbestos. Toxicol Pathol 1995; 23: 689-95.

[14]   Kamp DW, Panduri V, Weitzman SA *et al.* Asbestos-induced alveolar epithelial cell apoptosis: role of mitochondrial dysfunction caused by iron-derived free radicals. Mol Cell Biochem 2002; 235: 153-160.

[15]   Morris ER, Ellis R: Isolation of monoferric phytate from wheat bran and its biological value as an iron source to the rat. J Nutr 1976; 106: 753-60.

[16]   Ohlrogge JB and Kernan TP: Oxygen-dependent aging of seeds. Plant Physiol 1982; 70: 791-4.

[17]   Peng C, Wang X, Chen J, *et al.* Biology of aging and role of dietary antioxidants. Biomed Res Int 2014; 2014: 831841 Published online Apr 3, 2014. doi: 10.1155/2014/831841.

[18]   Amaral EG, Fagundes DJ, Marks G, *et al.* Study of superoxide dismutase's expression in the colon produced by azoxymethane and inositol hexaphosfate's paper, in mice. Acta Cir Bras 2006; 21 (Supplement 4): 27-31.

[19]   Cantu-Medellin N, Kelley EE. Xanthine oxidoreductase-catalyzed reactive species generation: A process in critical need of reevaluation. Redox Biol. 2013 Jun 10; 1(1): 353-8. eCollection 2013.

[20]   Granger DN, Rutili G, McCord JM. Superoxide radicals in feline intestinal ischemia. Gastroenterology 1981; 81: 22-9.

[21]   McCord J.M., Fridovich I. Superoxide dismutase. An enzymic function for erythrocuprein (hemocuprein) Journal of Biological Chemistry. 1969; 244: 6049-55.

[22]   Battelli, M.G., Corte, E.D., Stirpe, F., 1972. Xanthine oxidase type D (dehydrogenase) in the intestine and other organs of the rat. Biochemical Journal 126 (3), 747-9.

[23]   Muraoka S, Miura T. Inhibition of xanthine oxidase by phytic acid and its antioxidative action. Life Sci. 2004; 74: 1691-700.

[24]   Nascimento NR, Lessa LM, Kerntopf MR *et al.* Inositols prevent and reverse endothelial dysfunction in diabetic rat and rabbit vasculature metabolically and by scavenging superoxide. Proc National Acad Sci *USA* 2006; 103: 218-23.

[25]   Lee HJ, Lee SA, Choi H. Dietary administration of inositol and/or inositol-6-phosphate prevents chemically induced rat hepatocarcinognesis. Asian Pacific J Cancer Prev. 2005; 6: 41-47.

[26]   Raj DG, Ramakrishnan S, Devi CS. Myo-inositol and peroxidation -an *in vitro* study on human cataract lens and human erythrocytes. Indian J Biochem Biophs, 1995; 32: 109-11

[27]   Miyamoto SA, Kuwata G, Imai M, *et al.* Protective effect of phytic acid hydrolysis products on iron-induced lipid peroxidation of liposoaml membranes. Lipid*s* 2000; 35: 1411-3.

[28]   Shamsuddin AM. Metabolism and cellular functions of IP_6: a review. Anticancer Research 1999; 19: 3733-6.

[29]   Coppolino CA: Hexa-citrated phytate and process of preparation thereof. US7009067 (2006)

[30]   Coppolino CA and Shamsuddin AM.: Phytic citrate compounds and process for preparing the same. US7517868 (2009).

[31]   Coppolino CA and Shamsuddin AM.: Phytic citrate compounds and process for preparing the same. US7989435 (2011).

[32]   Abelson PH: A potential phosphate crisis. Science 1999; 283: 2015-6.

[33]   Rafacz-Livingston KA, Martinez-Amezcua C, Parsons CM, *et al.* Citric acid improves phytate phosphorus utilization in crossbred and commercial broiler chicks. Poult Sci 1999; 84: 1370-5.

[34]   Gilbert N. The Disappearing Nutrient. Nature 2009; 461: 716-8.

CHAPTER 17

Roles of Inositol & Inositol Phosphates in Immunity, Infection and Inflammation

Abstract: $InsP_6$ with or without inositol enhances NK cell activity and stimulates polymorphonuclear cell priming function and the macrophages to produce cytokines. $InsP_6$ can inhibit human immunodeficiency virus (HIV) replication; both $InsP_5$ and $InsP_6$ are critical in HIV particle assembly. $InsP_6$ plays significant roles in mucosal protection in the gastrointestinal tract, and inflammation.

Keywords: Asbestosis, gastric ulcer, granzymes, HIV, macrophage, natural killer, NK cells, perforin, polymorphonuclear cells.

INTRODUCTION

Plants and animals have to continually defend themselves from invaders - viruses, bacteria, fungi, parasites *etc.* The defense mechanisms are classified as either natural or acquired. The engulfing (phagocytosis) of bacteria by polymorphonuclear leukocytes (also known as neutrophils) and monocytes in our blood or by tissue macrophages is an example of natural defense or immunity. The engulfed bacteria are then digested by the enzymes that literally come in packets - lysosomal granules. Actual killing of the invading bacteria or fungi is done by activated oxygen species - hydroxyl radicals (•OH) that are produced through a series of activities that convert molecular oxygen (O_2) to the superoxide anion $•O_2$, these intracellular activities are known together as "respiratory burst". There is non-oxidative killing as well *via* a variety of enzyme systems. Thus, the process of inflammation and associated reaction by the army of cells standing by is our immediate natural defense. This natural immunity neither requires prior exposure to the offender or invader, nor is it enhanced by such experience. It is also non-specific, that is, it does not (or cannot) discriminate among the various offenders who are essentially foreign.

Acquired immunity, by contrast is specific, requires a sensitizing exposure to the invader or offender and subsequent encounters magnify the response. The cellular components of acquired immunity are the T lymphocytes, B lymphocytes, natural killer (NK) cells and mononuclear phagocytes.

A.K.M. Shamsuddin and Guang-Yu Yang

Since the immune modulation is not only an attribute but also a mechanism through which inositol and its phosphates mediate other functions, these topics are inseparable. This chapter therefore assimilates both the function and mechanism.

NATURAL KILLER (NK) CELLS

The thymus-derived T cells constitute 60-70% of the peripheral blood lymphocytes, B cells 10-20% and the natural killer (NK) cells 10-15%. NK cells are somewhat larger than small lymphocytes and have also been called large granular lymphocytes. The NK cells have an inherent ability to destroy virally infected cells and a variety of tumor cells, and some normal cells. This ability to lyse cells is innate and is not dependent on prior sensitization or experience; this ability is natural and forms the first line of defense against cancer or viral infections; however, they may be stimulated by various cytokines such as interferon (INFα/β and interleukin 2 (IL2). NK cells contain special weaponry - perforin and proteases known as granzymes. As the name suggests, perforin punches holes (perforations) on the target cells through which the granzymes and associated molecules enter and cause cell death (Fig. **17.1**).

Insofar as cancer is concerned, there is a correlation between neoplastic disease and depressed NK cell activity; and carcinogens inhibit NK cell activity [1]. NK cells have the ability to destroy tumor cells, and chemical agents that directly or indirectly increase NK cell activity have been shown to have antitumor activity *in vivo* [2, 3]. Thus, in order to investigate whether the observed anticancer action of inositol and InsP_6 was correlated with NK cell activity Shamsuddin Lab had embarked on investigating that.

Inositol and InsP_6 Enhancements of our Natural Defense

Natural Killer (NK) Cell Activity

YAC-1 a cell line of Maloney virus-induced lymphoma of A/Sn mice origin is sensitive to NK cell lysis and therefore these were used as the target cells on which to test the cytotoxicity of mouse spleen natural killer cells. CD-1 mice were injected with the carcinogen 1,2-dimethylhydrazine (DMH) to induce colon cancer. They were treated with inositol, InsP_6, inositol + InsP_6 or tap water (control). Mice who had developed carcinogen-induced tumors but had been treated with InsP_6 displayed significantly enhanced NK cell activity (compared to the untreated control animals). Treatment of mice with inositol or InsP_6 resulted

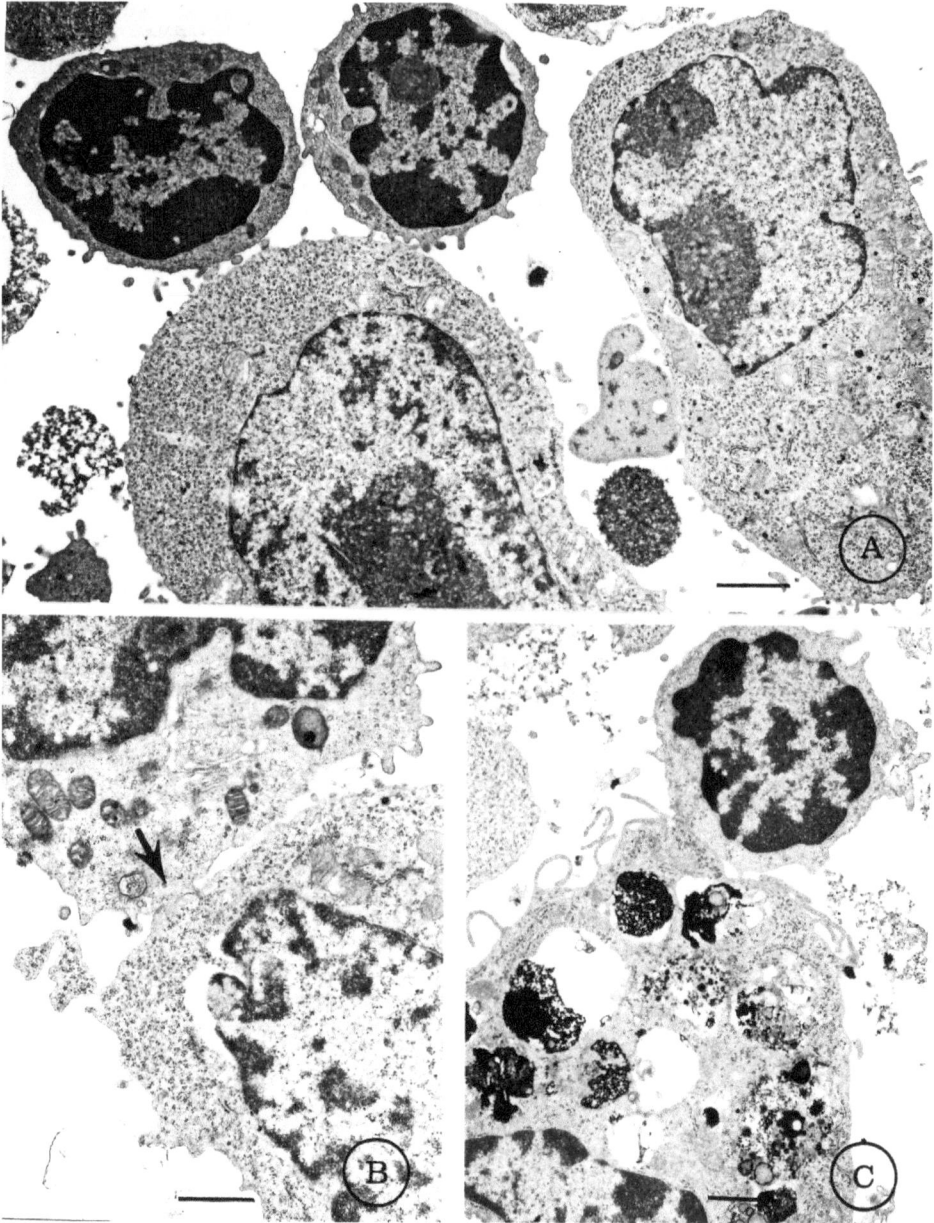

Fig. (17.1). Transmission electron micrographs of NK cell (small round cells with large nucleus and high nuclear-cytoplasmic ration) and target YAC-1 cancer cells. A. Two NK cells are seen in contact with a target YAC-1 cell, B. The NK cell (upper left) has established contact with the target cell as evidenced by fusion of the membranes (arrow) of NK cell and the target cell. C. The target YAC-1 lymphoma cell (lower large cell) is seen undergoing degeneration as evidenced by small nucleus with clumped and condensed chromatin, and numerous lysosomes within the cytoplasm.

in a 43% and 40% increase of NK cell activity respectively over the base line level in control mice that were not injected with the carcinogen DMH. In animals injected with the carcinogen and treated with inositol or $InsP_6$ the NK activity increased by 66% and 61% respectively; and inositol + $InsP_6$ treatment resulted in doubling (105% increase) in NK activity of the carcinogen-injected mice. Inositol + $InsP_6$ treatment not only increased the baseline activity of the normal mice by 58%, but also reversed the carcinogen-induced depression of NK activity (Fig. **17.2**) [4].

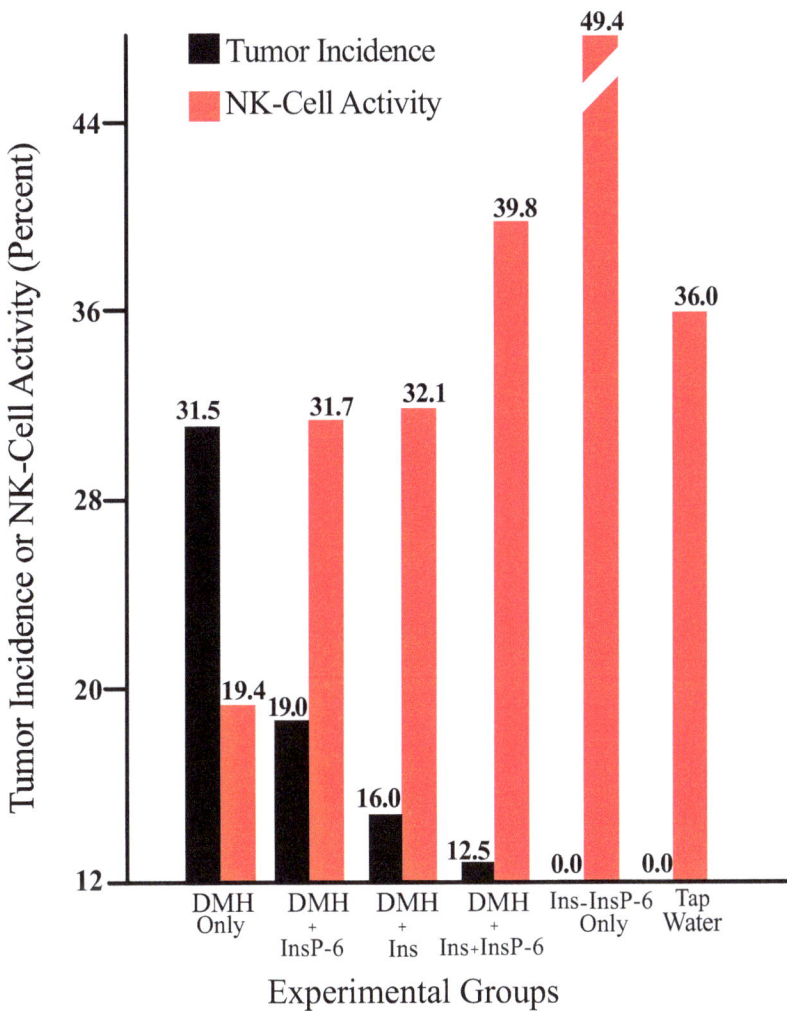

Fig. (17.2). DMH: 1, 2-dimethyl hydrazine, Ins: inositol. Note that the Inositol + $InsP_6$ ONLY group did not have induced tumor (0.0) and their NK activity was the highest (49.4%), more than tap water control.

The NK activity correlated positively with the extent of the tumor suppression - an increase in NK activity was associated with a lower incidence of cancer with a correlation coefficient of $r = -0.9811$ ($P < 0.0005$) (Fig. **17.3**) [4]. In order to investigate whether there was any correlation of *in vivo* enhancement of NK activity with *in vitro* system, splenic NK cells from healthy mice were treated *in vitro* with inositol, $InsP_6$ or inositol + $InsP_6$. Similar to the *in vivo* experiments, this resulted in an increase of NK activity by 66%-95% which was also dose-dependent [4].

CORRELATION BETWEEN NK- CELL ACTIVITY
& TUMOR INCIDENCE IN CD - I MICE

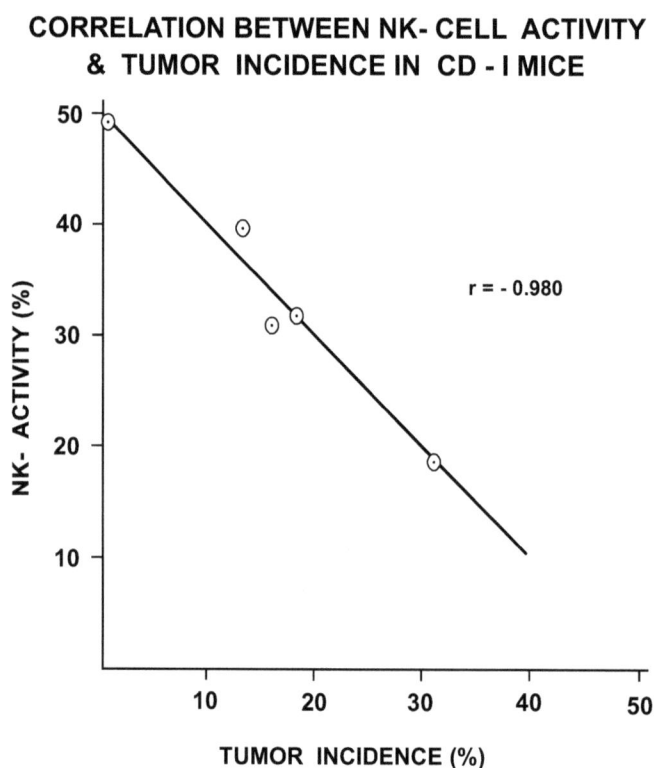

Fig. (17.3). Effect of inositol $\pm$ $InsP_6$ on DMH-induce tumor and NK activity is shown with a correlation coefficient of $r = -0.9811$ $P < 0.0005$.

Zhang *et al.,* reported a similar correlation in rats, concomitant to a decreased number and size of tumors as well as reduced metastasis [5]. But, the mechanism(s) exactly how inositol $\pm$ $InsP_6$ enhances NK cell activity is not understood. While $InsP_6$ may directly or indirectly *via* $InsP_3$ induce proliferation of NK cells *in vivo*, it is the level of activity and not the number of NK cells that is boosted by inositol $\pm$ $InsP_6$. It is possible that inositol $\pm$ $InsP_6$ triggers the release of NK cytotoxic factor (NKCF) important in cytolysis of target cells. However, that remains to be determined.

POLYMORPHONUCLEAR CELLS

The polymorphonuclear leukocytes (or neutrophils) are capable of destroying bacteria; however, as stated earlier, they first require stimulation in the form of a so-called respiratory burst - a series of chemical events that allow the neutrophils to generate free radicals capable of killing the bacteria. In connection with $InsP_6$, Eggleton and colleagues reported that human polymorphonuclear leukocytes - when pre-incubated with $InsP_6$ had a greatly augmented capacity to produce reactive oxygen species (when the leukocytes were stimulated either by chemicals or by the presence of bacteria). The presence of low levels of Ca^{2+} and Mg^{2+} (0.1 mM) during priming appeared to be an essential requirement. That the intracellular concentrations of $InsP_6$ in mammalian cells is in the 30-100 μM range suggest that the local release of this $InsP_6$ could have a physiologically important modulatory role on neutrophil functions [6]. Further studies on the release of IL-8, tumor necrosis factor (TNF-α) and IL-6 by neutrophils attached to either plastic or laminin for up to 6 hours in response to stimulation with lipopolysaccharide or N-formyl-Met-Leu-Phe (fMLP) showed an increase in IL-8 secretion by stimulated cells in the presence of $InsP_6$. Their data suggest a direct effect of $InsP_6$ was to trigger a sustained assembly of F-actin to modulate selective neutrophil functions [7].

Paradoxically, while $InsP_6$ acts as a rather potent antioxidant in most other instances, herein, it enhances the ability of polymorphonuclear leukocytes to produce more superoxides so that they can kill the bacteria, and more efficiently.

MACROPHAGE

Macrophages in the tissues are derived through differentiation of blood monocytes, and are active in both non-specific host defense (innate immunity) and specific adaptive immunity. One of their main functions is to engulf (phagocytose) foreign substances or invading agents such as bacteria.

Realizing the paucity of data on $InsP_6$'s ability to enhance the natural disease resistance of the body, Johnson and colleagues embarked on a study to investigate the effects of $InsP_6$ on the proliferation and viability of RAW 264.7 transformed macrophages and to investigate the role of $InsP_6$ as a free radical scavenger. They report that $InsP_6$ increases the rate of cell proliferation of the macrophages which was also dose-dependent; and it had an excitatory effect on the inflammatory cell secretions [8]. Once again, akin to the stimulation of respiratory burst and free radical generation in neutrophils, InsP_6 stimulates the macrophages to proliferate as opposed to cancer cells where it inhibits cell proliferation.

INTESTINAL BARRIER

We are fortunate to have many other innate defenses in our body; one of these is the intestinal lining epithelial cells. Aside from forming a mechanical barrier directly separating our interior from the luminal contents, they also are actively involved in the local immune response team. The cytokine IL-8 (interleukin-8) attracts the neutrophils through a process called chemotaxis and, causes them to release the deadly oxygen free radicals and enzymes to combat the intruder. IL-6 too is involved in inflammatory response in addition to being part of the normal immune response. Intestinal mucosal level of IL-8 and IL-6 are increased in various disease conditions of the intestine. Intestinal mucosal cells are continuously exposed to the bacterial products such as the endotoxin lipopolysaccharide (LPS) which plays an important role in the pathogenesis of bacterial infection especially by Gram negative bacteria which are the commonest in the intestinal lumen.

Since InsP_6 may be present in the intestinal lumen, Węglarz *et al.,* set out to study what if any effect InsP_6 in the intestinal milieu may have on the immune function of intestinal epithelial cells. The investigators looked into the *in vitro* influence of InsP_6 on the release of IL-8 and IL-6 by human colonic epithelial cells Caco-2 and the effect of InsP_6 treatment on secretion of these cytokines in cells stimulated with endotoxin LPS isolated from two strains of *Desulfovibrio desulfuricans*, wild intestinal type and soil strains, as well as to LPS from *E. coli.* Cells were also treated with IL-1β and with a combination of LPS and IL-1β. InsP_6 had a suppressive effect on IL-8 basal release and it dose-dependently reduced IL-8 secretion by Caco-2 cells stimulated with LPS and IL-1β. InsP_6 increased constitutive IL-6 as well as down-regulated IL-6 secretion stimulated by binary actions of LPS and IL-1β. Thus, InsP_6 in the intestinal milieu may exert immunoregulatory effects on colonic epithelium under physiological conditions or, during microbe-induced infection and/or inflammation in order to maintain the colonic mucosa healthy, or to counteract infection [9].

Further studies from Węglarz laboratory showed that the physiological intestinal concentrations of InsP_6 may have an inhibitory effect on IL-8 secretion by Caco-2 cells. One of the mechanisms of this may be the inhibition of protein tyrosine kinase signaling cascade [10].

Inflammation and Fibrosis

Gastric Ulcer

Studies on the effect of $InsP_6$ in experimental models of inflammation and gastric ulcer were reported by Kumar *et al.,* [11]. In the carrageenan-induced rat paw edema model they observed an anti-inflammatory activity of $InsP_6$, maximum at an oral dose of 150 mg/kg. $InsP_6$ showed the ability to prevent denaturation of proteins but it showed less anti-inflammatory activity than ibuprofen albeit, without sharing the side-effects of ibuprofen. The investigators believe that the ability of $InsP_6$ to bring down thermal denaturation of proteins might be a contributing factor in the mechanism of action against inflammation. $InsP_6$ at all the doses tested, showed significant protection from ulcers induced by ibuprofen, ethanol and cold stress. There was a significant increase in gastric tissue malondialdehyde levels in ethanol treated rats but these levels decreased following $InsP_6$ pretreatment. Moreover, pretreatment with $InsP_6$ significantly inhibited various effects of ethanol on gastric mucosa, such as, reduction in the concentration of non-protein sulfhydryl groups, necrosis, erosions, congestion and hemorrhage. Thus the gastro-protective effect of $InsP_6$ could be mediated by its antioxidant activity and cytoprotection of gastric mucosa [11].

Asbestosis

It is a well-established fact that exposure to asbestos can lead to fibrosis - the formation of fibrous (scar) tissue in response to an irritant - and to lung cancer. It is believed that oxidative damage (for instance caused by superoxide free radicals) is an early, key contributor in a sequence of chemical events that leads to asbestos-induced damage of the lungs. Several mechanisms appear to play a role, among them: Inflammation triggered by asbestos leads to a production of superoxides; and, the earlier discussed Fenton reaction may also play a role in that the iron present in the asbestos can catalyze the formation of superoxides.

What is $InsP_6$'s role in this? Given its capacity to chelate iron, as well as its ability to incapacitate free radicals, $InsP_6$ can diminish the tissue damage sustained from such inflammatory processes. In experiments where animals were exposed to asbestos have revealed that IP_6 can a) suppress the formation of superoxide free radicals, b) reduce DNA damage; and c) lower the degree of inflammation and the extent of subsequent lung fibrosis.

Kamp and associates demonstrated that $InsP_6$ reduces amosite asbestos-induced •OH generation, DNA strand break formation, and injury to cultured human pulmonary epithelial cells WI 26 cells *in vitro* [12]. In an *in vivo* rat model of asbestosis, asbestos exposure induced a severe inflammatory and fibrotic reaction in the animals' lungs after a single intra-tracheal instillation of amosite asbestos [13]. Sprague-Dawley rats were given either saline, amosite asbestos (5 mg; 1 ml saline), or amosite treated with $InsP_6$ (500 μM) for 24 hr. At various times after asbestos exposure, the rats were euthanized; the lungs were 'lavaged' and examined histologically [13]. A fibrosis score was determined by special staining. Compared to controls, asbestos elicited a significant pulmonary inflammatory response, as evidenced by an approximately 2-fold increase in broncho-alveolar lavage (BAL) cell counts and the percentage of BAL neutrophils and giant cells at 2 wk (0.1 v 6.5% and 1.3 v 6.1%, respectively); the results being statistically significant at $p < 0.05$ [13]. Compared to asbestos alone, $InsP_6$-treated asbestos elicited significantly less BAL PMNs (6.5 v 1.0%; $p < 0.05$) and giant cells (6.1 v 0.2%; $p < 0.05$) and caused significantly less fibrosis (5 v 0.8; $p < 0.05$) 2 weeks after exposure [13]. This study most convincingly demonstrates that a) asbestos cause pulmonary inflammation and fibrosis in rats and b) $InsP_6$ clearly reduces these effects [13].

VIRAL DISEASES

Human Immunodeficiency Virus (HIV)

"Human immunodeficiency virus is considered to be the cause of acquired immune deficiency disease (AIDS), which has a very unfavorable prognosis". Polyanionic substances such as heparin can inhibit replication of herpes simplex virus, and polysulfates can block the replication of HIV-1, HSV-1, HSV-2, CMV, togaviruses, arenaviruses, orthomyxoviruses and paramyxoviruses in cell culture [14]. It is believed that the polysulfates block the viral binding to the receptors on cells by covering the viral envelop protein. Otake *et al.,* embarked on investigating the potential of $InsP_6$ (inositol hexaphosphoric acid) and inositol hexasulfate (IHS) for anti-HIV-1 activity [15, 16]. The experiment on MT-4 cells - an HIV susceptible T-cell cell line demonstrates that $InsP_6$ and IHS inhibited the replication of HIV-1 [15, 16]. IHS completely inhibited the cytopathic effect of HIV and the HIV specific antigen expression at a concentration of 1.67 mg/ml. $InsP_6$ moderately inhibited both of HIV effects *i.e.,* replication and HIV specific antigen expression [15]. However, neither IHS nor InsP_6 inhibited HIV reverse transcriptase activity *in vitro* and there was no effect on late stage replication [16]. This work has not been expanded, but there are some intriguing reports on the role of InsP_6 in HIV viral particle assembly.

HIV-1 Gag protein, the principal structural component from which a retroviral viral particle is formed assembles into 100-120 nm sized particles in mammalian cells. Recombinant HIV-1 Gag protein assembles *in vitro* into particles that are far smaller than authentic viral particles - only 25-30 nm in diameter; and that differ significantly in other respects from authentic particles. However, addition of inositol phosphates or phosphatidylinositol phosphates results in particle size and properties analogous to the authentic HIV-1 particle indicating the necessity of these compounds in normal viral assembly [17]. The ability to assemble the correct virus like particles could be completely accounted for by the presence of InsP_5. Addition of only one InsP_5 molecule per 10 Gag molecules was sufficient to render the virus like particles completely resistant to 0.5 M NaCl, a property of authentic immature particles [17].

Further studies of the binding of InsP_6 to Gag show that basic regions at both ends of the protein contribute to InsP_6 binding [18]. Gag is in monomer-dimer equilibrium in solution *in vitro*, and mutation of the dimer drastically reduces Gag dimerization. However, addition of InsP_6 results in Gag that is in monomer-trimer rather than monomer-dimer equilibrium. These studies strongly point to the necessity of InsP_5 and InsP_6 in correct assembly of HIV-1 viral particle, at least *in vitro* [18].

CONCLUDING REMARKS

Inositol and InsP_6 enhance NK cell activity, stimulate polymorphonuclear cell priming function and the macrophages to produce cytokines. Additionally, they play significant roles in inflammation, viral assembly *etc*.

REFERENCES

[1]   Gorelik E, Herberman RB Inhibition of the activity of mouse natural killer cells by urethan. J Natl Cancer Inst. 1981; 66: 543-8.
[2]   Herberman RB, Holden HT. Natural killer cells as antitumor effector cells. J Natl Cancer Inst 1979; 62: 441-5.
[3]   Riccardi C, Santoni A, Barlozzari T, *et al*. *In vivo* natural reactivity of mice against tumor cells. Int J Cancer 1980; 25: 475-86.
[4]   Baten A, Ullah A, Tomazic VJ, *et al*. Inositol phosphate induced enhancement of natural killer cell activity correlates with tumor suppression. Carcinogenesis 1989; 10: 1595-8.

[5]   Zhang Z, Song Y, Wang X-L. Inositol hexaphosphate-induced enhancement of natural killer cell activity correlates with suppression of colon carcinogenesis in rats. World J Gastroenterol 2005; 11: 5044-6.

[6]   Eggleton P, Penhallow J, Crawford N. Priming action of inositol hexakisphosphate (InsP6) on the stimulated respiratory burst in human neutrophils. Biochim Biophys Acta 1991; 1094: 309-16.

[7]   Eggleton P: Effect of IP_6 on human neutrophil cytokine production and cell morphology. Anticancer Research1999; 19: 3711-5.

[8]   Johnson M, Tucci M, Benghuzzi H, *et al.* The effects of inositol hexaphosphate on the inflammatory response in transformed RAW 264.7 macrophages. Biomedl Sci Instrument 2002; 36: 21-6.

[9]   Węglarz, L.B., Wawszczyk, J., Orchel, A., *et al.* Phytic acid modulates *in vitro* IL-8 and IL-6 from colonic epithelial cells stimulated with LPS and IL-1β. Dig Dis Sci 2007; 52: 93-102.

[10]   Wawszczyk J, Orchel A, Kapral M, *et al.* Role of protein tyrosine kinase in the effect of IP_6 on IL-8 secretion in intestinal epithelial cells. Acta Pol Pharm. 2013; 70: 79-86.

[11]   Kumar S, Reddy S, Babu K, *et al.* Anti-inflammatory and antiulcer activities of phytic acid in rats. Ind J Exp Biol 2004; 42: 179-85.

[12]   Kamp DW Israbian VA Preusen SE *et al.* Asbestos causes DNA strand breaks in cultured pulmonary epithelial cells: role of iron-catalyzed free radicals. Am J Physiol. 1995; 268(Pt 1): L471-80.

[13]   Kamp DW, Israbian VA, Yeldandi AV, *et al.* Phytic acid, an iron chelator, attenuates pulmonary inflammation and fibrosis in rats after intratracheal instillation of asbestos. Toxicol Pathol 1995; 23: 689-95.

[14]   Witvrouw M, Desmyter J, De Clercq E. Antiviral portrait series: 4. Polysulfates as inhibitors of HIV and other enveloped viruses. Antiviral Chem Chemother 1994; 5: 345-59.

[15]   Otake T, Shimonaka H, Kanai M, *et al.* Inhibitory effect of inositol hexasulfate and inositol hexaphosphoric acid (phytic acid) on the proliferation of human immunodeficiency virus (HIV) *in vitro* [Japanese]. Kansensho-gaki Zasshi [Journal of Japanese Association of Infectious Diseases] 1989; 63: 676-83.

[16]   Otake T, Mori H, Morimoto M *et al.* Anti-HIV-1 activity of my*o*-inositol hexaphosphoric acid (IP6) and myo-inositol hexasulfate. Anticancer Res 1999; 19: 3723-6.

[17]   Campbell S, Fisher RJ, Towler EM, *et al.* Modulation of HIV-like particle assembly *in vitro* by inositol phosphates. Proc Natl Acad Sci USA 2001; 98: 10875-9.

[18]   Datta SA, Zhao Z., Clark PK, *et al.* Interactions between HIV-1 Gag molecules in solution: an inositol phosphate-mediated switch. J Mol Biol 2007; 365: 799-811.

CHAPTER 18

Mechanisms of Biological Actions of Inositol and InsP_6 I: Cell Survival, Proliferation and Differentiation

Abstract: A fundamental defect in cancer cells is the abnormal and uncontrolled cell proliferation. Along with that, there is de-differentiation to an immature or primitive phenotype. InsP_6 and inositol normalize the abnormal cell proliferation rate and induce increased differentiation so that the cancer cells begin to look and behave akin to normal cells. Examples of these are shown in colon cancer, erythroleukemia and mammary cancer, and rhabdomyosarcoma cells. In colon cancer cells that produce the cancer marker ß-D-Galactose-[1→3]-*N*-acetyl-D-galactosamine (Gal-GalNAc), InsP_6 inhibits the expression of the marker in the intracellular mucus without suppressing the production of mucus, a normal function of colon epithelial cells.

Keywords: Aberrant crypt, ACF, actin, anoikis, apoptosis, erythroleukemia, HeLa cells, HL-60 cells, lactalbumin, mucin, PCNA, rhabdomyosarcoma, tumor marker.

INTRODUCTION

It seems that classification or grouping people or actions is almost a constant phenomenon in life. People are grouped according to different identifiable characteristics such as race, ethnicity, religion, political affiliation *etc.*, and so are many other things. There are people who are "lumpers" - groups that do not wish to divide or classify and "splitters," who do. InsP_6 and inositol may bring about their actions in various ways; but we are often pressured to classify them. When there are divergent mechanisms, they are sure to overlap, therefore making a rigid classification difficult, if not irrelevant altogether. The mechanisms of anticancer action of inositol and its phosphates may be broadly looked at as biological, cellular, molecular, biochemical, chemical *etc.*, again with some overlap.

A fundamental defect in cancer cells is the abnormal and uncontrolled cell proliferation along with de-differentiation to an immature or primitive phenotype. Insofar as the mechanisms of anticancer action of inositol and InsP_6 are concerned, there are some clear-cut chemical reactions such as suppression of free radical generation. However, there are other mechanisms that cannot be categorized in any specific discipline for the pathways overlap across them. In this chapter we discuss the cellular mechanisms that involve normalization of

A.K.M. Shamsuddin and Guang-Yu Yang

abnormally elevated rate of cell proliferation, reversion of de-differentiated cells to normally differentiated one and cancer cell death.

CELL SURVIVAL, PROLIFERATION AND DIFFERENTIATION

Inositol and $InsP_6$ affect all the stages in a cell's life: survival (or death), cell division (or proliferation) and cell differentiation. In the body, most cells are associated with others forming tissues wherein they need mechanical strength for which not only they interact with each other, but also are interdependent. The extracellular matrix for example provides the epithelial cells with this mechanical support.

Apoptosis

It has been estimated that in every second approximately a million cells die in our body, mostly through pre-programmed death. In the skin and the gut lumen the dead cells simply slough off, or are quickly swallowed up by healthy neighbors or scavenger cells - the macrophages. This process of cell death is called apoptosis, as opposed to necrosis wherein cells die as a result of injury.

HeLa Cells

At concentrations up to 2-5 mM, $InsP_6$ inhibits cell proliferation of cancer cell lines with concomitant induced differentiation, but without a substantial increase in cell death in most cell lines. However, at higher dosage, or on prolonged treatment it induces apoptosis or programmed cell death. HeLa cells (derived from cancer of the uterine cervix of <u>H</u>enrietta <u>L</u>acks who died of her cancer in 4 October, 1951) on the other hand appear to be more sensitive, undergoing apoptosis at $InsP_6$ concentrations where very little apoptosis was observed in other cell lines. Treatment of HeLa cells with tumor necrosis factor (TNF) or insulin stimulated the Akt-nuclear factor κB (NFκB) pathway, a cell survival signal, which involves the phosphorylation of Akt and IκB, nuclear translocation of NFkB and NFkB-luciferase transcription activity. $InsP_6$ blocked all these cellular events. $InsP_6$ itself caused mitochondrial permeabilization, followed by cytochrome c release, which later caused activation of the apoptotic machinery, caspase 9, caspase 3 and poly (ADP-ribose) polymerase [1]. Exogenously applied $InsP_6$ directly activates the apoptotic machinery as well as inhibits the cell survival signaling, probably by the intracellular delivery followed by a dephosphorylation.

Prostate Cancer Model

In line with the broad-spectrum anti-cancer activity of $InsP_6$, induction of apoptosis is also seen in other cancers. Using advanced human prostate cancer cells DU145 to investigate the mechanisms of action of $InsP_6$, Singh *et al.,* demonstrated that at higher doses and longer treatment times, $InsP_6$ caused a marked increase in apoptosis, which was accompanied by increased levels of cleaved PARP and active caspase 3 [2]. Concomitantly, $InsP_6$ modulated CDKI-CDK-cyclin complex, and decreased CDK-cyclin kinase activity, possibly leading to hypophosphorylation of Rb-related proteins and an increased sequestration of E2F4. Higher doses of $InsP_6$ could induce apoptosis and that might involve caspases activation [2].

The investigators then expanded their work on transgenic adenocarcinoma of mouse prostate (TRAMP) cells, which reproduces the spectrum of benign, latent, aggressive and metastatic forms of human prostatic carcinoma. TRAMP-C1 cell line was treated with $InsP_6$ which resulted in cell death (apoptosis) with concurrent inhibition of cell growth in a dose- and time-dependent manner [3]. $InsP_6$ induced a moderate to strong (up to 14-fold over control) apoptotic cell death. Pretreatment of cells with caspases-inhibitor for 2 hours followed by 2 mM $InsP_6$ for 48 hours resulted in approximately 50% reversal in $InsP_6$-induced apoptosis suggesting a partial involvement of caspases activation in apoptosis caused by $InsP_6$. $InsP_6$ also showed a 6-fold induction in caspase-3 activity compared to control, again suggesting the involvement of caspases activation in $InsP_6$-induced apoptosis [3]. Their data also showed the involvement of both caspases-dependent and -independent mechanisms in $InsP_6$-induced apoptotic death of TRAMP-C1 cells.

Prostate Cancer Model In Vivo

Further work from Agarwal's lab show that apoptosis can be induced also *in vivo*: $InsP_6$ increased cyclin-dependent kinase inhibitors p21/Cip1 and p27/Kip1 protein levels in human prostate cancer DU145 cells that lacked functional p53. $InsP_6$-induced apoptosis occurred in a Cip/Kip-dependent manner in cell culture and xenograft [4, 5]. The investigators also provided evidence that p21 and p27 have a critical role in mediating the anticancer efficacy of $InsP_6$ both *in vitro* and *in vivo*.

Colon Cancer Model In Vivo

Additional evidence of the induction of apoptosis *in vivo* came from Jenab and Thompson [6]. Male Fischer 344 rats were injected with colon-carcinogen and fed a basal control diet or one supplemented with either wheat bran, or wheat bran stripped of $InsP_6$, or wheat bran stripped of $InsP_6$ + added $InsP_6$ to dissect out if the results are due to $InsP_6$. They observed that $InsP_6$ significantly increased the rate of apoptosis in the colonic epithelial crypts. The authors also concluded that aside from increased cell apoptosis, there was increased differentiation of colonic epithelial cells.

Skin Cancer Model In Vivo

Gupta *et al.,* [7] investigated the effect of topical application of $InsP_6$ on 7,12-dimethylbenzanthracene (DMBA)-induced carcinogenesis of mouse skin. $InsP_6$ induced DMBA-inhibited transglutaminase activity along with a significant inhibition of skin tumor development. $InsP_6$ suppressed DNA synthesis, as determined by ^{3}H-thymidine incorporation in a dose-dependent manner. $InsP_6$ also inhibited the enzyme thymidine kinase, which is responsible for ^{3}H-thymidine incorporation into DNA. They thus showed that topical application of $InsP_6$ inhibits DMBA-induced mouse skin tumor development and that $InsP_6$ exerts its tumor inhibitory effect probably by modulating apoptosis, proliferation and/or differentiation.

Liver Cancer Model In Vitro

Apoptosis was also studied in human hepatocellular carcinoma cells HepG2 by evaluating the expression of apoptosis-regulatory genes p53, Bcl-2, Bax, Caspase-3 and Caspase-9 by reverse transcriptase-PCR and DNA fragmentation assay [8]. The antioxidant activity of $InsP_6$ in Fe^{3+} reducing power assay was also investigated. Concomitant to $InsP_6$-inhibited growth of HepG2 cells in a concentration dependent manner, $InsP_6$-treated HepG2 cells showed up-regulation of p53, Bax, Caspase-3 and -9, and down- regulation of Bcl-2 gene. At the IC_{50} of 2.49 mM (concentration of the test substance causing 50% inhibition of cell growth) of $InsP_6$, the p53, Bax, Caspase-3 and-9 genes were up-regulated by 6.03, 7.37, 19.7 and 14.5 fold respectively. And the fragmented genomic DNA in $InsP_6$-treated cells provided evidence of apoptosis [8].

Inositol Pentaphosphate

The effect of different inositol polyphosphates on Akt activation showed that Inositol (1,3,4,5,6) pentakisphosphate [Ins(1,3,4,5,6)P_5] specifically inhibited Akt phosphorylation and kinase activity, whereas other inositol polyphosphates tested had no effect [9]. Ins(1,3,4,5,6)P_5 specifically promoted apoptosis in lung, ovarian, and breast cancer cell lines [9]. Fibroblast growth factor-2 (FGF-2) induced Akt phosphorylation in human umbilical vein endothelial cells (HUVEC) resulting in anti-apoptotic effect in serum-deprived cells and increase in cellular motility. Ins(1,3,4,5,6)P_5 blocked FGF-2-mediated Akt phosphorylation and inhibited survival of HUVEC [10]. Since, as described before, others have shown similar results with InsP_6, this study shows that perhaps one of the pathways for InsP_6-mediated apoptosis is *via* dephosphorylation to InsP_5 or lower phosphates.

Paradox

Enigmatically and to our advantage, in contrast to induction of apoptosis in cancer cells, InsP_6 can attenuate apoptosis where apoptosis is harmful. Aljandali *et al.,* show that InsP_6 reduces asbestos-induced apoptosis in alveolar epithelial cells in the lungs [11]. InsP_6 can protect against 6-hydroxydopamine- (6-OHDA-) induced apoptosis in immortalized rat mesencephalic dopaminergic cells under normal and iron-excess conditions, iron-excess being incriminated in the pathogenesis of neuronal cell degeneration of substantia nigra in Parkinson's disease. In a rat model of Parkinson's disease, Xu *et al.,* demonstrated that Caspase-3 activity was increased about 6-fold after 6-OHDA treatment (compared to control; $p < 0.001$) and InsP_6 pretreatment decreased it by 38%. Similarly, protection against 6-OHDA induced DNA fragmentation was observed with InsP_6 pretreatment. Under iron-excess condition, a 6-fold increase in caspase-3 activity and an increase in DNA fragmentation with 6-OHDA treatment were decreased by InsP_6 [12, 13]. More on this has been discussed in the context of oxidative damage and its modulation by InsP_6 as well as in the Chapter 13 (Parkinson's disease).

Anoikis

One of the most dangerous weapon in cancer cells' arsenal is their ability to detach from their habitat, invade surrounding tissues and travel to distant sites to cause havoc there (metastasis). But when otherwise normal cells are removed from natural surrounding such as loss of extracellular matrix attachment to cultured epithelial cells, they undergo self destruction or *anoikis* (derived from Greek to mean "homelessness"). Originally, it was presumed that anoikis is executed by apoptosis; but that has changed. Recent research show that blocking apoptosis does not prevent anoikis; detached cells die any way! Anoikic cells undergo a process whereby the cell digests parts of itself (autophagocytosis or autophagy). Autophagy may help the cell survive starvation, but if it runs its full course, it may cause cellular death through self-consumption. Since cancer cells have the ability to invade, naturally they escape anoikis. One may look at it in another manner and argue that anoikis prevents cancer! That remains to be seen and so is the role of InsP_6, if any in anoikis. Schafer *et al.,* [14] showed that oxidative stress induces anoikis and certain anti-oxidants prevent it; this would suggest that anti-oxidants may not have a beneficial effect in cancer treatment. That however remains to be proven.

Cell Proliferation

In most *in vitro* experiments, within the IC$_{50}$ dose InsP_6 causes inhibition of cell proliferation. This has also been observed *in vivo,* most conveniently in the colon cancer models where it is relatively easy to visualize and quantitate the rate of cellular mitosis, labeling index of cells *etc.* Studies in Shamsuddin Laboratory have shown that inositol, InsP_6 and InsP_6 + inositol cause a marked reduction in the number of mitosis, as opposed to tap-water control in animals induced to produce colon cancer [15].

In vitro models however allow for additional experiments that are not possible in the *in vivo* models, the limitations of each model systems notwithstanding. Thus, along with studies of cell proliferation, other parameters have also been tested. Sakamoto *et al.,* demonstrated the growth inhibition of HT-29 human colon cancer cells in a dose-dependent manner (Fig. **18.1**) [16]; studies of the differentiation were performed concomitantly, which will be described later in the chapter.

Plating efficiency is another method to determine the ability of cells to proliferate. InsP_6 has been shown to inhibit both the estrogen receptor positive MCF-7 and estrogen receptor negative MDA MB 231 cells (Fig. **18.2**) [17].

Since InsP_6 inhibits cell division, its effect on the cell cycle and the cell cycle regulatory proteins and their genes have been looked at. Reversible phosphorylation of specific intracellular proteins is known to be an important and

Growth Inhibition of HT-29 Human Colon Cancer Cells by InsP6

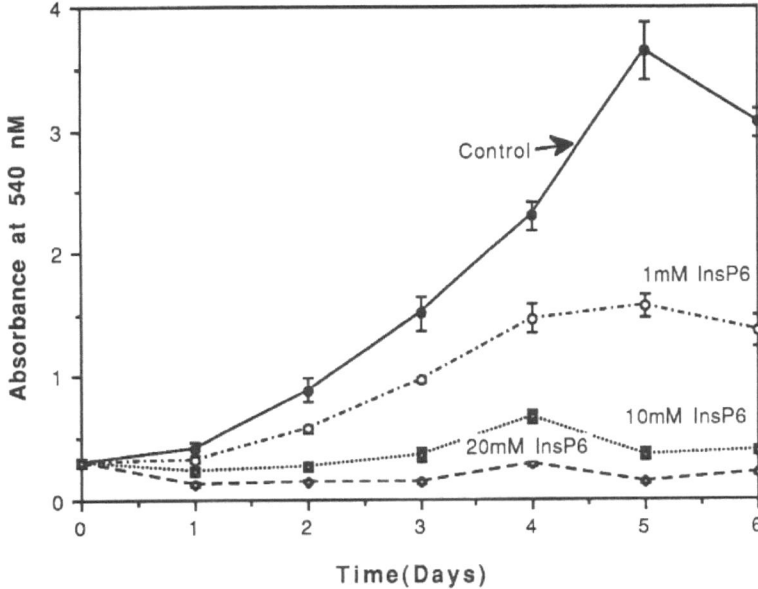

Fig. (18.1). Dose-response inhibition of HT-29 Human colon cancer cells by $InsP_6$. Note that at 10mM and 20mM concentrations there is near-total inhibition of cell growth.

versatile mechanism for regulating their biological activity. After rapid intake and dephosphorylation, $InsP_6$ enters inositol phosphate pool and controls variety of cellular functions, such as cell cycle regulation, differentiation *etc*.

While normal cells divide at a controlled and limited rate, malignant cells escape from the control mechanisms that regulate the frequency of cell multiplication, and usually have lost the checkpoint controls that prevent replication of defective cells. $InsP_6$ can regulate the cell cycle to block uncontrolled cell division and force malignant cells either to differentiate or go into apoptosis. Studies in Shamsuddin laboratory have shown that $InsP_6$ controls the progression of cells through the cycle by decreasing S- phase and arresting cells in the G_0/G_1-phase of the cell cycle. A significant decrease in the expression of proliferation markers indicated that $InsP_6$ disengaged cells from actively cycling [18].

$InsP_6$ can modulate cellular response at the level of receptor binding; $InsP_6$, after sterically blocking the heparin-binding domain of basic fibroblast growth factor (bFGF), disrupted further receptor interactions [19]. In addition to blocking of

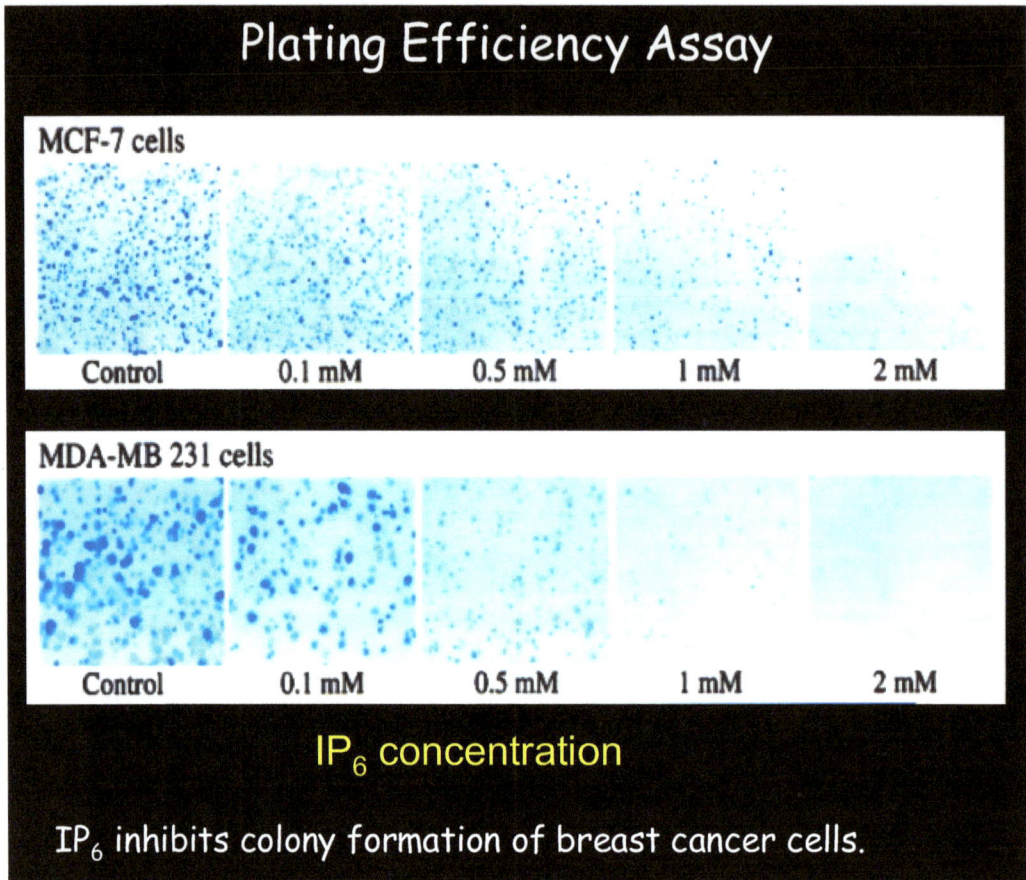

Fig. (18.2). Increasing concentration of InsP_6 results in decreased number of colonies (dark blue spots) irrespective of whether the cells are estrogen-receptor-positive (MCF-7 upper panel) or estrogen-receptor-negative (MDA-MB 237 lower panel).

phosphatidylinositol-3 kinase (PI3K) and activating protein-1 (AP-1) by InsP_6, protein kinase C (PKC) and mitogen-activated protein kinases (MAPK) are involved in InsP_6-mediated anticancer activity [20-22]. InsP_6 may also operate *via* a direct control of protein phosphorylation [23]. Interestingly, although inositol phosphate-regulated phosphorylation was shown for InsP_6, it was more effective than the lower inositol phosphate in particular InsP_3 or InsP_4, both known active signal transducers [23].

Using DU145 human prostate cancer cell line, Singh *et al.,* [2, 24] studied the cell cycle progression and apoptosis by flow cytometry. They also investigated the involvement of G_1 cell cycle regulators and their interplay and end point markers of apoptosis. A significant dose- and time-dependent growth inhibition of InsP_6-

treated cells was associated with an increase in cells in G_1. Except for a slight increase in cyclin D2, InsP_6 strongly increased the expression of cyclin-dependent kinase inhibitors (CDKIs) - Cip1/p21 and Kip1/p27, without any noticeable changes in G_1 CDKs and cyclins. InsP_6 also inhibited kinase activities associated with CDK2, 4 and 6, and cyclin E and D1. An increased binding of Kip1/p27 and Cip1/p21 with cyclin D1 and E was additionally demonstrated. In downstream of CDKI-CDK/cyclin cascade; InsP_6 increased hypophosphorylated levels of Rb-related proteins, pRb/p107 and pRb2/p130; and moderately decreased E2F4 but increased its binding to both pRb/p107 and pRb2/p130. At higher doses of InsP_6 and longer treatment times, InsP_6 caused a marked increase in apoptosis that was accompanied by increased levels of cleaved PARP and active caspase 3. InsP_6 modulated CDKI-CDK-cyclin complex, and decreases CDK-cyclin kinase activity, perhaps leading to hypophosphorylation of Rb-related proteins and an increased sequestration of E2F4 [24].

As mentioned before, TRAMP-C1 cell line was treated with InsP_6 which resulted in apoptosis with concomitant inhibition of cell growth in a dose- and time-dependent manner [3]. In the studies assessing whether cell growth inhibition by InsP_6 is associated with an alteration in cell cycle progression, InsP_6 treatment resulted in up to 92% cells in G_0-G_1 phase as compared to controls.

Wheat bran and its component InsP_6 have both been shown to decrease early biomarkers of colon carcinogenesis, *i.e.,* the PCNA labeling index of cell proliferation and certain aberrant crypt foci parameters [25]. InsP_6 lowered the labeling index and the position of the uppermost labeled cell in the distal colon. Dephytinization (removing InsP_6 from the diet) caused an increase in the overall labeling. Exogenous InsP_6 also reduced the number and size of aberrant crypt foci (ACF - a morphological marker for colonic precancerous lesion), the number of ACF per unit length colon as well as the number of sialomucin-producing (abnormal mucus produced by cancer and precancerous lesions of colon) ACF in various colon sections [25].

Differentiation

In general, cell differentiation depends on changes in gene expression resulting in synthesis and accumulation of different sets of RNA and consequently protein molecules; the latter often represent the differentiated features. Examples of such markers of differentiation include hemoglobin for mature red blood cells, prostate specific acid phosphatase for prostatic epithelial cells, lactalbumin for mammary cells, myoglobin for muscle cells *etc*. The various steps in the pathway leading from DNA to protein can be regulated, affecting gene expressions. These steps include: transcriptional control, RNA processing control, RNA transport and localization control, translational control, mRNA degradation control, or protein activity control. For most genes, transcriptional control is the critical one.

These alterations in gene expressions can be, and often are responses to external cues or stimuli; the signal(s) switching the regulatory regions of DNA near the site where transcription begins or by activating the gene regulatory proteins which turn genes on or off.

Erythroleukemia

$InsP_6$ has been demonstrated to induce differentiation of malignant cells of divergent origins to the normal phenotype. It was first demonstrated in K-562 human erythroleukemia cells, which showed increased hemoglobin production following $InsP_6$ treatment (Figs. **18.3** and **18.4**) [26].

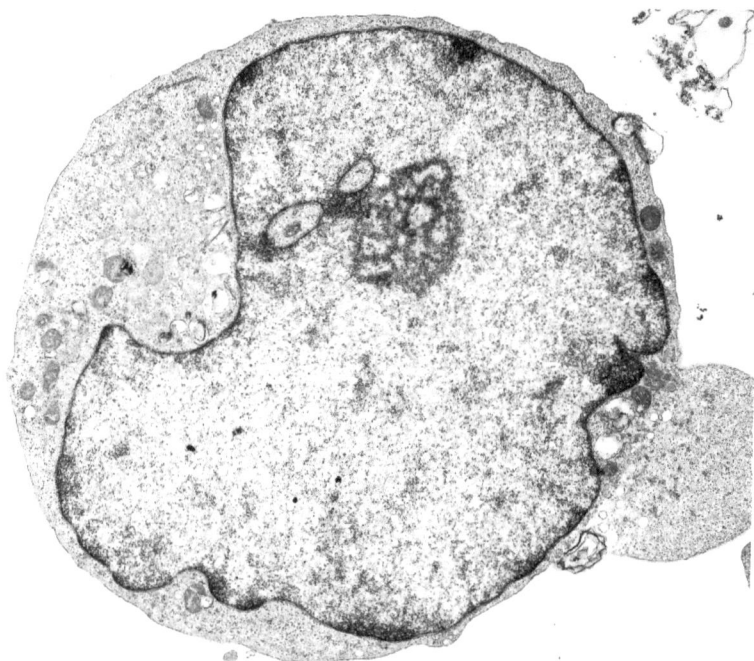

Fig. (18.3). Transmission electron micrograph of control untreated K-562 human erythroleukemia cell showing a very large nucleus with very high neuclear: cytoplasmic ratio. A few mitochondria are visible to the left side of the cytoplasm which is scanty.

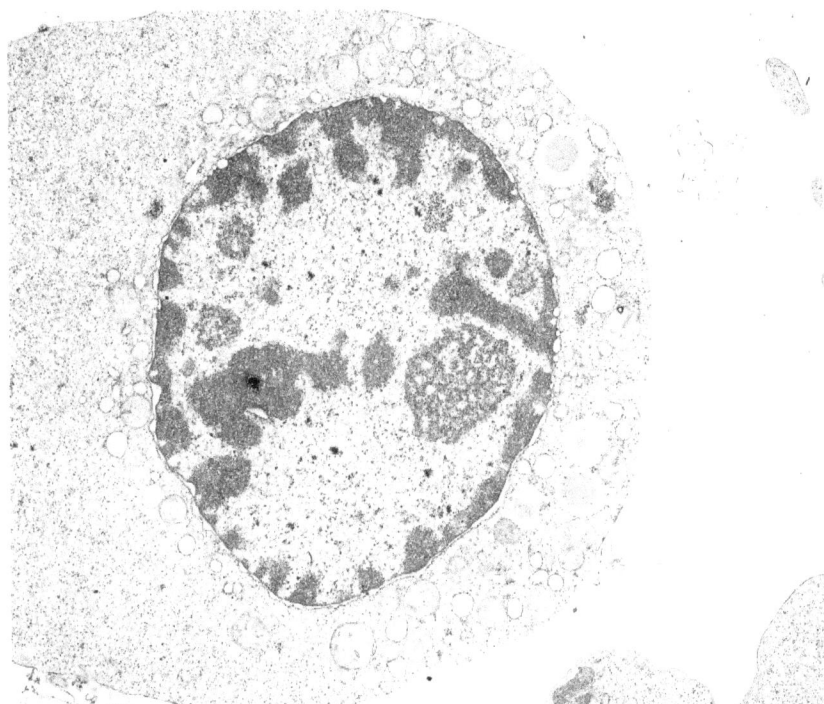

Fig. (18.4). K-562 cell treated with InsP_6 showing a relatively smaller nucleus with large cytoplasm - low nuclear: cytoplasmic ratio as compared to the untreated control above. Very few to no mitochondria or other organelles besides some vacuoles with probable hemoglobin are seen as evidence of differentiation.

Similar induction of tissue specific differentiation was reported for human colon carcinoma HT-29 cells, prostate cancer cells, breast cancer cells, and rhabdomyosarcoma cells.

Colon Cancer Cell Line

Along with a dose- and time-dependent growth inhibition as tested by MTT-incorporation assay, HT-29 human colon cancer cells also showed marked differentiation [16, 27]. The expression of cytokeratin and carcinoembryonic antigen (CEA) were both augmented by either InsP_6 or inositol at all concentrations tested, although the degree of augmentation was milder with inositol than with InsP_6. The combination of InsP_6 and inositol (both 0.66 mm) resulted in augmentation (P < 0.001) of cytokeratin expression, while that of CEA remained unchanged [16].

DNA-synthesis was suppressed by $InsP_6$ and significantly inhibited as early as 6 h after treatment at 1 mM concentration ($p < 0.05$) and continued to 48 h ($p < 0.01$). The expression of proliferation marker PCNA was down-regulated ($p < 0.05$) by $InsP_6$ treatment. The expression of a mucin antigen associated with goblet cell differentiation and defined by the monoclonal antibody CMU10 was augmented ($p < 0.0001$) by $InsP_6$. The tumor mucin marker ß-D-Galactose-[1→3]-N-acetyl-D-galactosamine (Gal-GalNAc), expressed by pre-cancer and cancer of colon, but not by the normal cells showed a time-dependent biphasic change by $InsP_6$; an increased expression after 1 day of treatment followed by suppression after 2 days suggest progression of mucin synthesis and differentiation of cancer cells with reversion to normal phenotype (Chapter 7, Fig. **7.4**). The significance of this action cannot be over emphasized, for a) it shows that $InsP_6$ perhaps suppresses the gene and or gene products responsible for expression of the tumor marker, and b) the marker can be used to monitor the outcome of cancer prophylactic drugs.

Involvement of Lower Inositol Phosphates

To investigate the mechanism of action of $InsP_6$ the intracellular phosphatases (including phytase) were inhibited by F to slow down the dephosphorylation of $InsP_6$. Ion-exchange chromatographic separation of intracellular inositol phosphates demonstrated an 84-98% decrease of *myo*-inositol, $InsP_1$ and $InsP_2$; and $InsP_3$ was reduced by 39% and, $InsP_4$ and $InsP_5$ by 21% and 13% respectively, whereas intracellular $InsP_6$ was increased by 24.6% at 5 min following ^{3}H- $InsP_6$. Since neither the rate of uptake of ^{3}H-$InsP_6$ was unaffected, nor was the efficacy of growth inhibition altered by F inhibition of phytase, data suggest that dephosphorylation by phytase plays no role in influencing the anti-neoplastic action of $InsP_6$. Alkaline phosphatase activity, brush border enzyme, associated with absorptive cell differentiation was increased following 1 and 5 mM $InsP_6$ treatment for 1-6 days [27].

Mammary Cancer Cell Lines

Estrogen is a hormone believed to be of great significance in the genesis and treatment of breast cancer. Some breast cancers show a growth response when specific hormones are present. In order for a tumor to respond to a certain hormone, it must have a receptor that is specific to that hormone. Breast cancers are commonly tested for the presence (or absence) of estrogen or progesterone receptors (ER or PR). For instance, MCF-7 human breast cancer cells are estrogen receptor-positive. In contrast, MDA MB-231 breast cancer cells are receptor-negative. InsP_6 has been shown to be equally successful in curtailing the growth of both of these cell lines as shown in Fig. (**18.2**) before. Concomitant to this reduction in cell proliferation, it also showed differentiation as evidenced by increased lactalbumin production (Fig. **18.5**) which is also dose-dependent (Fig. **18.6**) [17].

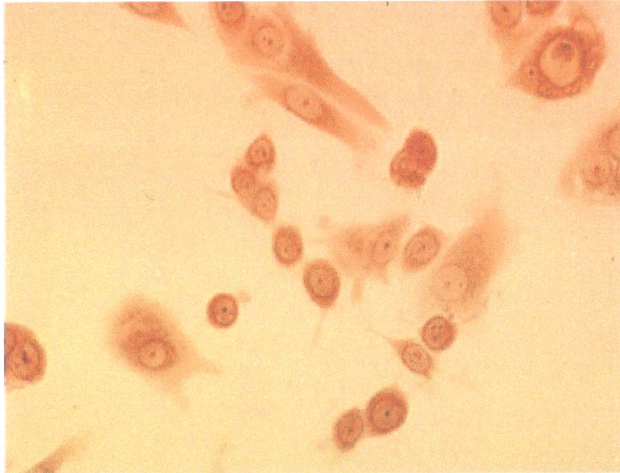

5 mM InsP6

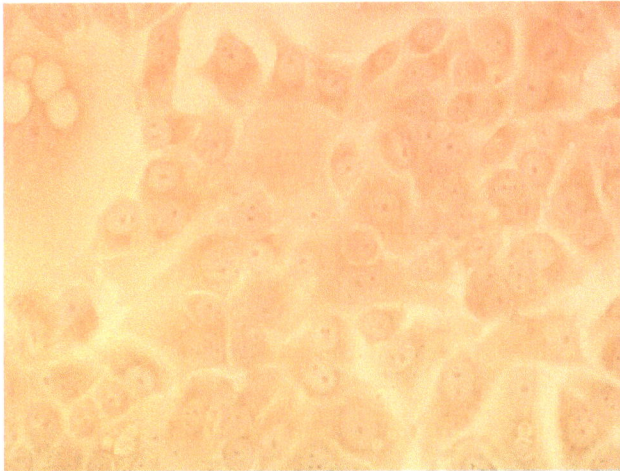

control

Fig. (18.5). Shows the effect of InsP_6 on lactalbumin expression by human breast cancer cell line. Note that InsP_6 treatment results in fewer cells (decreased proliferation - upper panel) and increased differentiation. Upper panel shows increased lactalbumin as dark brown with associated decrease in cell number [17].

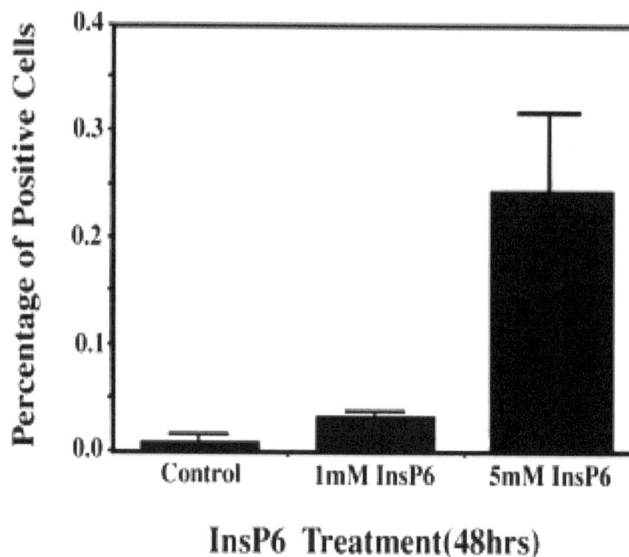

Fig. (18.6). The bar-gram shows that with increased doses of InsP_6 there is increased lactalbumin expression (dose-response relationship) [17].

Rhabdomyosarcoma

Studies in Shamsuddin Laboratory have shown that human rhabdomyosarcoma RD cells also undergo highly conspicuous morphological changes (*i.e.*, altered phenotype) following InsP_6 treatment *in vitro* [28]. Typically, untreated RD cancer cells are small and spindle-shaped. They also tend to attach themselves to the bottom of the culture dish, where they form patchy patterns; one normally does not see any evidence of skeletal muscle differentiation (maturation). As early as after the second or the third three-day treatment of InsP_6 these cells showed clear signs of differentiation: the treated cells grew larger and produced higher levels of the muscle-specific actin (a protein found explicitly in normal, differentiated skeletal muscle cells); cancerous untreated RD cancer cells cannot produce muscle-specific actin.

The molecular mechanisms involving these InsP_6 - induced differentiation will be fascinating to study. It is however known that PI 3-K (phosphoinositide 3-kinase) plays an important role in granulocytic differentiation of HL-60 leukemia cells. Interestingly, the intracellular concentration of InsP_6 and InsP_5 is elevated by about two orders of magnitude during chemotactic stimulation of HL-60 cells [29]. Clearly an elevated level of InsP_6 as described above also for HT29 human colon cancer cells [23] plays a yet to be determined role in these differentiated functions.

CONCLUDING REMARKS

Inositol and its phosphates normalizes abnormal rate of cell proliferation and induces differentiation in a wide variety of malignancies - epithelial and mesenchymal. These are mediated *via* activation or suppressor of various genes, and various inositol polyphosphates are involved, though the exact mechanism(s) are unclear; however an increase in intracellular InsP_6 in colon cancer cells and perhaps InsP_5 (in HL-60 cells) seems to be important.

REFERENCES

[1]   Ferry S, Matsuda M, Yoshida H, *et al*. Inositol hexakisphosphate blocks tumor cell growth by activating apoptotic machinery as well as by inhibiting the Akt/NFkappaB-mediated cell survival pathway. [Erratum appears in Carcinogenesis. 2003 Jan; 24(1): 149]. Carcinogenesis 2002; 23: 2031-41.

[2]   Singh RP, Agarwal C, Agarwal, R. Inositol hexaphosphate inhibits growth, and induces G1 arrest and apoptotic death of prostate carcinoma DU145 cells: modulation of CDKI-CDK-cyclin and pRb-related protein-E2F complexes. Carcinogenesis 2003; 24: 555-63.

[3]   Sharma G, Singh RP, Agarwal R. Growth inhibitory and apoptotic effects of inositol hexaphosphate in transgenic adenocarcinoma of mouse prostate (TRAMP-C1) cells. Int J Oncol 2003; 23: 1413-8.

[4]   Roy S, Gu M, Ramasamy K, *et al*. p21/Cip1 and p27/Kip1 Are essential molecular targets of inositol hexaphosphate for its antitumor efficacy against prostate cancer. Cancer Res. 2009; 69: 1166-73.

[5]   Roy S, Singh RP, Agarwal C, *et al*. Downregulation of both p21/Cip1 and p27/Kip1 produces a more aggressive prostate cancer phenotype. Cell Cycle 2008; 7: 1828-35.

[6]   Jenab M, Thompson LU. Phytic acid in wheat bran affects colon morphology, cell differentiation and apoptosis. Carcinogenesis 2000; 21: 1547-52.

[7]   Gupta KP, Singh J, Bharathi R. Suppression of DMBA-induced mouse skin tumor development by inositol hexaphosphate and its mode of action. Nutrition & Cancer 2003; 46: 66-72.

[8]   Al-Fatlawi AA, Al-Fatlawi AA, Irshad M, *et al*. Rice Bran Phytic Acid Induced Apoptosis Through Regulation of Bcl-2/Bax and p53 Genes in HepG2 Human Hepatocellular Carcinoma Cells. Asian Pac J Cancer Prev. 2014; 15: 3731-6.

[9]   Piccolo E, Vignati S, Maffucci T, *et al*. Inositol pentakisphosphate promotes apoptosis through the PI 3-K/Akt pathway. Oncogene 2004; 23: 1754-65.

[10]   Maffucci T, Piccolo E, Cumashi A, *et al*. Inhibition of the phosphatadylinositol 3-kinase/Akt pathway by inositol pentakisphosphate results in anti-angiogenic and antitumor effects. Cancer Research 2005; 65: 8339-49.

[11]   Xu Q, Kanthasamy AG, Reddy MB. Neuroprotective effect of natural iron chelator phytic acid in a cell culture model Parkinson's disease. *Toxicology* 245: 101-108, 2008.

[12]   Xu Q, Kanthasamy AG, Reddy MB. Phytic acid protects against 6-hydroxy dopamine-induced dopaminergic neuron apoptosis in normal and iron excess conditions in a cell culture model. Parkinson's Disease. 2011: 431068, Published online 7 February 2011. 2011 Feb 7; 2011: 431068. doi: 10.4061/2011/431068.

[13]   Aljandali A, Pollack H, Yeldandi A, *et al*. Asbestos causes apoptosis in alveolar epithelial cells: role of iron-induced free radicals. J Lab Clin Med 2001; 137: 330-9.

[14]   Schafer ZT, Grassian AR, Song L, *et al*. Antioxidant and oncogene rescue of metabolic defects caused by loss of matrix attachment. Nature 2009; 461: 109-13.

[15]   Shamsuddin AM, Ullah A, Chakravarthy AK. Inositol and inositol hexaphosphate suppress cell proliferation and tumor formation in CD-1 mice. Carcinogenesis 1989; 10: 1461-3.

[16]   Sakamoto K, Venkatraman G, Shamsuddin AM. Growth inhibition and differentiation of HT-29 cells *in vitro* by inositol hexaphosphate (phytic acid). Carcinogenesis 1993; 14: 1815-9.

[17]   Shamsuddin AM, Yang G-Y, Vucenik I. Novel anti-cancer functions of IP_6: Growth inhibition and differentiation of human mammary cancer cell lines *in vitro*. Anticancer Research 1996; 16: 3287-92.

[18]   El-Sherbiny YM, Cox MC, Ismail ZA *et al*. G_0/G_1 arrest and S phase inhibition of human cancer cell lines by inositol hexaphosphate (IP_6). Anticancer Res 2001; 21: 2393-403.

[19]   Morisson RS, Shi E, Kan M, Yamaguchi F, *et al*. Inositol hexaphosphate ($InsP_6$): An antagonist of fibroblast growth factor receptor binding and activity. *In vitro* Cell Dev Biol 1994; 30A: 783-9.

[20]   Huang C, Ma W-Y, Hecht SS, *et al*. Inositol hexaphosphate inhibits cell transformation and activator protein 1 activation by targeting phosphatidylinositol-3' kinase. Cancer Res 1997; 57: 2873-8.

[21].   Nickel KP, Belury MA: Inositol hexaphosphate reduces 12-*O*-tetradecanoylphorbol-13-acetate-induced ornithine decarboxylase independent of protein kinase C isoform expression in keratinocytes. Cancer Lett 1999; 140: 105-11.

[22].   Vucenik I, Ramakrishna G, Tantivejkul K, *et al*. Inositol hexaphosphate (IP_6) differentially modulates the expression of $PKC\delta$ in MCF-7 and MDA-MB 231 cells. Proc Amer Assoc Cancer Res 1999; 40: 653.

[23]   Solyakov L, Cain K, Tracey BM, *et al*. Regulation of casein kinase-2 (CK-2) activity by inositol phosphates. J Biol Chem 2004; 279: 43403-10.

[24]   Singh R P, Sharma G, Mallikarjuna GU, *et al. In vivo* suppression of hormone-refractory prostate cancer growth by inositol hexaphosphate: induction of insulin-like growth factor binding protein-3 and inhibition of vascular endothelial growth factor. Clinical Cancer Research 2004; 10: 244-50.

[25]   Jenab M, Thompson LU. The influence of phytic acid in wheat bran on early biomarkers of colon carcinogenesis. Carcinogenesis. 1998; 19: 1087-92.

[26]   Shamsuddin AM, Baten A, Lalwani ND. Effects of inositol hexaphosphate on growth and differentiation in K-562 erythroleukemia cell line. Cancer Letters 1992; 64: 195-202.

[27]   Yang GY, Shamsuddin AM. IP_6-induced growth inhibition and differentiation of HT-29 human colon cancer cells: involvement of intracellular inositol phosphates. Anticancer Research 1995; 15: 2479-87.

[28]   Vucenik I, Kalebic T, Tantivejkul K, *et al*. Novel anticancer function of inositol hexaphosphate: inhibition of human rhabdomyosarcoma *in vitro* and *in vivo*. Anticancer Research 1998; 18: 1377-84.

[29]   Pittet D, Schlegel W, Lew DP *et al*. Mass changes in inositol tetrakis- and pentakis phosphate isomers induced by chemotactic peptide stimulation of HL-60 cells. J Biol Chem 1989; 264: 18489-93.

CHAPTER 19

Mechanisms of Biological Actions of Inositol and InsP_6 II: DNA Damage Repair

Abstract: DNA damage is one of the earliest changes in cancer formation. Within the cell and the nucleus, inositol and its phosphates are found wherein they are involved in DNA damage repair machinery, chromatin remodeling, RNA editing and mRNA transport.

Keywords: Chromatin remodeling, mRNA transport, nuclear pore, NHEJ, NPC, RNA editing, RNA interference.

INTRODUCTION

In the previous chapter we have discussed the general effects of inositol and its phosphates, especially InsP_6 on cellular level. Here we discuss the effects on the cellular genome; some of the mechanistic pathways have already been considered in Chapter 3 in relation to intracellular signaling. Since there is considerable overlap between the various mechanisms, there may be some repetitions which are unavoidable.

Controlled Cell Growth

When unimpeded and functioning properly, cells divide under well regulated conditions. Unrestrained cell division is a harbinger of cancer.

During the cell division process, most cells create an exact copy of their contents. At this point, the most vital prerequisite is an authentic and accurate replication of a cell's genetic material - deoxyribonucleic acid (DNA). Once the cell replication process has been completed, DNA is distributed to each of the daughter cells making them genetically identical copies of their parents' cells. In order to allow this genetic replication process the body must make fresh DNA.

DNA

DNA's building blocks consist of a chain of Adenine (A) - Guanine (G) - Cytosine (C) and Thymine (T). The two strands are wound together with A paired with T (A = T) and G paired with C (G = C). This complementary base-pairing

A.K.M. Shamsuddin and Guang-Yu Yang

enables the base-pairs (A = T or G = C) to be packed in the most favorable arrangement in the interior of the double helix. When the time has come for a cell to produce a copy of its DNA, the double helix uncoil and the new bases line up in such a way that they correspond to those on the original strand.

Specific bases within a group of cells can be radioactively tagged. This procedure makes it possible to determine the exact moment when the DNA synthesis occurs. The rate at which the DNA synthesis takes place can also be evaluated by measuring the quantity of the tagged bases that are being processed. Thymidine (which combines the thiamine T base with ribose) is used exclusively during the DNA synthesis process. Radioactively marked thymidine (^{3}H-thymidine) is incorporated into the DNA allowing measurement of how much of the substance had been taken up.

Studies first in Shamsuddin's laboratory showed a suppression of DNA synthesis as measured by ^{3}H-thymidine incorporation and down-regulation of proliferation marker PCNA (proliferating cell nuclear antigen) by InsP_6 [1-4]. A marked decrease in the expression of proliferation markers indicated that InsP_6 disengaged cells from active cycling. Using dual parameter flow cytometry and combined analysis of the expression of cell cycle-related proteins, it was also demonstrated that InsP_6 controls the progression of the cells through the cell cycle. As stated in the previous chapter, InsP_6 treatment significantly decreased the S-phase and arrested the human colon and breast cancer cells in the G_0/G_1 phase [2-4]. Interestingly, the intracellular levels of InsP_6 were high in G_1 and G_2/M phases of cell cycle, but dropped by 50-75% during the S phase [2-4].

InsP_6-treated leukemia cells accumulate in G_2M phase of cell cycle (as opposed to G_0/G_1 phase in breast cancer cells); once again arrest of cells in the cycle, albeit in a different phase [5]. Further investigation using cDNA microarray analysis showed an extensive down-modulation of genes involved in transcription and cell cycle regulation (c-myc, HPTPCAAX1, FUSE, and cyclin H) and an up-regulation of cell cycle inhibitors such as CKS2, p57 and Id-2. Genes such as STAT-6 and MAPKAP, involved in important signal transduction pathways were also down-regulated [5].

DNA Damage

In order to continue our life form, all species must pass on the genetic information to the progeny, unchanged or unaltered. However, our DNA, is under constant assault from physical and chemical agents both inside and outside our body. As if to counter that constant onslaught, life has evolved to not only detect, but also repair (within limits) the damages inflicted on the DNA. And every day, each of the approximately 10^{13} cells of our body ends up having tens of thousands of lesions in our DNA. These lesions can potentially affect the genome copies during DNA replication and then transcription. Of course, if these errors in the message are not corrected, or improperly corrected, they may result in gene mutation which may be catastrophically detrimental to the health of the cell, and even the organism. But fortunately, these damages are mostly not permanent and we have evolved to repair these damages, by and large. The double helical structure of the DNA is ideally suited for repair as it carries two separate copies of all the genetic information; when one strand is damaged, the complementary strand retaining an intact copy of the same information is used to restore the correct nucleotide sequence and hence, repair the damage. Single strand damages are less dangerous than double strand damage; the latter may be fatal to the cell. Ionizing radiation is a cause for double stranded DNA breaks. Aside from accidental (Chernobyl, Fukushima Daiichi), war-time (Hiroshima and Nagasaki) or terrorist-activated [dreaded] nuclear blast, one may find ionizing radiation in some homes as radioactive radon gas (from uranium decay) that contributes to lung cancer. We are also exposed to ionizing radiation from diagnostic and therapeutic agents such as technetium-99m (^{99m}Tc) iodine-131 (^{131}I) or cancer radiotherapy.

Fortunately, each cell contains many DNA repair systems, every one with its own enzymes; most of these systems use the undamaged strand of the double helix containing a copy of the original information as a template for repairing the damaged strand. Basically, for DNA with single strand damage the damaged sequence is excised, the original sequence is restored by using the undamaged strand as template by the enzyme DNA polymerase and finally the break in the DNA is sealed by another enzyme, appropriately called DNA ligase. The various cancer-causing agents (carcinogens, such as those in tobacco products, fungal toxins as aflatoxins, *etc.*) as well as many of the cancer-treatment drugs (!) also damage DNA by forming adducts. The most pervasive of the environmental DNA damaging agents is ultraviolet (UV) light. Notwithstanding the increasing size of the hole, thanks to the ozone layer, the most dangerous part of the solar ultraviolet spectrum - UVC does not reach us. However, the residual UVA and UVB can induce approximately 100,000 DNA lesions/cell/day [6]. A common DNA lesion by UV radiation from sunlight is TT dimers produced by covalent linkage between two adjacent thymidine molecules in the DNA strand. If the lesion is not excised, the misinformation owing to abnormal base-pairing would lead to its substitution (mutation) in the daughter DNA chain during DNA replication. This mutation will then be propagated throughout subsequent generations of cells.

DNA Repair

As one would expect, to combat the various types of DNA damages there exists a rather elaborate maintenance/repair system for each category. For example, there is base-excision repair (BER) which removes the subtle modifications of DNA such as small base adduct. Bulkier single-strand lesions that distort the DNA helical structure (*e.g.,* caused by UV light) are processed by nucleotide excision repair (NER). The double stranded breaks (DSBs) are handled through homologous recombination and non-homologous end joining (NHEJ); the latter - an imperfect process, simply brings two ends together wherein bases may be lost or added. Genomic stability depends upon the NHEJ mechanism and deficiencies in this pathway can lead to events that initiate or propagate tumorigenesis [7]. NHEJ rejoins DSBs during G_0, G_1, and early S-phases of the cell cycle; it is the dominant mechanism used for DSB repair in humans and other multicellular organisms. End joining is carried out by DNA ligase IV (ligase IV), which acts as part of the XLF/XRCC4/ligase IV complex [8]. Acting mainly during the S and G_2 phases of the cell cycle, homologous recombination on the other hand tends to restore the original DNA sequence.

InsP₆ in DNA Repair

As mentioned before, double strand breaks are potentially dangerous; they may cause the chromosomes to break into small fragments and eventually leading to cell death. This type of damage is caused by ionizing radiation, oxidizing agents, *etc.*

A serine/threonine kinase DNA-dependent protein kinase (DNA-PK) composed of the phosphoinositide-3-kinase-related protein kinase (PIKK), DNA-PK catalytic subunit (DNA-PKcs), and the heterodimeric Ku70/80 regulatory subunit is required for NHEJ [8].

$InsP_6$ has been demonstrated to stimulate non-homologous end-joining. It has been proposed to be brought about by the binding of $InsP_6$ to the DNA-PK_{cs} [8]. A detailed study shows that it is not DNA-PK_{cs} (a large protein of ~3500 amino acids, M_w ~465 kDa), but the DNA end binding protein Ku (consists of Ku70 - 70 kDa, and Ku86 - 83 kDa) that binds to IP₆ [9]. $InsP_6$ has a potent effect on mobility and dynamics of Ku through a region of Ku70 [10]. Recent studies show that Ku is precisely regulated by binding to $InsP_6$ and, *via* activation of DNA-PK, plays a key role in non-homologous end-joining repair [11].

Once the assault on the cell has gone past the scope of DNA repair, the otherwise heretofore normal cell is likely to transform to a malignant (cancer) cell. Insofar as the transformation of cells from normal to malignant is concerned, there are various models and pathways; one of these pathways is the activation of transcription factors activating protein-1 (AP-1) and nuclear factor NFκB *via* phosphatidylinositol 3-kinase (PI-3 kinase). Using tumor promoter-induced cell transformation of human skin JB6 cells, Huang *et al.,* [12] have demonstrated that $InsP_6$ blocks epidermal growth factor-induced PI-3 kinase and AP-1 activity. That $InsP_6$ also acts as an anti-mutagenic agent has been recently demonstrated by Ra Yoon *et al.,* [13].

DNA damage is also caused by reactive oxygen compounds arising as by-products of oxidative respiration or from environmental toxic agents. This has been discussed in Chapter 16.

Chromatin Remodeling, RNA Editing and mRNA Export

Chromatin Remodeling

DNA in eukaryotic nucleus is packaged into chromatin. This limits the access of DNA-binding proteins to DNA; hence the regulatory transcription machinery proteins are unable to function and express the genes properly.

Chromatin remodeling is a dynamic process that allows the access of condensed genomic DNA to the regulatory protein by altering the nucleosome architecture such that genes are exposed to or hidden from the transcriptional machinery. The nucleosome can be restructured by two mechanisms: the movement of nucleosomes along DNA by ATP-dependent chromatin remodeling complexes; and the modification of core histones by histone acetyltransferases, deactylases, methyltransferases, and kinases.

The SWI2/SNF2 family of ATP-dependent chromatin-remodeling complexes is widely used to regulate DNA accessibility for transcription. Four related classes of protein complexes (SWI2/SNF2, ISWI, Mi2, and INO80) use the energy of ATP hydrolysis to alter nucleosome architecture.

Since these chromatin remodeling proteins play an essential role in transcriptional regulation, they have been linked to cancer. "Besides actively regulating gene expression, chromatin remodeling imparts an epigenetic regulatory role in several key biological processes, *e.g.*, DNA replication and repair; apoptosis; chromosome *etc.*". Thus it is not surprising that targeting chromatin remodeling pathways is evolving as a major therapeutic strategy in the treatment of several cancers.

Shen *et al.,* [14] and Steger *et al.,* [15] have shown that significant role of inositol polyphosphates in chromatin remodeling and gene expression. Production of InsP_4 and/or InsP_5 modulates the ability of the SWI/SNF and INO80 chromatin remodeling complexes to induce transcription of some phosphate-responsive genes; it is thought that this effect is mediated perhaps by affecting the ability of these complexes to interact with Pho4, Pho2, and/or chromatin [15].

As mentioned above, SWI2/SNF2, ISWI, Mi2, and INO80 use the energy of ATP hydrolysis to alter nucleosome architecture. InsP_6 inhibits nucleosome mobilization by NURF, ISW2, and INO80 complexes. In contrast, nucleosome mobilization by the yeast SWI/SNF complex is stimulated by InsP_4 and InsP_5. Mutations in genes encoding inositol polyphosphate kinases that produce InsP_4, InsP_5 and InsP_6 impair transcription *in vivo* [14]. It is possible that the amounts or ratios of inositol polyphosphates are altered under certain physiological conditions which may signal for global regulation of mRNA export and transcription within the cell [15]. Results from these investigators (published simultaneously) provide a link between inositol polyphosphates, chromatin remodeling, and gene expression [14, 15].

RNA Editing

Nucleotide sequences of the RNA transcripts are altered by RNA editing whereby one or more U (uracil) nucleotides are inserted or removed from selected regions of a transcript. This results in major modifications of the original reading frame and sequence, hence changing the meaning of the message. For some genes the editing can be extensive enough that over one-half of the nucleotides are U nucleotides. Discovered first in RNA transcripts that code for proteins in the mitochondria of trypanosomes, extensive editing of mRNA has also been found in the mitochondria of many plants. RNA editing takes place in the nucleus, cytosol, mitochondria or plastids; and takes place in vertebrates, albeit rare.

Most cellular proteins and the translation machinery confuse between inosine and guanosine, recognizing the former as the latter; inosine pairs most stably with cytidine. Therefore, editing of RNA can alter a codon, create splice sites, and change its structure. One form of RNA editing is catalyzed by adenosine deaminases that act on RNA (ADARs) is a family of enzymes that deaminates adenosine to form inosine in double-stranded RNA (dsRNA). ADARs are important for proper neuronal function and are also implicated in the regulation of RNA interference (RNAi) [16].

The crystal structure of the catalytic domain of human ADAR2 that $InsP_6$ is buried within the enzyme core [16]. Macbeth *et al.,* show that $InsP_6$ is required for the enzymatic activity of ADAR2. $InsP_6$ is also essential for *in vivo* and *in vitro* deamination of adenosine 37 of tRNA[ala] by ADAT1 (adenosine deaminases that act on tRNA) [16].

mRNA Export

Gene expression involves distinct cellular processes that include transcription, mRNA processing, mRNA export from the nucleus to the cytosol through the nuclear pore complex, and translation in the cytosol. It has been estimated that only about $1/20^{th}$ of the total mass of RNA made in the nucleus ever leaves it. RNA export through the nuclear pores is an active process as each mRNA precursor molecule remains tethered to sites inside the nucleus until all of the spliceosome components have been dissociated from it. During the process of synthesis of mRNAs, proteins associate with the RNA to form messenger ribonucleoprotein particles (mRNPs). RNA-binding protein composition of these mRNPs is dynamic, changing as the mRNP moves through the different steps of gene expression. The DEAD-box proteins (DBPs) are intricately involved in these mechanisms and act in nucleotide-dependent processes such as RNA duplex unwinding and mRNP remodeling to alter the protein composition of an mRNP. DBPs are a family of enzymes that acts as an ATPase, bind with ATP to convert it to ADP and inorganic phosphate in an RNA-dependent manner. One DBP, Dbp5 (aka DBX19 in humans) is required for mRNP export out of the nucleus. Dbp5 ATPase activity is enhanced by a specific protein binding partner, Gle1, bound to $InsP_6$. At the cytoplasmic face of the nuclear pore complex (NPC) the export protein Gle1 bound to $InsP_6$ spatially activates the ATP-hydrolysis and mRNP-remodeling activity of the DEAD-box protein Dbp5 [17]. The interaction between Dbp5 and Gle1 is stabilized by $InsP_6$ however $InsP_6$ by itself has no effect on Dbp5 activity [18].

Targeting of the mRNP to the nuclear face of the NPC and through its central channel is dependent upon the mRNA export factors Mex67 and Mtr2. By interacting with both the mRNP and Nups, Mex67 directly facilitates the translocation. The essential sequence of events that occur at the NPC cytoplasmic face are linked to precise, localized activation of Dbp5 ATPase activity by Gle1-InsP_6 for the final step of export to the cytosol [17].

CONCLUDING REMARKS

InsP_6 is involved in diverse pathways within the cell wherein it is an integral part of the DNA damage repair machinery, chromatin remodeling, RNA editing and transport. While the biological and physiological relevance of these are being revealed, exactly how some of these roles of InsP_6 fit into pathogenesis of diseases would be fascinating to investigate.

REFERENCES

[1]   Shamsuddin, A. M., and G. Y. Yang.: Inositol hexaphosphate inhibits growth and induces differentiation of PC-3 human prostate cancer cells. Carcinogenesis 1995; 16: 1975-9.

[2]   Yang GY, Shamsuddin AM. IP$_6$-induced growth inhibition and differentiation of HT-29 human colon cancer cells: involvement of intracellular inositol phosphates. Anticancer Res 1995; 15: 2479-87.

[3]   Shamsuddin AM, Yang G-Y, Vucenik I. Novel anti-cancer functions of IP$_6$: Growth inhibition and differentiation of human mammary cancer cell lines *in vitro*. Anticancer Research 1996; 16: 3287-92.

[4]   El-Sherbiny YM, Cox MC, Ismail ZA, *et al.* G0/G1 arrest and S phase inhibition of human cancer cell lines by inositol hexaphosphate (IP$_6$). Anticancer Research 2001; 21: 2393-403.

[5]   Deliliers, G. L., Servida, F., Fracchiolla, N. S., Ricci, C., Borsotti, C., Colombo, G., and Soligo, D.: Effect of inositol hexaphosphate (IP$_6$) on human normal and leukaemic haematopoietic cells. Brit J Haematol 2002; 117: 577-87.

[6]   Lord CJ, Ashworth A. The DNA damage response and cancer therapy. Nature 2012; 481: 287-94.

[7]   Ferguson DO, Alt FW. DNA double strand break repair and chromosomal translocation: lessons from animal models. Oncogene 2001; 20: 5572-9.

[8]   Hanakahi LA, Bartlet-Jones M, Chappell C *et al.* Binding of inositol phosphate to DNA-PK and stimulation of double-strand break repair. Cell 2000; 102: 721-9.

[9]   Ma Y, Lieber MR. Binding of inositol hexakisphosphate (IP6) to Ku but not to DNA-PKcs. Journal of Biological Chemistry 2002; 277: 10756-9.

[10] Byrum J, Jordan S, Safrany ST *et al.* Visualization of inositol phosphate-dependent mobility of Ku: depletion of the DNA-PK cofactor InsP6 inhibits Ku mobility. Nucleic Acids Res 2004; 32: 2776-84.

[11] Cheung JC, Salerno B, Hanakahi LA. Evidence for an inositol hexakisphosphate-dependent role for Ku in mammalian nonhomologous end joining that is independent of its role in the DNA-dependent protein kinase. Nucleic Acids Res 2008; 36: 5713-26.

[12] Huang C, Ma WY, Hecht SS *et al.* Inositol hexaphosphate inhibits cell transformation and activator protein 1 activation by targeting phosphatidylinositol-3' kinase. [Erratum appears in Cancer Research 1997 Nov 15; 57(22): 5198]. Cancer Research 1997; 57: 2873-8.

[13] Ra Yoon M, Hyun Nam S, Young Kang M. Antioxidative and antimutagenic activities of 70% ethanolic extracts from four fungal mycelia-fermented specialty rices. J Clin Biochem Nutr. 2008; 43: 118-25.

[14] Shen X, Xiao H, Ranallo R *et al.* Modulation of ATP-dependent chromatin-remodeling complexes by inositol polyphosphates. Science 2003; 299: 112-4.

[15] Steger DJ, Haswell ES, Miller AL *et al*. Regulation of chromatin remodeling by inositol polyphosphates. Science 2003; 299: 114-6.

[16] Macbeth MR, Schubert HL, Vandemark AP *et al*. Inositol hexakisphosphate is bound to ADAR2 core and required for RNA editing. Science 2005; 309: 1534-9.

[17] Folkmann A, Noble KN, Cole CN *et al*. Dbp5, Gle1-IP6 and NUP159: a working model for mRNP export. Nucleus 2011; 2: 540-8. doi: 10.4161/nucl.2.6.17881. Epub 2011 Nov 1.

[18] Alćzar-Rom'n AR, Bolger TA, Wente SR. Control of mRNA export and translation termination by inositol hexakisphosphate requires specific interaction with Gle1. J Biol Chem 2010; 285: 16683-92. doi: 10.1074/jbc.M109.082370.

CHAPTER 20

Mechanisms of Biological Actions of Inositol and InsP_6 III: Epigenetics, Telomerase, Angiogenesis

Abstract: Epigenetic changes such as DNA methylation, histone modifications *etc.*, are also important in cancer formation. InsP_6 reverses the carcinogen-induced epigenetic changes in at least two different carcinogen-induced lung cancer models. InsP_6 has also been found to repress telomerase activity in prostate and brain cancers. For cancer to metastasize, the cells need access to blood vessels to disseminate. Formation of new blood vessels - angiogenesis facilitates the process. InsP_6 has been demonstrated to inhibit angiogenesis both *in vitro* and *in vivo*. This is mediated by suppression of pro-angiogenic factors such as vascular endothelial growth factor (VEGF).

Keywords: Angiopoietin, calreticulin, FGF, histone modification, hypermethylation, iNOS, matrix metalloproteinase, microRNA, microvessel density, MMP, PDGF, TERT, telomerase reverse transcriptase, TRAMP, vasostatin, VEGF.

INTRODUCTION

In the preceding chapter we have discussed the various pathways by which inositol and its phosphates, most importantly InsP_6 exert their function by affecting DNA repair. Also crucial are chromatin remodeling, RNA editing and mRNA export from the nucleus. However, epigenetic changes are no less important in carcinogenesis. We therefore discuss these in the context of how InsP_6 modulates them and affects the biological outcome, in this chapter.

In general, mortality from cancers is the result of dissemination of cancer cells to distant sites, called metastasis. Cancer cells must gain access to lymphatic and blood vessels for them to metastasize. Formation of new blood vessels - angiogenesis is associated with tumor growth and spread to distant organs. Thus, it has been an area of interest insofar as cancer control. We will examine the role of InsP_6 in angiogenesis.

EPIGENETIC CHANGES

The early steps in carcinogenesis involving gene expression without affecting DNA sequence modification are no less important. Aberrations in DNA methylation and histone modifications are correlated with tumorigenesis. DNA methyltransferases (DNMTs), methyl CpG binding proteins, methyl CpG DNA binding domain protein, and histone deacetylases (HDACs) are the major molecules involved in epigenetics. There is an increased activity of DNMTs during initiation and progression of lung cancer. DNMTs have been associated with tumor suppressor gene hypermethylation, cell proliferation and blocking of normal differentiation [1]. It appears that microRNAs (miR) may be both targets as well as effectors in aberrant DNA hypermethylation. One of the modifiers of DNMTs is miR-29b which is down-regulated in non-small cell lung cancer, wherein miR-29 family targets DNMT3a and 3b, causing down-regulation of these genes and re-expression of the DNA hypermethylated and silenced tumor suppressor genes [2].

Pandey & Gupta report that, in a ethylnitrosourea (ENU) induced mouse-lung tumorigenesis model, while the carcinogen ENU up-regulated the epigenetic events such as the expressions of DNMT1, MeCP2, MBD1, and HDAC1, these alterations were reduced by InsP_6 administration [3]. After 3 months of ENU exposure, there was hyperplasia and lymphocytic infiltration in the lungs with concomitant up-regulation of inflammation and DNA damage repair enzymes COX-2 and MLH1, and down-regulation of tumor suppressor gene p16. ENU exposure also up-regulated the epigenetic events such as the expressions of DNMT1, MeCP2, MBD1, and HDAC1, which were reduced by InsP_6 administration, indicting the regulation of gene expression by InsP_6 before the onset of ENU-induced lung tumors [3].

In a follow-up study, the investigators used a different carcinogen urethane, which also induces lung tumors in mice. They examined the tumor development, status of DNMTs, HDACs and MBDs, DNA methylation and expression of microRNA-29b during the period before and after the appearance of the tumors (1-36 weeks). Well-defined tumors appeared after 12 weeks and larger tumors appeared at 36 weeks, which were prevented by InsP_6. DNMT1, DNMT3a and DNMT3b were up-regulated following urethane exposure at the time of no tumor till the tumor developed and microRNA-29b was down-regulated. HDAC - the histone modifier also showed progressive up-regulation. Periodic increase in methyl binding protein MBD2 supported the expression of gene silencing pathways in terms of the down-regulation of tumor suppressor genes, p16 and MLH1. All of these alterations were protected by InsP_6 [4].

TELOMERASE

The ends of the DNA strands are identified as 3' or 5'. At the 3' end of all vertebrate DNA exists nucleotide repeat sequences TTAGGG, the region is called telomere. This segment of DNA is non-coding and essentially caps (protects the DNA from losing important coding sequences) the coding segments. With each cell division, there is shortening of the telomere. Telomerase elongates the shortened telomere without which important DNA message will be lost. Since the activation of telomerase is crucial for cells to gain immortality and proliferation ability, Jagadeesh & Banerjee [5] examined the role of InsP_6 in the regulation of telomerase activity in mouse and human prostate cancer cells. They demonstrated that InsP_6 repressed telomerase activity in prostate cancer cells in a dose-dependent manner. In addition, they showed that InsP_6 prevented the translocation of TERT (telomerase reverse transcriptase) to the nucleus, and inhibited phosphorylation of Akt and PKCα. Thus, InsP_6 represses telomerase activity in prostate cancer cells by post-translational modification of TERT *via* the deactivation of Akt and PKCα. Using human glioblastoma (brain cancer) T98G cell line Karmakar *et al.* [6] also demonstrated the repression of telomerase activity by InsP_6. An important point to keep in mind is that the telomerase activity may have conflicting roles - truncated telomeres can be seen as harbinger of both good and bad; further research is needed to confirm, validate and fully understand these.

ANGIOGENESIS

Angiogenesis is the formation of new blood vessels. It is an essential normal physiological process in growth and development of organs or tissues, and in wound healing. Granulation tissue is the initial response of the wound to start the repair or healing process. It consists of new capillaries, fibroblasts and inflammatory cells, starting at 48-72 hours after an injury and lasting for several days. Endothelial cells in the vicinity of the injury divide and form solid sprouts from preexisting blood vessels; intracytoplasmic vacuoles develop and coalesce producing lumen; the sprouts arborize and anastomose to establish the new capillary bed. This is essential for bringing in the nutrient, oxygen and other factors essential to the healing process. The sprouting capillaries have a tendency to protrude from the surface as tiny granules, hence the term "granulation" tissue; those who have had the misfortune to injuring these by rubbing or pulling the surgical dressing have the firsthand experience of how profusely vascular these are. Many of the new capillaries never develop a definitive blood flow; they are reabsorbed.

The formation of new blood vessels is dependent on an ever-increasing number of factors, some acting directly and others indirectly. They include endothelial growth factor, placental growth factor, angiopoietin (Ang1 and Ang2). Fibroblast growth factors (FGF), vascular endothelial growth factor (VEGF), platelet-derived growth factor (PDGF), interleukins (IL) 1, 2 and 8, *etc.*, also assist, albeit indirectly. FGF promotes endothelial cell proliferation and physical organization of endothelial cells into tube like structures - the future capillaries. VEGF increases the number of capillaries; Ang1 and Ang2 are important for formation of mature blood vessels; matrix metalloproteinase (MMP) help degrade the proteins to keep the blood-vessel wall solid.

Insofar as cancer, Folkman had hypothesized that tumor growth is dependent on angiogenesis [7]. For cancer cells to spread locally or to distant sites (metastasis), they need to invade the vasculature - lymphatic vessels and capillaries. Thus, formation of new vasculature facilitates and enhances the ability of the cancer cells to metastasize, at the least. Because angiogenesis depends on the interaction between endothelial and tumor cells studies in Shamsuddin laboratory investigated the effect of $InsP_6$ on both the cell types. $InsP_6$ inhibited the proliferation and induced the differentiation of endothelial cells *in vitro*; the growth of bovine aortic endothelial cells (BAECs) as evaluated by MTT proliferation assay was inhibited in a dose-dependent manner ($IC_{50} = 0.74$ mM; Fig. **20.1**). The combination of $InsP_6$ and vasostatin, a calreticulin fragment with anti-angiogenic activity, was synergistically superior in growth inhibition than either compound alone (Fig. **20.2**). $InsP_6$ inhibited human umbilical vein endothelial cell (HUVEC) tube formation (*in vitro* capillary differentiation) on a reconstituted extracellular matrix Matrigel, and disrupted preformed tubes (Fig. **20.3**). $InsP_6$ significantly reduced basic fibroblast growth factor (bFGF)-induced vessel formation ($p < 0.01$) *in vivo* in Matrigel plug assay (Fig. **20.4**) [8].

For experiments on the other component - the cancer cells, human hepatoma cell line HepG2 was used. Exposure of HepG2 cells to $InsP_6$ for 8 hours resulted in a dose-dependent decrease in the mRNA levels of VEGF. $InsP_6$ treatment of HepG2 cells for 24 hours also significantly reduced the VEGF protein levels in conditioned medium, in a concentration-dependent manner ($p = 0.012$; Fig. **20.5**). These data show that $InsP_6$'s anti-angiogenesis action is directed towards both the endothelial cells and the cancer cells [8].

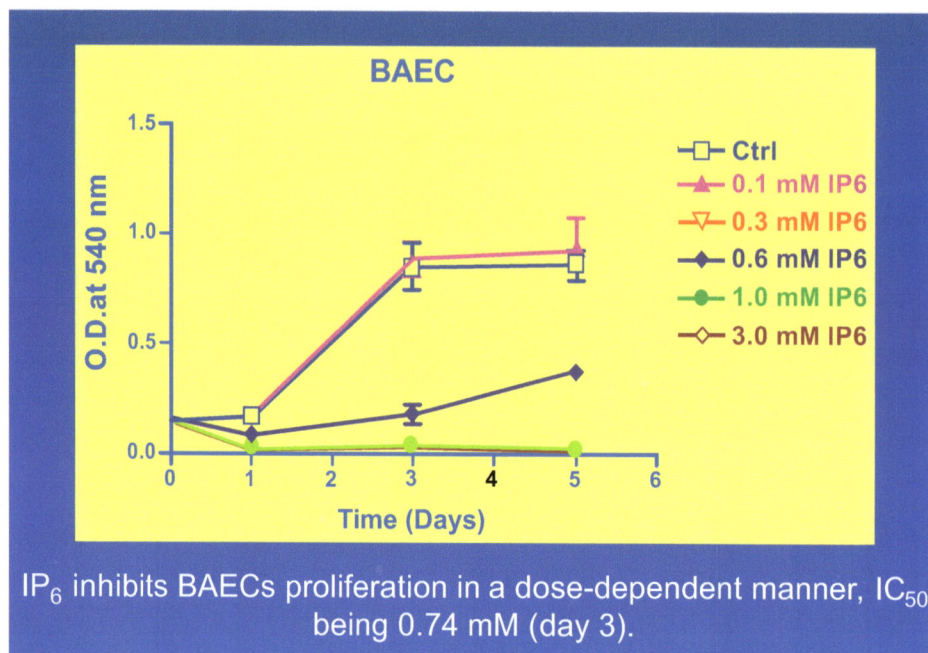

Fig. (20.1). Inhibition of bovine aortic endothelial cells by $InsP_6$ in a dose dependent manner.

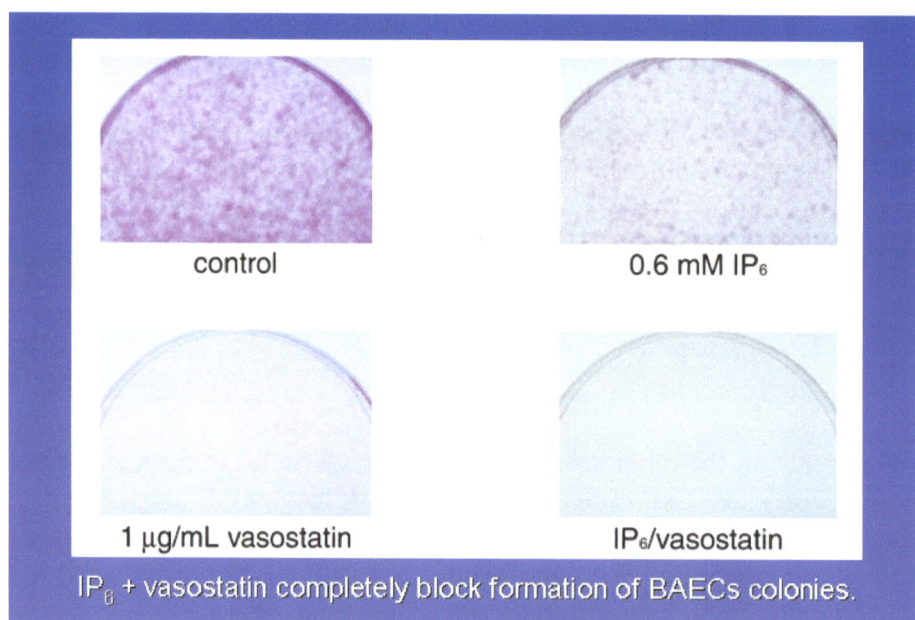

Fig. (20.2). Synergism of $InsP_6$ and vasostatin in inhibition of colony formation of endothelial cells.

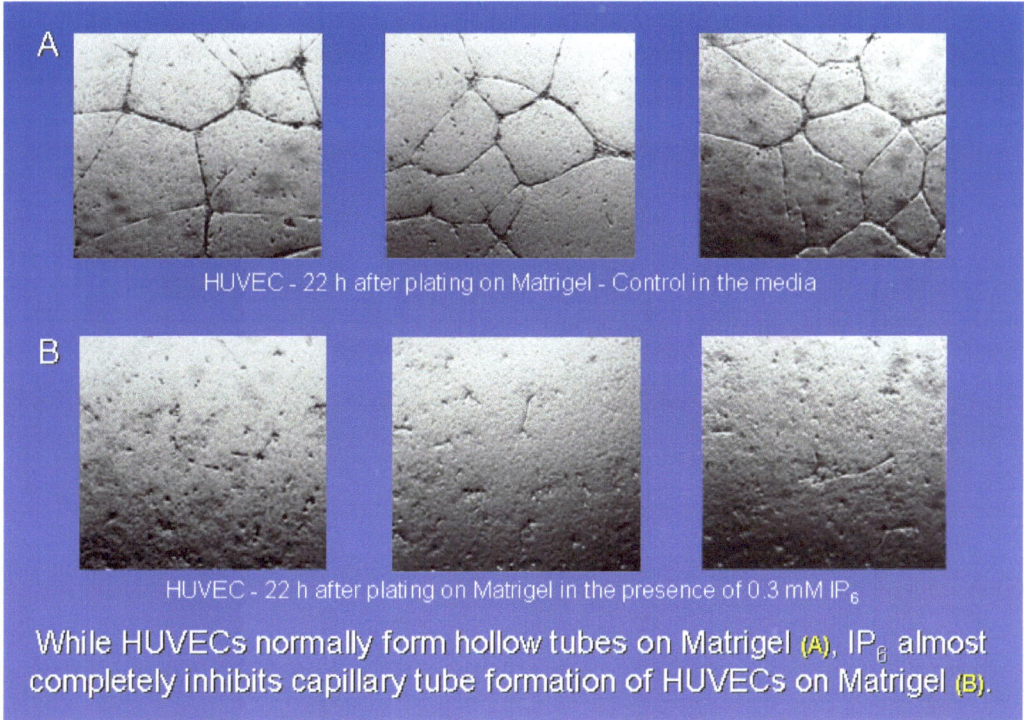

Fig. (20.3). InsP_6 inhibits endothelial capillary tube formation in *in vitro* Matrigel assay.

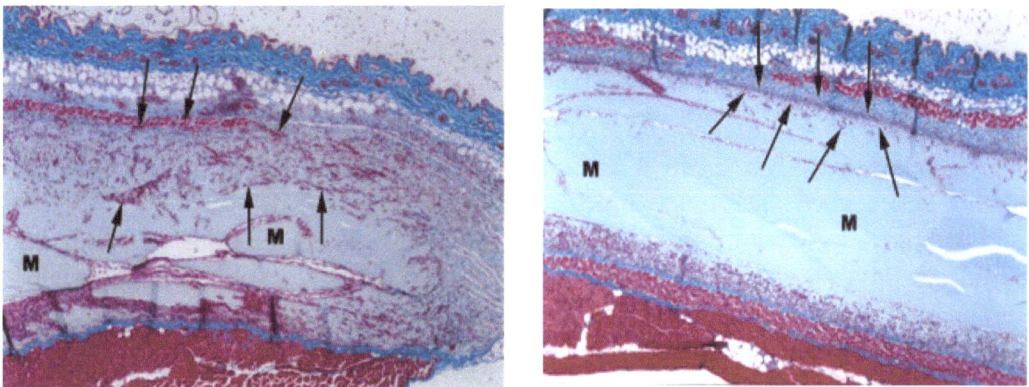

Fig. (20.4). Left panel: Mouse skin (control) showing wide area of neovascularization between arrows; Note that InsP_6 treatment resulted in near complete inhibition of angiogenesis (right panel, between arrows).

Agarwal's laboratory has also investigated the anti-angiogenesis action of InsP_6 on prostate cancer. DU145 prostate cancer cells were injected into nude mice, and animals were fed normal drinking water, or 1 or 2% InsP_6 in drinking water for 12 weeks, and the tumors were analyzed for proliferating cell nuclear antigen PCNA), terminal deoxy-nucleotidyl transferase-mediated nick end labeling, and CD31.

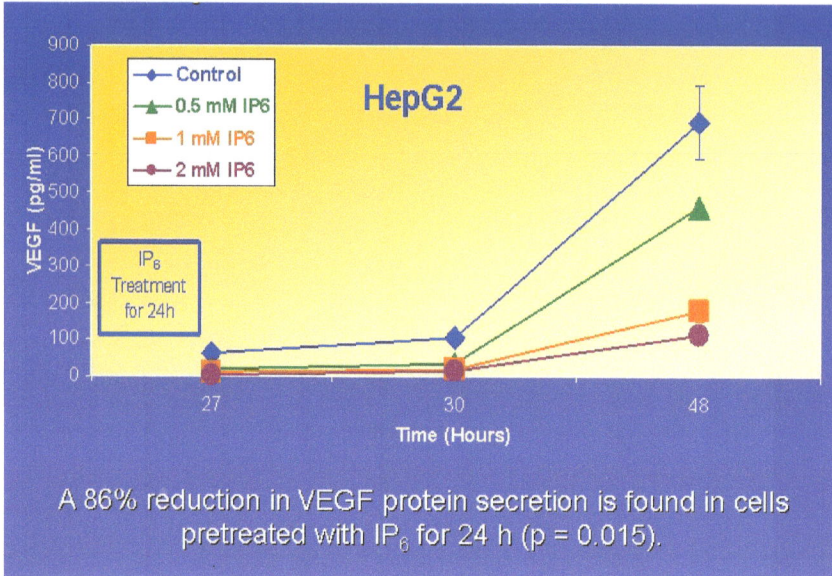

Fig. (20.5). InsP_6 inhibits VEGF protein from human hepatocellular carcinoma HepG2 cells in a dose-dependent manner.

Tumor-secreted insulin-like growth factor binding protein (IGF-BP)-3 and VEGF were quantified in plasma. InsP_6 feeding of the animals resulted in suppression of hormone-refractory human prostate tumor growth. The investigators also reported that there was no adverse effect on body weight gain, diet, and water consumption during entire study. There was a dose-response inhibition of tumor growth: at the end of study, tumor growth inhibition by 1 and 2% InsP_6 feeding was 47% and 66% ($p = 0.049$-0.012) respectively, in terms of tumor volume/mouse. As regards tumor weight/mouse, the reduction by 1 and 2% InsP_6 feeding was 40 and 66% ($p = 0.08$-0.003) respectively. Tumor xenografts from InsP_6-fed mice showed significantly ($p < 0.001$) decreased proliferating cell nuclear antigen-positive cells, but increased number of apoptotic cells. Tumor-secreted IGFBP-3 levels were also increased up to 1.7-fold in InsP_6-fed groups. Furthermore, InsP_6 strongly decreased tumor micro-vessel density and inhibited tumor-secreted VEGF levels [9]. In a subsequent study with transgenic adenocarcinoma of the mouse prostate (TRAMP) model, the investigators confirmed that indeed InsP_6 inhibited angiogenesis in that model as well. In the TRAMP model, with increasing tumor grade there was higher expression of pro-angiogenic factors resulting in increased microvessel density (MVD); MVD further promoted the progression of the tumor to invasive stages.

InsP_6 feeding significantly decreased tumor perfusion/permeability and MVD. InsP_6 treatment also led to a dose-dependent decrease in tumor perfusion. These results suggest that InsP_6 exerts its anti-angiogenic effect by affecting the expression of pro-angiogenic factor; support for this was found in a significant decrease in the expression of pro-angiogenic factor VEGF in InsP_6-fed mice as compared to controls ($P < 0.001$). Studies of the expression of inducible nitric oxide synthase (iNOS), an enzyme involved in the production of nitric oxide (NO) which facilitates neo-vascularization and invasion, show that InsP_6 significantly decreased (31-34%, $P < 0.001$) iNOS immunoreactivity scores. InsP_6 also inhibited NF-κB activity as evidenced by a significant decrease in the nuclear expression of phospho NF-κB/p65 and phospho AKT $^{ser\,473}$ levels in the TRAMP prostate. Their data show that InsP_6 feeding inhibited the recruitment of new vascular network during angiogenesis, by down regulating the expression of pro-angiogenic factors [10].

Fig. (**20.6**) summarizes the various mechanistic pathways of the actions of inositol and its phosphates.

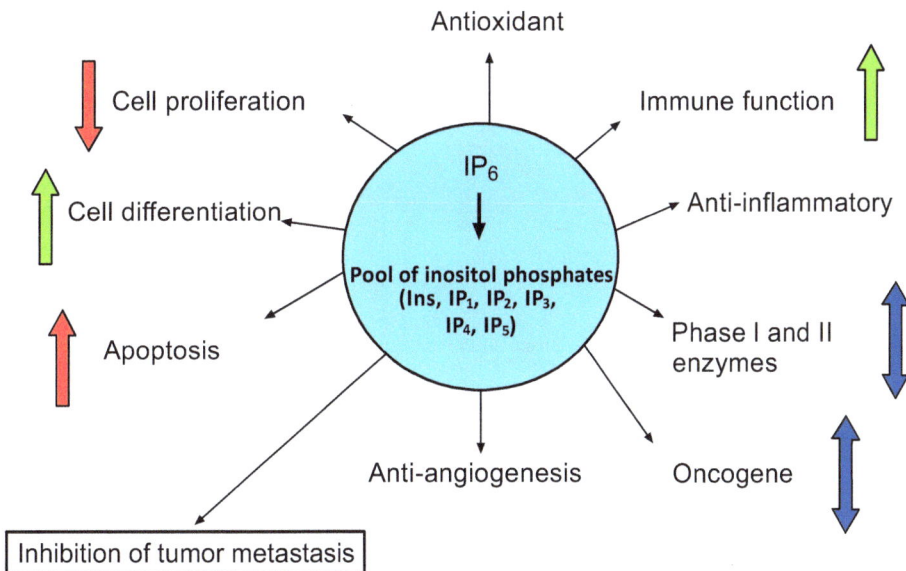

Fig. (20.6). Schematic representation of the various chemical, biochemical, cell biological, immunological actions of inositol (Ins) and it's phosphates (IP).

CONCLUDING REMARKS

Aside from normalizing cell proliferation and inducing differentiation, and repair DNA damages, inositol phosphates also affect the initiation and progression of cancer cells by affecting telomerase, epigenetic events and tumor angiogenesis.

REFERENCES

[1]   Tang M, Xu W, Wang Q *et al*. Potential of DNMT and its epigenetic regulation for lung cancer therapy Curr. Genomics 2009; 10: 336-52.

[2]   Fabbri M, Garzon R, Cimmino A *et al*. MicroRNA-29 family reverts aberrant methylation in lung cancer by targeting DNA methyltransferases 3A and 3B Proc Natl Acad Sci USA 2007; 104: 15805-10.

[3]   Pandey M, Gupta KP. Epigenetics, an early event in the modification of gene expression by inositol hexaphosphate in ethynitrosurea exposed mouse lungs. Nutr Cancer 2011; 63: 89-99. doi: 10.1080/01635581.2010.516868.

[4]   Pandey M, Sultana S, Gupta KP. Involvement of epigenetics and microRNA-29b in the urethane induced inception and establishment of mouse lung tumors. Exp Mol Pathol. 2014; 96: 61-70. doi: 10.1016/j.yexmp.2013.12.001. Epub 2013 Dec 19.

[5]   Jagadeesh S, Banerjee PP. Inositol hexaphosphate represses telomerase activity and translocates TERT from the nucleus in mouse and human prostate cancer cells *via* the deactivation of Akt and PKC alpha. Biochem Biophy Res Com 2006; 349: 1361-7.

[6]   Karmakar S, Banik NL, Ray SK. Molecular mechanism of inositol hexaphosphate-mediated apoptosis in human malignant glioblastoma T98G cells. Neurochemical Research 2007; 32: 2094-102.

[7]   Folkman J. Tumor angiogenesis: therapeutic implications. N Engl J Med 1971; 285: 1182-6.

[8]   Vucenik I, Passaniti A, Vitolo MI *et al*. Anti-angiogenic activity of inositol hexaphosphate (IP$_6$). Carcinogenesis 2004; 25: 2115-23.

[9]   Singh RP, Sharma G, Mallikarjuna GU *et al*. *In vivo* suppression of hormone-refractory prostate cancer growth by inositol hexaphosphate: induction of insulin-like growth factor binding protein-3 and inhibition of vascular endothelial growth factor. Clin Cancer Res 2004; 10: 244-50.

[10]   Raina K, Ravichandran K, Rajamanickam S *et al*. Inositol hexaphosphate inhibits tumor growth, vascularity and metabolism in TRAMP mice: a multiparametric magnetic resonance study. Can Prev Res 2013; 6: 40-50. doi: 10.1158/1940-6207.CAPR-12-0387. Epub 2012 Dec 4.

CHAPTER 21

Industrial Applications, Technology & Environmental Sciences

Abstract: There are emerging new uses of $InsP_6$ in both the existing as well as in developing technologies, and in protective roles in our environment; these are in addition to the vast and wide applications in healthcare. Discussed here is its use in preservation of food and wine, improvement of the taste of wine, application in dentistry, lithium ion battery, diagnostic imaging, mass spectrometry, fuel cell and nanotechnology, *etc*.

Keywords: AFB1, aflatoxin, chitosan, dental cement, dental etchant, diagnostic imaging, drug delivery, food safety, fuel cell, Li-ion battery, mass spectroscopy, nanotechnology, plasmon, Raman spectroscopy, SERS, technetium, toxicology.

INTRODUCTION

Aside from the health benefits of inositol and $InsP_6$, there are numerous other divergent uses of $InsP_6$, most of which are owing to the chelating property and antioxidant action. These range from preserving food and wine, through protecting tires, use in cosmetic industry, batteries, spectroscopy, and even in fuel cell and nanotechnology.

FOOD & WINE INDUSTRY

Owing to the ability to chelate cations, $InsP_6$ has been in use in food and wine industry since the mid twentieth century [1] and The International Organization of Vine and Wine (OIV) recommends use of calcium $InsP_6$ for removal of iron from wine.

Wine Preservation and Taste

While some amount of oxidation ("let the wine breathe") brings out the aroma and the taste of red wine, oxidation often results in browning reactions, loss of characteristic aromatic compounds, and the production of carbonyls associated with undesirable aromas. Non-enzymatic wine oxidation is thought to be catalyzed by trace quantities of transition metals, specifically, iron and copper [2].

A.K.M. Shamsuddin and Guang-Yu Yang

In the presence of Fe^{2+}, oxygen is reduced by a sequential one-electron reduction to yield a superoxide anion radical. The free radical is then quickly converted to a hydroperoxyl radical (•OOH) under acidic wine conditions. It is then converted to H_2O_2 which reacts quickly with either bisulfite or reduced transition metals *e.g.*, Fe^{2+} or Cu^+. This metal-catalyzed reduction of H_2O_2 through Fenton reaction - as discussed in Chapter 16, yields highly oxidizing hydroxyl radicals •OH which is capable of reacting with organic components in wine. Since ethanol is the major organic component in wine, it is therefore the principal target for these radicals yielding 2-hydroxyethyl radicals and 1-hydroxyethyl radicals, the latter can be further oxidized to acetaldehyde. Because the transition metals are responsible for catalyzing various reactions in wine, often leading to many undesirable effects, the most effective way to prevent metal-catalyzed oxidation processes would be to remove all trace iron and copper from the juice, must, or wine. Thus investigators have been attempting to remove iron using various chelators including InsP_6 with success [2].

"Fishy aftertaste" is sometimes perceived in wine consumed with seafood and is undesirable. Iron in wine (2.8-16 mg L^{-1}) mostly from soil, dust and processing equipment, has been reported to be a key compound that produces fishy aftertaste. While excessive iron concentrations can be reduced by use of specific fining agents or cooling to induce precipitation, the very low levels of iron in wine that cause fishy aftertaste cannot be removed by conventional methods such as ion exchange, chemical precipitation or adsorption using activated carbon, or chelating resin, without affecting the other compounds and the process is prohibitively expensive. Likewise, while chemical precipitation is a relatively inexpensive process, its utility is limited by safety, rather lack of it. Thus, Tsuji *et al.,* [3] developed a safe and cost-effective method of removing iron from red and white wines consisting of alcohol-treated yeast cells; InsP_6 (inositol hexaphosphoric acid a.k.a. phytic acid) enhanced the ability to remove the iron in a synergistic manner [3].

Food Safety

Heterocyclic Aromatic Amines

Heterocyclic aromatic amines are potent mutagens and carcinogens generated during the heat processing of meat. These are produced during frying, and their levels increase with increasing frying time and temperature. Zhang *et al.,* [4] demonstrated that pork patties had the highest concentration of heterocyclic aromatic amines compared with pork meatballs and pork strips. The addition of

InsP_6 (or other antioxidants) to pork before frying had an inhibitory effect on heterocyclic aromatic amines generation [4].

Non-enzymatic chemical reaction between an amino acid and reducing sugar in the presence of heat (Maillard reaction) results in browning of meat when it is roasted or seared. The carbonyl group of the sugar reacts with the nucleophilic amino group of the amino acid forming a complex mixture of molecules that are responsible for the color and flavor; and the type of amino acid determines the resultant flavor; acrylamide, a widely used chemical that is also a known lethal neurotoxin and animal carcinogen, is formed at high temperatures. Wang *et al.,* [5] tested the effects of InsP_6 on the Maillard reaction and the formation of acrylamide. InsP_6 enhanced browning in glucose/β-alanine system. Browning was suppressed by the addition of calcium and magnesium ions, but an additive effect was observed for ferrous ions and InsP_6 in glucose/β-alanine solution at pH 8.0. The kinetics of Maillard reaction was first-ordered reaction in the presence of InsP_6. When potato slices (as in potato chips or crisps) were treated with sodium InsP_6 and calcium chloride successively, the formation of acrylamide was greatly suppressed [5].

Mycotoxins

Mycotoxins are toxic secondary metabolites produced by fungi that may contaminate our food, both of animals and humans at all stages of the food chain. Deoxynivalenol is a type of β-trichothecene mainly produced by the fungi *Fusarium graminearum* and *Fusarium culmorum,* which are found naturally, worldwide. Deoxynivalenol disrupts the functions of cellular membranes and alters the intercellular communication. Pacheco *et al.,* [6] used deoxynivalenol on a porcine intestinal epithelial cell line IPEC-1 to determine the effects of InsP_6 on intestinal epithelial integrity.

Pretreatments of cell monolayer for 24 h with 0.5 mM or 1.0 mM of InsP_6 were able to partially restore ($p < 0.001$) the decreased transepithelial electrical resistance values induced by deoxynivalenol indicating that InsP_6 may prevent its toxic effect [6]. InsP_6 at a concentration of 0.5 mM or 1.0 mM protected the membranes of the IPEC-1 intestinal epithelial cell line against cell damage induced by the mycotoxin. Thus addition of InsP_6 in food or feed may protect cellular systems against mycotoxins thereby reducing the losses in animal production and improving animal and human health [6].

Fungi such as *Aspergilli, Fusaria* and *Peniállia* are common contaminants of cereal grains including corn; and mycotoxins in animal feed pose most danger to the animal as well as human health and economy. Of the mycotoxins, aflatoxin B1 (AFB1) produced by *Aspergillus flavus* and *Aspergillus parasiticus* pose the most health hazard. Being a food contaminant, it causes hepatotoxicity and cancer. Fundamental to the mechanism of AFB1-induced carcinogenesis is formation adducts with the DNA and protein; free radical generation is also considered vital to its toxicity [7]. Thus, Abu Elsaad & Mahmmod [8] investigated whether InsP_6 would modify AFB1-induced toxicity. In a male Albino rat model, they show a decrease in sex hormone levels, an increase in testicular lipid peroxidation product levels and a significant decrease in testicular glutathione content, catalase and total peroxidase and superoxide dismutase activities in control AFB1-exposed rats. There was associated degeneration and high rate of mitotic division within the spermatogenic nuclei, as well as karyomegaly and pyknosis of the nuclei. InsP_6 caused a reduction in the toxicity of free radicals, marked improvement of AFB1-induced testosterone level, testicular malondialdehyde along with noted improvement histopathologically [8].

NANOTECHNOLOGY

Drug Delivery

Studies of the use of InsP_6 in nanotechnology were started by investigating the potential for drug delivery. Assuming that InsP_6 and perhaps other Ins$P$s do not travel across the plasma membrane, initially investigators tested them in liposomes perhaps with the aim of delivering Ins$P$s inside the cell [9]. Thin layer chromatography of liposomes show that those containing InsP_6 migrate completely differently than the ones containing other inositol phosphates (Ins$P$s). Another unique feature of InsP_6-containing liposomes is that unlike the other Ins$P$s, liposome-entrapped InsP_6 elicits dose-dependent contractions of the isolated rat aorta suggesting that liposomes loaded with InsP_6 undergo physicochemical alterations that eventually change their drug- delivery capacity.

Oral administration of drugs is inherently more acceptable by the recipients than other routes such as injections - intramuscular, intravenous, *etc.* However, there are drugs which are not suitable for oral intake especially protein and enzymes, as they are destroyed by the digestive processes in the stomach. Despite development of several different methods such as microencapsulation, microemulsion, liposomes *etc.*, to circumvent the problem, there still is a need to deliver crucial life-saving drugs taken on a daily basis such as insulin, to ensure patient compliance.

Owing to the non-toxic nature of chitosan, a linear polysaccharide derived from the shells of shrimp and other crustaceans, it has been used as a drug carrier in pharmaceutical industry, besides other industries. Chitosan interacts with insulin and enhances intestinal permeation of the latter. In the process of preparation of chitosan capsules, tripolyphosphate (TPP) is used as a cross-linking agent. Because of the number of anions in InsP_6 capable of reacting with the cations of chitosan is two-fold higher than TPP, Lee *et al.,* [10] have developed InsP_6-chitosan capsules for oral delivery of insulin to preserve its pharmacological activity and to enhance its bioavailability. Thus developed capsules were tested *in vitro* for controlled release of insulin in gastrointestinal fluids. They also performed *in vivo* study to evaluate the bioactivity of encapsulated insulin after oral administration to diabetic mice. InsP_6-chitosan capsules significantly decreased blood glucose levels while TPP-chitosan capsules was less effective. The relative pharmacological bioactivity of InsP_6-chitosan capsules prepared was 6.4% while that of TPP-chitosan capsules was 1.1% [10]. Thus, InsP_6-chitosan capsules have good potential for use in oral delivery of insulin for sustained control of the blood glucose level thereby alleviating the need for injections that are uncomfortable if not painful, and hence reason for non-compliance.

Higdon *et al.,* [11] developed a "ceramic delivery system" for releasing sustained levels of InsP_6 + inositol in a tissue culture setting and assessed the proliferation rate and viability of HTB 122 intraductal breast cancer cells exposed to sustained levels of InsP_6 + inositol, and compared to conventional means to deliver drugs dissolved in media. They then evaluated the morphological changes associated with these treatment processes. Compared to the treatment by conventional means and to the sham group (empty capsule), cells treated with sustained delivery resulted in cellular atrophy, as well as fragmentation. The investigators claim that the ceramic delivery systems in tissue culture "gives breakthrough information for basic research on limiting and eliminating contamination and the logistical problems associated with intermittent dosing in tissue culture" [11].

Surface-Enhanced Raman Spectroscopy

Surface-enhanced Raman spectroscopy or scattering (SERS) is a surface-sensitive technique that enhances Raman scattering by molecules adsorbed onto rough metallic surface or, by nanostructures such as plasmonic-magnetic (plasmon is a quantum of plasma oscillation just as photons and phonons which are quantization of electromagnetic and mechanical vibrations, respectively) silica nanotubes which may detect single molecules. Using a combination of Au and Ag nanoparticles, and Raman active dyes, SERS can be used to target specific DNA or RNA sequences.

By direct grafting of Au nanoparticles onto the surface of magnetic network nanostructure with the help of InsP_6 Yang *et al.,* [12] developed a novel magnetically responsive and surface-enhanced Raman spectroscopy active nanocomposite. InsP_6 acted as a stabilizer and a bridging agent to weave Fe_3O_4 nanoparticles into magnetic network nanostructure, which is easily dotted with Au nanoparticles. Au-magnetic network nanostructure presenting the large surface and high detection sensitivity enabled it to exhibit multifunctional applications involving sufficient adsorption of dissolved chemical species for enrichment, separation, as well as a Raman amplifier for the analysis of trace pesticide residues at femtomolar level by a portable Raman spectrometer. This will have potential usefulness in effective on-site assessments of agricultural and environmental safety [12].

The presence of InsP_6 micelles allowed N Wang *et al.,* [13] to obtain stable small Ag seeds (size diameter < 10 nm). Ag-Au bimetallic nanoparticles were then synthesized through a replacement reaction with the rapid inter-diffusion process between such small Ag seeds in nanoclusters and $HAuCl_4$. Adjusting the dosage of $HAuCl_4$ resulted in different products, which possessed unique surface plasmon resonances. The Ag-Au alloy nanoparticles with the cauliflower-like structure had a suitable surface plasmon resonance for highly sensitive Raman detection application as a SERS substrate with a long-term stability of six months [13].

InsP_6 has also been found to act as stabilizer of electrochemical formation of cobalt nanoparticles for the fabrication of graphene-cobalt nanocomposite [14].

ENVIRONMENTAL TOXICOLOGY

Contaminated Soils and Groundwater

One of the methods of treatment of contaminated soils and groundwater is *in situ* chemical oxidation which includes permanganate, persulfate, ozone and catalyzed H_2O_2 propagations. Catalyzed H_2O_2 propagations are considered to be most efficient for the destruction of environmental contaminants, and is based on the standard Fenton reaction. However, while catalyzed H_2O_2 propagations provides robust treatment chemistry capable of effectively destroying nearly all contaminants, the rapid decomposition of hydrogen peroxide is a limiting factor often vastly reducing its effectiveness. Thus there is a need for stabilizing H_2O_2 in the subsurface for *in situ* chemical oxidation.

In their quest for improving the *in situ* chemical oxidation through stabilizing H_2O_2, Schmidt *et al.,* [15] investigated the use of Na-citrate or Na-InsP_6 for effectiveness in one-dimensional columns of iron oxide-coated and manganese oxide-coated sand. Hydrogen peroxide (5%) with and without 25 mM citrate or Na-InsP_6 was applied to the columns. While citrate was not an effective stabilizer for hydrogen peroxide in iron-coated sand, InsP_6 was highly effective, increasing H_2O_2 residuals by two orders of magnitude over unstabilized H_2O_2 [15]. Both citrate and InsP_6 were effective stabilizers for manganese-coated sand, increasing H_2O_2 residuals by four-fold over unstabilized H_2O_2. InsP_6 and citrate did not degrade and were not retarded in the sand columns; and the addition of InsP_6 increased column flow rates relative to unstabilized columns. Thus, InsP_6 and citrate are effective stabilizers of H_2O_2 under the dynamic conditions, and that citrate and InsP_6 can be added to H_2O_2 before injection to the subsurface as an effective means for increasing the radius of influence of catalyzed H_2O_2 propagations in *in situ* chemical oxidation, conclude the authors [15]. It would be interesting to see how the new InsP_6-citrate molecule performs in this regard.

Cadmium (Cd)

Cadmium is one of the most dangerous environmental pollutants; its toxicity in humans is expressed through damages and diseases of various organs systems including the brain where it may impair brain development and increase brain deiodinase activity with resultant decrease in the serum level of the thyroid hormone T4. Mohammed *et al.,* [16] investigated the protective effect of Na-InsP_6 added to the diet in a rat model intoxicated with Cd. Compared to control rats, serum calcium, iron, and total iron-binding capacity, and serum T_3 and T_4 in Cd-treated rats were decreased; in contrast, Cd-intoxicated animals treated with InsP_6 showed a significant improvement when compared with the Cd-treated rats only. Serum thyroid stimulating hormone level was significantly increased in Cd-treated rats compared with the control group; again, the addition of InsP_6 in diet decreased the high levels of thyroid stimulating hormone supporting a prophylactic effect of InsP_6 against Cd-induced toxicity [16].

MATERIAL SCIENCE

Bioactive Glass

Bioactive glass is known to develop an interfacial bond between the implant and surrounding tissues and the first material was prepared by melt-quenching method at temperatures between 1,300 °C and 1,450 °C. The bioactivity of thus developed glass is less than optimal. Since phosphate materials offer good biocompatibility, Li & Qiu [17] investigated the application of $InsP_6$ as a precursor to synthesize CaO-P_2O_5-SiO_2 glasses by sol-gel method. They found that a wide range of compositions of gel-glasses could be prepared with $InsP_6$ as compared to other phosphorus precursors, or the melt-quenching method. Additionally, $InsP_6$ assisted calcium incorporation into glass networks. *In vitro* tests in simulated body fluid showed that they were bioactive over a much broader compositional range especially at high phosphate content, thus enabling one to design bioactive materials with various degradation rates by adjusting the phosphate content [17].

Immobilized $InsP_6$

In 1976 Scheiner and Breitenbach described a method for immobilizing $InsP_6$ on cross-linked agarose gels (Sepharose-4B) *via* (6 + 3) carbon spacer [18, 19]. The resulting gel is appropriate for utilization in affinity chromatography of enzymes of $InsP_6$ metabolism such as wheat bran phytase, affinity chromatography of hemoglobin, as well as potentials for use on surgical wound dressings as an antioxidant and bacteriostatic agent [19].

Dental Application

Cementing Agent

Cementing agents are materials that offer cohesion and adhesion, and include various substances aside from the commonly known Portland cement in brick-laying. They comprise of a powder and a liquid, capable of incorporating fillers and, harden to form solid mass. As most of us may know, cementing is widely used in dentistry for management of various dental ailments, not least of which is dental cavities; they are either resin cements or acid-base cements. Prosser *et al.,* [20] described the formation of $InsP_6$ dental cement by combining with zinc oxide or calcium fluoroaluminosilicate glasses, and evaluated its properties. Thus developed $InsP_6$ cements are fast-setting, more resistant to erosion by water and acids. However, the cements adhere to enamel but not to the dentin of teeth [20].

Etchant

Dental etchants which are acidic materials prepare enamel for the attachment of a bonded restoration. Etchants remove the outermost layer of the tooth surface - enamel, and expose a rough, porous layer providing greater surface area for the bonding process. Extreme care must however be taken to correctly use dental etchants so that the acid is not left on the enamel surface for too long. Historically, application of 30-40% phosphoric acid for about 15 seconds has been the 'gold standard' etchant. Although more effective enamel bonding is achieved through etching with phosphoric acid, it is now considered too harsh, as it results in the exposure of collagen fibrils that are devoid of hydroxyapatite. These fragile collagen fibrils are susceptible to collapse, resulting in incomplete infiltration of resin. This poor impregnation of collagen could give unsatisfactory results jeopardize the long-term durability of the bond [21].

Owing to the Ca^{2+} chelating property of $InsP_6$ Nassar *et al.,* [21] used $InsP_6$ (inositol hexaphosphoric *acid*) as an etchant and compared the results with the 'gold standard' phosphoric acid. They report that 1% $InsP_6$ solution produced resin-dentin bond-strength values that were significantly higher than that of the phosphoric acid-etched dentin. $InsP_6$ effectively removed the smear layer and plugs, thus exposing the collagen network. Furthermore, $InsP_6$ at 0.01% concentration had a minimal effect on pulpal cells, whereas phosphoric acid resulted in a marked decrease in their viability [21]. Since $InsP_6$ also acts as a cross-linker [10], it mechanically strengthens the collagen and prevents its collapse [21].

LI-ION BATTERY

Li-ion batteries are one of the most popular types of rechargeable battery for portable electronics. The energy density of lithium-ion is typically twice that of the standard nickel-cadmium with potential for even higher energy densities. Lithium-ion is a low maintenance battery; there is no memory and no scheduled cycling is required to prolong the battery's life; and the self-discharge is less than half compared to nickel-cadmium.

It is however fragile and requires a protection circuit to maintain safe operation. Aging is another major concern, some capacity deterioration is noticeable after one year irrespective of whether the battery is in use or not. Thus, there is ongoing research to improve, some of it being focused on overcoming the poor cycling stability issue associated with its large volume changes during charging and discharging processes, mostly through nanostructured material design. Silicon has a high-specific capacity as an anode material for Li-ion batteries. Wu *et al.,* [22] incorporated a conducting polymer hydrogel into Si-based anodes wherein the hydrogel is polymerized *in-situ*, resulting in a well-connected three-dimensional network structure consisting of Si nanoparticles coated by the conducting polymer. This offers multiple advantages such as a continuous electrically conductive polyaniline network, binding with the Si surface through either the cross-linker hydrogen bonding with InsP_6 or electrostatic interaction with the positively charged polymer, and porous space for volume expansion of Si particles. With this anode, the investigators demonstrate a cycle life of 5,000 cycles with over 90% capacity retention at current density of 6.0 A g^{-1} [22].

FUEL CELL

Fuel cells convert the chemical energy from a fuel to electrical energy through reaction with oxygen or other oxidizing agents. Hydrogen is the most common fuel. Irrespective of the types of fuel cells and there are many, they all contain an anode, a cathode and an electrolyte; the type of fuel cell is determined by the type of electrolyte. In the standard proton exchange membrane fuel cells (PEMFC), a proton conducting polymer membrane (electrolyte) separates the anode from the cathode. Nafion membrane is the most widely used; however its proton conductivity drops off sharply in high temperature as a result of dehydration. PEMFCs suffer from catalyst poisoning, low catalyst activity, *etc.* Introduction of phosphate groups to the membrane may alleviate some of these challenges, and owing to the multitude of chemical properties, InsP_6 was the chosen by Li *et al.,* [23]. Inositol hexaphosphoric *acid* was first immobilized by MIL101 which was then utilized as a novel filler to incorporate into Nafion to fabricate hybrid proton exchange membrane for application in PEMFC. High loading and uniform dispersion of InsP_6 in MIL 101(Cr) were achieved. The Nafion/InsP_6@MIL hybrid membranes showed high proton conductivity at different relative humidities. Furthermore, the mechanical property of Nafion/InsP_6@MIL hybrid membranes was substantially enhanced and the thermal stability of membranes was well preserved [23].

DIAGNOSTIC IMAGING

^{99m}Tc Scintigraphy

Researchers have been constantly striving for better and more accurate methods of disease detection and evaluate the extent of the disease for better management.

Cancers are obvious ones; and they metastasize to the regional lymph nodes, liver and eventually to the bone marrow at late stages. Advances in diagnostic ultrasound, computerized tomography (CT), nuclear magnetic imaging, PET scan, *etc.,* have been very helpful towards evaluating the diseases, radionuclide scan still remains a reliable method for investigating a variety of liver diseases such as cirrhosis, infections, trauma, and of course, neoplasms [24]. Then there are diseases of the lymph nodes. Lymph nodes and liver are organs of the reticuloendothelial system that also includes the spleen. The ability of reticuloendothelial cells to phagocytose foreign particles has been exploited to develop imaging techniques that use radioactive colloids.

Using stannous chloride as a complexing/reducing agent, Subramanian *et al.,* formed a radiocolloid by binding ^{99m}technitium (^{99m}Tc) with Na-InsP_6 (commonly referred to in the nuclear medicine literature as ^{99m}Tc-Phytate) [25]. This was a landmark development as compared to other available imaging agents of the time (early 1970's), it can be easily formulated into an "instant labeling" kit with long shelf-life and high labeling efficiency [24]. Furthermore, not only it is free from stabilizing proteins and heavy metals common in others, but upon injection it also reacts with calcium ions in the plasma to form an insoluble radiocolloid which can then be imaged with a gamma camera [24]. The biodistribution of ^{99m}Tc-calcium InsP_6 can be modulated by changing the molar ratio of calcium: InsP_6; excellent splenic uptake is achieved with increased ratio [24].

Since then, ^{99m}Tc-InsP_6 has been used for liver/spleen scan, bone marrow and lymph node imaging [24]. There are other indirect uses of this technology. For instance it may be of help in evaluating the extent of portal hypertension. Portal hypertension is the primary cause of massive hemorrhage of the upper gastrointestinal tract as a sequel of cirrhosis of the liver; and it may be lethal. Thus, measuring the pressure in the portal vein is crucial for confirming therapy and predicting prognosis. ^{99m}Tc-phytate when injected into the splenic parenchyma passes through the splenic and portal vein into the liver. Because of the elevated portal vein pressure porto-systemic shunting takes place owing to increased pressure in the portal venous system, a portion of the colloid particles cannot enter the liver and instead enter the heart *via* compensatory circulation, which leads to simultaneous liver imaging and heart imaging. Thus the radioactivity ratio between heart and liver reflects the extent of the porto-systemic shunt and can be used to evaluate the extent of the elevation in portal vein pressure [26].

Since *in vivo* optical imaging is becoming more useful and hence attractive, Kondakov *et al.,* have investigated the application of optical imaging using ^{99m}Tc-InsP_6 [27]. They studied *in vitro* radioluminescence of ^{99m}Tc-InsP_6 with IVIS Spectrum CT™ optical imaging system. The distribution of ^{99m}Tc-InsP_6 colloid was also studied *in vivo* with and without scintillating materials. The investigators report that the radioluminescence of fluids induced by ^{99m}Tc-based tracers could be detected using charge-coupled device optical imaging systems [27].

OTHER INDUSTRIAL APPLICATIONS

Mass Spectrometry

Mass spectrometry is a well-established technique for the detection of trace amounts of biological materials. However, there are limitations of mass spectrometry, especially in the detection of polar and small analytes because of the matrix-induced background noise caused by poor ionization efficiencies or ionization suppression. Matrix-assisted laser desorption/ionization time-of-flight imaging mass spectrometry (MALDI-IMS) allows the visualization of lipids and proteins in biological tissues; biological compounds with molecular weights of 1-50 k Da have been successfully visualized by MALDI-IMS, however the ability to detect smaller compounds (< 500 Da) is less than satisfactory. Because of the six phosphate groups of InsP_6 Hong *et al.,* [28] hypothesized that it could be a new matrix additive for the efficient detection and visualization of tissue-distributed bioactive small dipeptides. Thus, exploiting the ability of InsP_6 (*acid*) to chelate ionization-interfering salts in the tissue, Hong *et al.,* achieved enhanced visualization of small peptides absorbed through a rat intestinal membrane by using InsP_6 as a matrix additive in laser desorption/ionization time-of-flight imaging mass spectrometry (MALDI-IMS) [28]. The investigators compared various additives with a chelating ability such as EDTA-2Na, tartaric acid, citric acid, di-AC, phosphoric acid and NTMP with InsP_6; only InsP_6 enhanced the dipeptide detection, while the others did not show any effect at the same concentration of additives [28]. Thus, here is yet another practical application of InsP_6 in analytical techniques.

Miscellaneous

InsP_6 has been investigated for application in various other industries with a varying degree of enthusiasm in different countries. It has been studied as a replacement for the highly toxic cyanide in etching solutions for offset printing [29]. Additional uses include as an anti-corrosive in antifreeze, paints, surface rust-proofing treatment especially for the interior surface of soda cans, water conditioning, metal electroplating, copper in chloride solution, *etc.* [29-31].

CONCLUDING REMARKS

As if the medical uses of $InsP_6$ are not broad enough, a very wide spectrum of use and application in other fields makes one wonder about the versatility of the inositol compounds. Much of these functions may be attributed to the chemical structure of the molecule and its ability to chelate divalent cations. Certainly, the preceding examples are not the end of the list of other potential application of this fascinating molecule.

CONFLICT OF INTEREST

Professor Shamsuddin is the inventor of several patents related to inositol & $InsP_6$, and the hexacitrated form of $InsP_6$.

REFERENCES

[1]   Deibner L, Bouzigues H. Action of iron-removing agents employed in oenology Ind Agric Aliment 1954; 71: 833-7.
[2]   Kreitman GY, Cantu A, Waterhouse AL *et al.* Effect of metal chelators on the oxidative stability of model wine. J Agric Food Chem. 2013; 61: 9480-7. doi: 10.1021/jf4024504. Epub 2013 Sep 23
[3]   Tsuji T, Kanai K, Yokoyama A *et al.* Novel method to reduce fishy aftertaste in wine and seafood pairing using alcohol-treated yeast cells. J Agric Food Chem 2012; 60: 6197-203. doi: 10.1021/jf300265x. Epub 2012 Jun 6.
[4]   Zhang Y, Yu C, Mei J *et al.* Formation and mitigation of heterocyclic aromatic amines in fried pork. Food Addit Contam Part A Chem Anal Control Expo Risk Assess 2013; 30: 1501-7. doi: 10.1080/19440049.2013.809627. Epub 2013 Jul 17.
[5]   Wang H, Zhou Y, Ma J, *et al.* The effects of phytic acid on the Maillard reaction and the formation of acrylamide. Food Chem. 2013; 141: 18-22. doi: 10.1016/j.foodchem.2013.02.107. Epub 2013 Mar 7.
[6]   Pacheco GD, Silva CA, Pinton P, *et al.* Phytic acid protects porcine intestinal epithelial cells from deoxynivalenol (DON) cytotoxicity. Exp Toxicol Pathol 2012; 64: 345-7. doi: 10.1016/j.etp.2010.09.008. Epub 2010 Oct 23.
[7]   Towner R, Mason R, Reinke L. *In vivo* detection of aflatoxin-induced lipid free radicals in rat bile. Biochim Biophys Acta 2002; 1573: 556-2.
[8]   Abu Elsaad AS, Mahmmod HM. Phytic acid exposure alters AFB1-induced reproductive and oxidative toxicity in Albino rats (*Rattus norvegicus*). Evid Based Complement Alternat Med 2009; 6: 331-41. doi: 10.1093/ecam/nem137. Epub 2007 Oct 17.
[9]   Brailoiu E, Huhurez G, Slatineanu S, *et al.* TLC characterization of liposomes containing D-myo-inositol derivatives. Biomed Chromatogr. 1995; 9: 175-8.
[10]   Lee H, Jeong C, Ghafoor K *et al.* Oral delivery of insulin using chitosan capsules cross-linked with phytic acid. Biomed Mater Eng. 2011; 21: 25-36. doi: 10.3233/BME-2011-0654.
[11]   Higdon KK, Scott A, Benghuzzi H, *et al.* Development of sustained delivery system as a novel technique for tissue culture. Biomed Sci Instrum. 2000; 36: 117-22.
[12]   Yang T, Guo X, Wang, H *et al.* Au dotted magnetic network nanostructure and its application for on-site monitoring femtomolar level pesticide. Small. 2013 Oct 16. doi: 10.1002/smll.201302604. [Epub ahead of print].

[13]   Wang N, Wen Y, Wang Y, *et al.* The IP$_6$ micelle-stabilized small Ag cluster for synthesizing Ag-Au alloy nanoparticles and the tunable surface plasmon resonance effect. Nanotechnology. 2012; 23: 145702. doi: 10.1088/0957-4484/23/14/145702. Epub 2012 Mar 21.

[14]   Guo SX, Liu Y, Bond AM, *et al.* Facile electrochemical co-deposition of a graphene-cobalt nanocomposite for highly efficient water oxidation in alkaline media: direct detection of underlying electron transfer reactions under catalytic turnover conditions. Phys Chem Chem Phys 2014; 16: 19035-45. doi: 10.1039/c4cp01608d

[15]   Schmidt JT, Ahmad M, Teel AL, *et al.* Hydrogen peroxide stabilization in one-dimensional flow columns. J Contam Hydrol. 2011; 126: 1-7. doi: 10.1016/j.jconhyd.2011.05.008. Epub 2011 Jun 7.

[16]   Mohammed TM, Salama AF, El Nimr TM, *et al.* Effects of phytate on thyroid gland of rats intoxicated with cadmium. Toxicol Ind Health. 2013 Jun 24. [Epub ahead of print]

[17]   Li A, Qiu D.Phytic acid derived bioactive CaO-P$_2$O$_5$-SiO$_2$ gel-glasses. J Mater Sci Mater Med. 2011; 22: 2685-91. doi: 10.1007/s10856-011-4464-7. Epub 2011 Nov 1.

[18]   Scheiner O, Breitenbach M. Herstellung von Sepharose-derivaten zur Affinitätschromatographie von Enzyme des *myo*-Inostiolphosphatstoffwechsels. Monatshefte f Chemie 1976; 107: 581-6.

[19]   Breitenbach M, Scheiner O. Synthesis and applications of immobilized phytic acid. In: Graf E Ed. Phytic acid chemistry & applications. Minneapolis, Pilatus Press 1986; pp. 127-130.

[20]   Prosser HJ, Brant PJ, Scott RP, *et al.* The cement-forming properties of phytic acid. J Dent Res. 1983; 62: 598-600.

[21]   Nassar M, Hiraishi N, Islam MS, *et al.* Effect of phytic acid used as etchant on bond strength, smear layer and pulpal cells. Eur J Oral Sci 2013; 121: 482-7.

[22]   Wu H, Yu G, Pan L, *et al.* Stable Li-ion battery anodes by in-situ polymerization of conducting hydrogel to conformally coat silicon nanoparticles. Nat Commun 2013; 4: 1943. doi: 10.1038/ncomms2941.

[23]   Li Z, He G, Zhang B, *et al.* Enhanced proton conductivity of Nafion hybrid membrane under different humidities by incorporating metal-organic frameworks with high phytic acid loading. http://www.ncbi.nlm.nih.gov/pubmed/24892655 ACS Appl Mater Interfaces 2014 Jun 3. [Epub ahead of print]

[24]   Baker RJ. Biological properties of phytate-containing radiopharmaceuticals. In: Graf E Ed. Phytic acid chemistry & applications. Minneapolis, Pilatus Press 1986; pp. 137-149.

[25]   Subramanian G, McAfee JG, Mehter A *et al.* ^{99m}Tc-stannous phytate: a new *in vivo* colloid for imaging the reticuloendothelial system. J Nucl Med 1973; 14: 459.

[26]   Gao L, Yang F, Ren C, *et al.* Diagnosis of cirrhotic portal hypertension and compensatory circulation using transsplenic portal scintigraphy with ^{99m}Tc-phytate. J Nucl Med 2010; 51: 152-6 doi: 10.2967/jnumed.109.067983

[27]   Kondakov AK, Gubskiy IL, Znamenskiy IA *et al.* Possibilities of optical imaging of the ^{99m}Tc-based radiopharmaceuticals. J Biomed Opt. 2014; 19(4): 046014. doi: 10.1117/1.JBO.19.4.046014

[28]   Hong SM, Tanaka M, Yoshii S *et al.* Enhanced visualization of small peptides absorbed in rat small intestine by phytic-acid-aided matrix-assisted laser desorption/ionization-imaging mass spectrometry. Anal Chem. 2013; 21: 10033-9. doi: 10.1021/ac402252j. Epub 2013 Oct 10.

[29].   Sands HS, Biscobing SJ, Olson RM. Commercial aspects of phytic acid: an overview. In: Graf E Ed. Phytic acid chemistry & applications. Minneapolis, Pilatus Press 1986; pp. 119-25.

[30]   Oda N, Ogino T, Terada H. Surface treatment of tin-coated steel sheet or can. Japan Patent 79-68733 (1979).

[31]   Peca D, Pihlar B, Ingrid M. Protection of copper surface with phytic acid against corrosion in chloride solution. Acta Chim Slov 2014; 61: 457-67.

CHAPTER 22

Safety of Inositol and Inositol Phosphates

Abstract: Owing to the various health benefits of inositol and inositol phosphates, especially InsP_6, and their availability as nutritional supplement for human consumption, it is imperative that their safety be addressed. The safety of *myo*-inositol has not been an issue, however, there have been lingering doubts in some quarters about the safety of InsP_6. There had been some unsubstantiated concerns in the past regarding the intake of foods high in InsP_6 that might reduce mineral bioavailability; recent studies using extracted salts of InsP_6 have refuted those. Both inositol and InsP_6 have been listed as "generally recognized as safe" by the US FDA. There should therefore be no concern for their consumption as appropriate.

Keywords: Bioavailability, cooking methods, culture media, GRAS, infant formula, low phytate mutant, lpa, Mediterranean diet.

INTRODUCTION

As has been stated before, inositol and InsP_6 are natural compounds abundantly present in cereals and legumes. Inositol has been generally considered to be a member of B vitamin family, however because it is endogenously produced in the body from glucose it is not classified as an essential nutrient. Since the landmark publication of Eagle *et al.*, [1] inositol has been in use as an essential component of cell and tissue culture media for a very long time; and so is its use in infant formulas. The United States Food and Drug Administration (FDA) have accorded GRAS (generally recognized as safe) status to both inositol and calcium-inositol hexaphosphate aka calcium phytate [2]. We shall therefore not discuss the safety of inositol for human consumption since this is unquestioned.

InsP_6 AND MINERAL BIOAVAILABILITY

In the past some concerns had been expressed regarding intake of foods high in InsP_6 that might reduce the bioavailability of dietary minerals. However, studies by Henneman as early as in 1958, in humans using pure form did not show any side-effects [3]. More recent studies demonstrate that this so-called anti-nutrient effect of InsP_6 can be observed only when large quantities of InsP_6 are consumed in combination with a diet poor in trace elements [4-11].

Claims of the adverse effect of InsP_6 on mineral bioavailability first appeared in papers published in the early 1900's. Sir Edward Mellanby reported in 1919-1921

A.K.M. Shamsuddin and Guang-Yu Yang

that a high cereal-diet in the absence or deficiency of a fat-soluble "calcifying" vitamin had a rickets-producing effect on puppies; the greater the amount eaten the more intense the resulting disease. It was later found that the more severe rickets developed when the diet consisted mainly of oatmeal, maize or whole wheat flour as opposed to white flour or rice. The rickets-producing effect of oats could be reduced by either boiling them with mineral acid or by subjecting them to a malting process suggesting that cereals contained a substance that interfered with calcium metabolism which could be reversed by chemical or enzymatic method; this was the era when the term "anti-vitamin" was added to the lexicon. So the hunt was on, and phytic acid or inositol hexaphosphoric acid was identified as the culprit. "...I[i]t appeared *probable* that, when oatmeal or oats were boiled with dilute HCR, the cereal phytate was hydrolysed to inositol and phosphate, and when oats and other cereals were malted, a similar change was effected by a phytase [italics added]..." wrote Sir Mellanby [12]. While his studies conclusively demonstrated that rickets was a result of dietary deficiency of fat-soluble vitamin (D), it nevertheless emboldened the negative publicity for $InsP_6$ for nearly a century; since then, several authors have attributed "anti-nutritional" properties to $InsP_6$.

However, more recent studies not only contradict these findings, but since the late 1980's, important physiological functions and health benefits of $InsP_6$ have also been discovered. The apparent discrepancy between the useful and presumed harmful effects of $InsP_6$ is partly due to the difficulty of the analytical determination of minute amounts of $InsP_6$ in biological samples, as already mentioned in previous chapters. Actually, knowledge of the beneficial properties of $InsP_6$ has become evident with the development of analytical methodologies for the determination of low levels of $InsP_6$ in biological fluids and tissues. As stated above, the "anti-nutritional" effect of $InsP_6$ has been attributed mostly to its interference with mineral bioavailability. However, a variety of factors need to be taken into account when this effect of $InsP_6$ is evaluated, one of the most important is that all these previous studies were based on 'diets rich in $InsP_6$' as opposed to experimenting by administering pure $InsP_6$; obviously 'diets rich in $InsP_6$' have numerous confounding factors other than $InsP_6$ that could affect the mineral bioavailability.

Epidemiological Data

"Safety" depends not only on the ratio of $InsP_6$ to overall mineral content of the diet, but also upon food processing and cooking methods [13-15]. It has been demonstrated that if essential minerals are present in the proper ratio with respect to $InsP_6$, there is no reason for a modification of mineral balance. Furthermore, no negative effect on element bioavailability was observed in individuals on a Mediterranean diet who ingested 1-2 g of $InsP_6$ per day [8, 10, 11]. Most importantly, studies of long-term intake of $InsP_6$ in food [5, 6] did not cause a mineral deficiency in humans. Henneman *et al.,* did not report of any mineral deficiency in 10 individuals given pure form (as opposed to the then usual practice of giving high-$InsP_6$ food) of Na-$InsP_6$ as much as 8.8 g per day and followed for an average of 24 months [3]. On the contrary, there was therapeutic benefit: 9 of these 10 patients with Idiopathic Hypercalciuria and kidney stones enjoyed rather prolonged suppression of their hypercalcuria for which there was no other treatments available. In addition, 8 of those 10 patients did not have any further growth of their kidney stones - another therapeutic bonanza [3].

Experimental Animals

When high $InsP_6$ dosages and diets with low mineral content, *e.g.,* some soy-based diet, were used, a deficit in mineral absorption was detected in some cases. Studies in experimental animals fed pure form of $InsP_6$ (as opposed to high $InsP_6$-diet) showed no significant toxic effects on body weight, serum, or bone minerals or any pathological changes in either male F344 or female Sprague-Dawley rats for their lifetime [16-19]. Analysis of serum Ca^{2+}, Fe^{2+}, Mg^{2+} and Zn^{2+} levels in rats given 15mM inositol or 15mM Na- $InsP_6$ or 15mM inositol + 15mM Na- $InsP_6$ orally for 17 weeks showed no significant decrease in the levels of these minerals (Table **22.1**) [17]. Since serum levels may not necessarily reflect the mineral status most accurately, their bone levels were also tested. Table **22.2** shows the data from bone Ca^{2+}, Mg^{2+} and Zn^{2+} levels in the bone of rats given 15mM inositol or 15mM Na- $InsP_6$ or 15mM inositol + 15mM Na- $InsP_6$ orally for 47 weeks [18]. Femoral bone samples were collected from 3-6 rats/group and analyzed by atomic absorption spectroscopy following wet and dry ashing. As can be clearly seen (Table **22.2**), there were no significant differences in the levels of bone Ca^{2+}, Mg^{2+} and Zn^{2+} in the animals receiving inositol $\pm$ $InsP_6$ as opposed to the tap water control [18]. Grases *et al.,* [4] not only confirmed these findings but also reported that abnormal calcification was prevented in rats given $InsP_6$. Moreover, in a more recent study Grases *et al.,* [11] monitored the effect of $InsP_6$ on mineral status during a long time period through a second generation of rats to evaluate any possible effects related to the pregnancy and lactation. Except for a lower zinc concentration in bone, they did not observe any decrease in overall mineral bioavailability. Interestingly enough, compared to the control, animals that were fed with equilibrated purified diet with or without InsP$_6$ had approximately 10-fold higher zinc levels in bone [11]. Even using high-InsP$_6$ diet, Barbara Harland demonstrated that rat plasma copper and zinc concentrations were not negatively affected by the high InsP$_6$ content of Black Seeds which are indigenous to the Mediterranean, the Arabian Peninsula, Asia and Africa, wherein they are used as a natural remedy [20]. And, it was further demonstrated that zinc deficiency was unrelated to InsP$_6$ content when a vegetarian diet was compared with meat-based diets with equal InsP$_6$ content [21]. It is important to note, that most if not all the studies reporting the so-called anti-nutrient effect were done with phytic acid (inositol hexaphosphoric acid), which can chelate; we and others, however, used the salt form - inositol hexaphosphate (phytate) which already has minerals bound to the molecule, as it exists in nature.

Table 22.1. Effects of Inositol ± InsP$_6$ on serum minerals [17].

Treatment (*n*)	Ca^{2+} (mg/dL)	Fe^{2+} (mg/dL)	Mg^{2+} (mg/dL)	Zn^{2+} (mg/dL)
Tap water (n = 6)	10.2 ± 0.7[d]	305.0 ± 61.9	3.1 ± 0.6	82.3 ± 20.0
InsP$_6$ (n = 3)[a]	11.1 ± 1.1	376.0 ± 35.4	4.2 ± 1.2	107.0 ± 38.2
Inositol (n = 4)[b]	9.7 ± 0.5	430.5 ± 9.2	3.0 ± 0.9	74.5 ± 14.8
Inositol + Na- InsP$_6$[c] (n = 5)	9.9 ± 0.2	316.8 ± 37.4	2.7 ± 0.4	88.8 ± 3.5

[a,b]15mM inositol and InsP$_6$, [c]15mM inositol + 15mM Na-InsP$_6$. [d]Values are mean ± SD.

Table 22.2. Effects of inositol ± InsP$_6$ on bone minerals [18].

Treatment (*n*)	Ca^{2+} (mg/g)	Mg^{2+} (mg/g)	Zn^{2+} (µg/g)
Tap water (n = 6)	116.9 ± 13.9[d]	1.13 ± 0.14	109.2 ± 14.9
InsP$_6$ (n = 3)[a]	124.8 ± 11.3	1.19 ± 0.14	127.4 ± 11.5
Inositol (n = 4)[b]	117.4 ± 14.2	1.10 ± 0.16	116.7 ± 14.5
Inositol + Na- InsP$_6$[c] (n = 5)	125.9 ± 9.0	1.14 ± 0.06	115.1 ± 9.9

[a,b]15 mM inositol and InsP$_6$, [c]15 mM inositol + 15 mM Na-InsP$_6$. [d]Values are mean ± SD.

We would be remiss if we did not refer to an early study suggesting a positive correlation between ingestion of InsP$_6$ and the incidence of urinary bladder papillomas in a rat model of carcinogenesis [22]. Using a wide-spectrum model of carcinogenesis, Hirose *et al.*, tested various naturally occurring anti-oxidants including InsP$_6$ to test their efficacy in prevention of various cancers. They found InsP$_6$ to be marginally effective against liver and colon carcinogenesis, with an increased incidence of papillomas - a preneoplastic lesion in urinary bladder [22]. However, subsequent studies showed that the sodium salt, but not the potassium or the magnesium salt, could encourage development of pre-neoplastic lesions of the urinary bladder [23]. Hiasa *et al.*, [24] reported that sodium salt of $InsP_6$ causes some necrosis and calcification of renal papilla in experimental rats. However, as mentioned before, Henneman *et al.*, [3] did not report any such events in their investigations in humans who were followed for an average of 24 months with high doses (8.8 g/day) of Na-$InsP_6$.

Studies in mice done at the US National Cancer Institute found it to be safe at dosage of 10 g/kg in mice [25], several fold higher than the maximum therapeutic dose recommended for humans. Further toxicology studies showed that rats tolerated up to 62.6 mg/kg i.v. $InsP_6$ as a single dose and between 250-350 mg/kg intraperitoneal (i.p.) as a single dose; mice tolerated up to 50 mg/kg i.v. and 400 mg/kg i.p. as single doses [25]. The reasons for this discrepancy among salt forms amongst different species would be interesting studies. Be that as it may, the $InsP_6$ formulation suggested for human use is the naturally occurring calcium and magnesium, and not sodium salt.

Genetic Engineering to Produce Low Phytic Acid (lpa) Mutant

Despite the extensive data in laboratory animals and in humans from epidemiological studies, there are still overzealous and persistent efforts to reduce the $InsP_6$ content in our food globally to alleviate the presumed mineral deficiency in developing countries and prevent eutrophication in the developed ones. One approach has been selective plant breeding to yield low phytic acid (*lpa*) mutant species of maize, rice, *etc.*, [26]. Raboy *et al.*, [27] initially described two non-lethal maize *low*-$InsP_6$ mutants, *lpa1-1* and *lpa2-1*. Seed $InsP_6$ phosphorous was reduced in these mutants by 50% to 66% but seed total phosphorous was unaltered. And the decrease in $InsP_6$ phosphorous in mature *lpa1-1* seeds was accompanied by a corresponding increase in inorganic phosphate [27]. Pilot studies have demonstrated that zinc absorption from low-$InsP_6$ maize is higher than from their near isohybrid wild-type controls, and the increase in absorption is inversely related to the $InsP_6$ content of the maize [28].

As intriguing and attractive as these *lpa* mutants may be from a technological point of view, they are either not as satisfactory or, worse they have their own problems. Associated with strong pleiotropic effects on the whole plant, *lpa1-241* mutation causes a reduction of up to 90% of $InsP_6$ in maize. Badone *et al.*, [29] reported an interaction between the accumulation of anthocyanin pigments in the kernel and the *lpa* mutations. The *lpa1-241* mutant accumulated a higher level of anthocyanins as compared to wild type either in the embryo or in the aleurone layer in a genotype that is able to accumulate anthocyanin. These pigments were abnormally localized in the cytoplasm, conferring a blue pigmentation of the scutellum [29]. Mazariegos *et al.,* [30] used a randomized controlled, doubly masked trial to determine the effect of substituting low-InsP_6 maize, a daily 5-mg zinc supplement, or both, in infants between ages 6-12 months on impaired linear growth velocity, a common feature of zinc deficiency in the Western Highlands of Guatemala. The authors expected a significant increase in linear growth velocity from reduction of InsP_6 in the diet of older infants whose complementary food was plant-based with high InsP_6 intake. Dietary InsP_6 reduction was hypothesized to improve zinc absorption in this population. A total of 412 infants were randomized to receive low-InsP_6 maize or control maize. Within each maize group, infants were further randomized to receive a zinc supplement or placebo. Length, weight, and head circumference were measured at 6, 9, and 12 months of age. The investigators did not find any significant differences between the two maize groups or between the Zn supplement and placebo groups and no treatment interaction was observed for length-for-age, weight-for-length or head circumference Z-scores. Low linear growth in older Guatemalan infants was not improved with either low- InsP_6 maize or a daily 5-mg zinc supplement. "Low contribution of maize to the complementary food of the infants negated any potential advantage of feeding low-phytate (InsP_6) maize" concluded the authors [30]. Thus, reducing the InsP_6 content in our diet is not only unnecessary but also may be harmful, paradoxically even from the nutritional point of view! That removal or destruction of InsP_6 from our diet is dangerous has been articulated as early as in 1976 by Morris & Ellis who wrote "Destruction of phytate [InsP_6], whether by enzymatic or by chemical hydrolysis, during baking may have a deleterious effect on the biological availability of iron as monoferric phytate [InsP_6] in bread" [31], but those words of wisdom seem to have been largely ignored, as several others [32-34].

CONCLUDING REMARKS

In conclusion, both inositol and InsP_6 are safe for human consumption.

REFERENCES

[1]   Eagle H, Omaya, VI, Levy M *et al. myo*-inositol as an essential growth factor for normal and malignant human cells in tissue culture. J Biol Chem 1957; 226: 191-205.

[2]   http://www.fda.gov/Food/IngredientsPackagingLabeling/GRAS/SCOGS/ucm084104.htm

[3]   Henneman PH, Benedict PH, Forbes AP *et al.* Idiopathic hypercalciuria N Engl J Med 1958; 17: 802-7.

[4]   Grases F, García-Gonsales R, Torres JJ *et al.* Effects of phytic acid on renal stone formation in rats. Scand J Urol Nephrol 1988; 32: 261-5.

[5]   Walker ARP, Fox FW, Irving JT. Studies in human mineral metabolism. I. The effect of bread rich in phytate phosphorous on the metabolism of certain mineral salts with special reference to calcium. Biochem J 1948; 42: 452-62.

[6]   Cullumbine H, Basnayake V, and Lemottee J. Mineral metabolism on rice diets. Brit J Nutr 1950; 4: 101-1.

[7]   Kelsay JL: Effect of fiber, phytic acid, and oxalic acid in the diet on mineral bioavailability. Am J Gastroent 1987; 278: 983-6.

[8]   Sandstrom B, Bugel S, McGaw BA *et al.* A high oat-bran intakes does not impair zinc absorption in human when added to a low-fiber animal protein-based diet. J Nutr 2000; 130: 594-9.

[9]   Manary MJ, Hotz C, Krebs NF *et al.* Dietary phytate reduction improves absorption in Malawian children recovering from tuberculosis but not in well children. J Nutr 2000; 130: 2959-64.

[10]   Siqueira EM, Arruda SF, de Sousa LM *et al.* Phytate from an alternative dietary supplement has no effect on the calcium, iron and zinc status in undernourished rats. Arch Latinoam Nutr 2001; 51: 250-7.

[11]   Grases F, Simonet BM, Perelló J *et al.* Effect of phytate on element bioavalability in the second generation of rats. J Trace Elem Med Biol 2004; 17: 229-34.

[12]   Mellanby, E: The rickets-producing and anti-calcifying action of phytate. J Physiology 1949; 109: 488-533.

[13]   Ma G, Jin Y, Piao J *et al.* Phytate, calcium, iron, and zinc contents and molar ratios in foods commonly consumed in China. J Agric Food Chem 2005; 53: 10285-90.

[14]   Szkudelski T: Phytic acid-induced metabolic changes in rat. J Anim Physiol Anim Nutr (Berl) 2005; 89: 397-402.

[15]   Yangklang C, Wensing T, Lemmens AG *et al.* Effect of sodium phytate supplementation on fat digestion and cholesterol metabolism in female rat. J Anim Physiol Anim Nutr (Berl) 2005; 89: 373-8.

[16]   Ullah A, Shamsuddin AM. Dose-dependent inhibition of large intestinal cancer by inositol hexaphosphate in F344 rats. Carcinogenesis 1990; 11: 2219-22.

[17]   Vucenik I, Sakamoto K, Bansal M *et al.* Inhibition of rat mammary carcinogenesis by inositol hexaphosphate (phytic acid). A pilot study. Cancer Letters 1993; 75: 95-102.

[18]   Vucenik I, Yang G-Y, Shamsuddin AM. Inositol hexaphosphate and inositol inhibit DMBA induced rat mammary cancer. Carcinogenesis 1995; 16: 1055-8.

[19]   Vucenik I, Yang G, Shamsuddin AM. Comparison of pure inositol hexaphosphate (IP$_6$) and high-bran diet in the prevention of DMBA-induced rat mammary carcinogenesis. Nutr Cancer 1997; 28: 7-13.

[20]   Harland B. Rat plasma cooper and zinc concentrations were not negatively affected by high phytate of "Black Seed". The FASEB J 2006; 20: A196.

[21]   Kristensen MB, Hels O, Morberg CM *et al.* Total zinc absorption in young women, but not fractional zinc absorption, differs between vegetarian and meat-based diets with equal phytic acid content. Br J Nutr 2006; 95: 963-7.

[22]   Hirose M, Ozaki K, Takaba K *et al.* Modifying effects of the naturally occurring antioxidants γ-oryzanol, phytic acid, tannic acid and *n*-tritriacontane-16,18-dione in rat a wide-spectrum organ carcinogenesis model. Carcinogenesis 1991; 12: 1917-21.

[23]   Takaba K, Hirose M, Ogawa K *et al.* Modification of *N*-butyl-*N*-(4-hydroxylbutl) nitrosamine-initiated urinary bladder carcinogenesis in rats by phytic acid and salts. Food Chem Toxicol 1994; 32: 499-503.

[24]   Hiasa Y, Kitahori Y, Morimoto J *et al.* Carcinogenicity study in rats of phytic acid "Daiichi", a natural food additive. Food Chem Toxic 1992; 30: 117-25.

[25]   Sausville E: Personal communication, National Cancer Institute, Bethesda, Maryland USA, July 1, 1999.

[26]   Raboy V: The ABCs of low-phytate crops. Nat Biotechnol. 2007; 25: 874-5.

[27]   Raboy V, Gerbasi PF, Young KA *et.* Origin and seed phenotype of maize low phytic acid 1-1 and low phytic acid 2-1. Plant Physiol 2000; 124: 355-68.

[28]   Hambidge KM, Huffer JW, Raboy V *et al.* Zinc absorption from low-phytate hybrids of maize and their wild-type isohybrids. Am J Clin Nutr 2004; 79: 1053-9.

[29]   Badone FC, Cassani E, Landoni M *et al.* The low phytic acid1-241 (lpa1-241) maize mutation alters the accumulation of anthocyanin pigment in the kernel. Planta 2010; 231: 1189-99. doi: 10.1007/s00425-010-1123-z. Epub 2010 Feb 27.

[30]   Mazariegos M, Hambidge KM, Westcott JE *et.* Neither a zinc supplement nor phytate-reduced maize nor their combination enhance growth of 6- to 12-month-old Guatemalan infants. J Nutr 2010; 140: 1041-8,. doi: 10.3945/jn.109.115154. Epub 2010 Mar 24.

[31]   Morris ER, Ellis R: Isolation of monoferric phytate from wheat bran and its biological value as an iron source to the rat. J Nutr 1976; 106: 753-60.

[32]   Shamsuddin AM.: Phytate and Colon Cancer Risk. Am J Clin Nutr 1992; 55: 478.

[33]   Shamsuddin AM.: Demonizing phytate. Nature Biotechnology 2008; 26: 496-7.

[34]   Murgia I, Arosio P, Tarantino D *et al.* Biofortification for combating 'hidden hunger' for iron. Trends Plant Sci 2012; 17: 47-55. doi: 10.1016/j.tplants.2011.10.003. Epub 2011 Nov 16.

Glossary

ADP: Adenosine 5'-diphosphate is a nucleotide produced by hydrolysis of the terminal phosphate of ATP (adenosine 5'-triphosphate).

Akt: A serine/threonine specific protein kinase a.k.a. protein kinase B (PKB) which is responsible for a multitude of cellular functions including apoptosis, cell survival, cell proliferation, cell migration, *etc.*

Alkaptonuria: is a genetic disease with inborn error of catabolism of the amino acids phenylalanine and tyrosine leading to the accumulation of homogentisic acid and resultant tissue damage in various organs.

Anchorage dependence: Dependence of cell growth on attachment to a substratum as *in vitro.*

Angiogenesis: Growth of new blood vessels that sprout from existing ones.

Anoikis: Greek for homelessness, self-destruction of cells when removed from normal/natural surroundings.

Apoptosis is a normal phenomenon in multi-cellular organisms. Also known as programmed cell death wherein a 'suicide' program is activated resulting in DNA fragmentation, shrinkage of the cytoplasm and cell death.

Atheroma: Accumulation and swelling (-oma) of artery wall due to cells (mostly macrophages - see below) and cell debris, lipid, calcium and fibrous connective tissue.

Atherosclerosis: Hardening of the arteries due to formation of atheroma (see above).

ATP: Adenosine 5'-triphosphateis the principal carrier of chemical energy in the cells and is composed of adenine + ribose + 3 phosphate groups. The terminal phosphate groups are highly reactive; a large amount of free energy is released when the terminal phosphates are removed or transferred to another molecule.

Autophagy/Autophagocytosis: Digestion of cell's own organelle; packets of enzymes contained in lysosomes carry out this activity.

A.K.M. Shamsuddin and Guang-Yu Yang

Benign: Non-invasive tumors those are self-limiting.

Bioavailability: A subcategory of absorption or the fraction of an administered dose of unchanged drug that reaches the systemic circulation, one of the principal pharmacokinetic properties of drugs.

Body mass index (BMI): A measure of relative weight based on an individual's mass and height.

Calcineurin: A protein phosphatase, AKA protein phosphatase 3, and calcium-dependent serine-threonine phosphatase.

Calcification: The accumulation of calcium salts in a body tissue; it may be normal physiological, related to aging, or pathological in disease conditions.

Calcinosis cutis or cutaneous calcification: A type of calcification wherein calcium deposits form in the skin.

Carcinogen: Any agent that causes cancer; they may be physical, chemical or biological.

Carcinogenesis: Generation or formation of cancer.

Carcinoma: The most common types of cancers in human; arise from the epithelial cells such as from the lungs, colon, liver etc, as opposed to sarcoma that are from non-epithelial cells as muscle or bone.

Caspases are cysteine-aspartic proteases or cysteine-dependent aspartate-directed proteases - a family of cysteine proteases that are involved in the process of cell apoptosis, necrosis, and inflammation.

Cell adhesion: The binding of a cell to a surface or substrate, such as an extracellular matrix or another cell.

Cell cycle is the series of events that take place in a cell leading to its division and duplication.

Cell division: Separation of cell into two daughter cells

Cell line: Population of cells which are capable of dividing indefinitely as opposed to normal cells which do not divide forever.

Chelate: To combine reversibly, usually with high affinity, with a metal ion such as calcium, magnesium, iron, zinc, *etc.*

Chemotaxis: Motile response of a cell or an organism guided by chemical gradients.

Chromatin: It is the complex of DNA and proteins that makes up the contents of the nucleus of a cell.

Chromatography: A technique by which a mixture of substances is separated by size, electrical charge or other properties.

Clathrin: A protein that plays a major role in the formation of coated vesicles (*vide infra*).

Coated Pit: It is an invagination of plasma membrane with bristle-like layer of proteins on its cytoplasmic surface. It pinches off to form a coated vesicle during endocytosis.

Coated vesicle: Small membrane-bound organelle formed by the pinching off of a coated pit.

Cyclins: A family of proteins that control the progression of cells through the cell cycle by activating cyclin-dependent kinases (Cdks).

Cyclin-dependent kinases are a family of protein kinases that regulate the cell cycle.

Diabetes mellitus, **simply diabetes:** A group of metabolic diseases in which there high blood sugar levels over a prolonged period.

Differentiation A process by which a cell changes to a clearly recognizable specialized type, often with specialized function.

Dynamin I: A substrate for calcineurin that is responsible for membrane fission during endocytosis.

Endocytosis: It is an energy-using process by which cells absorb molecules (such as proteins) by engulfing them.

Endothelium: Single layer of flattened cells (endothelial) that lines the inner surface of blood vessels.

Epimerization: Reaction that alters the steric arrangement around an atom.

Epithelium: Cell layer or layers that cover an external surface as in skin, or lining of cavities such as the gastrointestinal tract, urinary bladder, *etc.*

Eukaryote (eukaryote): Unicellular or multi-cellular organisms that have distinct nucleus and cytoplasm. All forms of life except viruses, bacteria and archea fall in this category.

Eutrophication is an increase in the concentration of chemical nutrients in an ecosystem. This increased amount of nutrient may at the beginning increase the productivity of the primary ecosystem, such as increased phosphorous from fertilizers result in increased crop production. However, depending on the degree of eutrophication, at a later time period, there can be severe negative impact on the environment, such as reduction in water quality and fish population, overgrowth of algae, anoxia in lakes, *etc.*

Exocytosis: A process by which most molecules are secreted from eukaryotic cells *via* packaging into membrane-bound vesicles or sacs those fuse with the plasma membrane and finally the contents being released outside the cell.

Extracellular matrix: Structural (lattice) element in tissues, composed of complex network of polysaccharides and proteins.

Fibroblast: Most common cell type found in connective tissue, migrates during wound healing to repair the damage, secretes collagen and other molecules; may be responsible for scar formation.

Free radical: An atom, molecule, or ion that has unpaired valence electrons or an open electron shell, and therefore may be seen as having one or more "dangling" covalent bonds.

G proteins: a.k.a. **guanosine nucleotide-binding proteins**, A family of proteins that are important signal transducing molecules in cells.

Growth Factor: A naturally occurring substance capable of stimulating cellular growth.

Hemoglobin S: Mutant hemoglobin that a single amino acid substitution in normal hemoglobin A by a valine replacing a glutamine in the 6th position of the beta chain of globin.

Hepatocyte: Liver cell.

Hepatoma: Cancer of the liver cells.

HLA: Human leukocyte antigen system is the locus of genes that encode for major histocompatibility complex (MHC). This group of genes resides on chromosome 6 encoding cell surface antigen presenting proteins; and has many functions. These proteins are unique to each person; the immune cells use the HLAs to distinguish self-cells from the non-self-cells.

Inositol synthase: Also called inositol 1-phosphate synthase is an enzyme converting glucose-6-phosphate to *myo*-inositol 1-phosphate as first step of *myo*-inositol synthesis intracellularly

Inositol oxygenase: A non-heme di-iron enzyme oxidizing *myo*-inositol to glucuronic acid.

Inositol pyrophosphates, IP_7 (or PP-IP_5; 5-diphosphoinositol pentakisphosphate) and IP_8 [or $(PP)_2IP_4$; bisdiphosphoinositol tetrakisphosphate], as the name indicates there are seven or eight phosphate groups attached to the six-carbon inositol ring, thus possessing one and two pyrophosphate moieties respectively.

Integrin: A member of the large family of transmembrane proteins involved in the adhesion of cells to the extracellular matrix.

Intima: The innermost layer of the blood vessel lined by endothelial cells.

Ionizing radiation: Radiation that carries enough energy to liberate electrons from atoms or molecules, thereby ionizing them.

IP3: Inositol triphosphates or *myo*-inositol (1,4,5)-trisphosphate, functions as a second messenger in calcium release; also written more accurately as InsP_3.

IP6 or InsP6 *myo*-inositol (1,2,3,4,5,6)-hexakisphosphate a.k.a. "phytate".

In vitro: A process taking place outside the body as in cell culture in the laboratory or biochemical reaction in cell-free systems.

In vivo: In an intact organism.

Isomers Molecules that are formed from the same atoms in the same chemical linkages but having different 3-dimensional conformations.

Keratin: Epithelial (mainly) cell proteins that form filaments, examples include hair, nails, feather, *etc*.

Kinase: The enzyme that transfers phosphate groups from high-energy donor molecules, such as ATP, to specific substrates through the process called phosphorylation.

Lectin: Protein found in plant seeds that bind tightly to a specific sugar.

Leukemia: Cancer of the white blood cells.

Lipid peroxidation: The oxidative degradation of lipids.

Lymphocytes are broadly divided into 2 types - B and T cells. These white blood cells are responsible for specific immune response.

Lysosome: Membrane bounded organelle containing digestive enzymes.

Macrophage: Scavenger cell derived from blood monocytes; also have a role in immune response. The monocyte leaves the bloodstream to mature or differentiate into macrophage in tissues.

Malignant: Invasive tumors/tumor cells that can travel to distant sites (metastasis).

Metastasis: Spread, most often referred to cancer cells from their site of origin to distant areas of the body.

Molar (1 M) describes a solution with a concentration of 1 mole of a substance dissolved in 1 liter of solution.

Mole: Consists of 6×10^{23} molecules of a substance.

Mutation: Heritable change in the nucleotide sequence within a chromosome.

Natural Killer (NK) cell: Cytotoxic lymphocyte that can kill virus-infected cells or cancer cells.

Neutrophil: Specialized white blood cell that constitute an early stage in the defense against invading microorganisms. It enters the tissues during infection or inflammation and take-up or engulf particles such as bacteria.

NF-κB: Nuclear factor kappa-light-chain-enhancer of activated B cells) is a protein complex or transcription factor that controls transcription of DNA.

Nuclear pore: Channel through the nuclear envelope (a double membrane surrounding the nucleus) that allows selected molecules to move between the nucleus and the cytoplasm.

Oncogene: An altered (mutant) gene whose product can help in the process of converting a normal cell to cancer.

Osteoblast: Cell that forms bone by secreting bone matrix.

Osteoclast: Cell that erodes bone.

Osteoporosis: A group of diseases characterized by reduced bone mass per unit of bone volume resulting in weak and fragile bone with increased risks of fractures of hip, wrist and spine.

Oxidation: Loss of electrons from an atom; happens with addition of oxygen to a molecule or with removal of hydrogen.

p53: Tumor suppressor gene found to be mutated in ~50% of human cancers.

Phagocytosis: Process by which particulate materials are engulfed or "eaten" by cells.

Pharmacokinetics: Abbreviated as PK, part of pharmacology dedicated to determining the fate of substances administered externally to a living organism, including the process of the administered substance from liberation, absorption, distribution, metabolism to excretion.

Phenotype: The observable character of a cell or an organism.

Phosphatase: Enzyme that removes phosphate groups from a molecule.

Phospholipase A: Enzyme that releases arachidonic acid from phosphatidylinositol.

Phospholipase C: Enzyme that cleaves the phospholipid phosphatidylinositol 4,5-bisphosphate (PIP_2) into diacylglycerol (DAG) and inositol 1,4,5-trisphosphate (InsP_3).

Phosphorylation: Reaction in which a phosphate group becomes covalently bound to another molecule.

Phosphoinositide kinases: The enzymes that phosphorylate phosphatidylinositol and its derivatives.

Phosphoinositide phosphatases: The enzymes that remove phosphates from phosphatidylinositol and belong to the protein tyrosine phosphatase (PTP) superfamily.

Phytase also called myo-inositol hexakisphosphate phosphohydrolase Phosphatase enzymes that catalyze the hydrolysis of phytic acid - InsP_6.

Pinocytosis: "Cell drinking" is a form of endocytosis in which soluble materials are taken up by the cell.

Prokaryotes: A group of organisms whose cells lack a membrane-bound nucleus (karyon).

Protein kinase C: A family of protein kinase enzymes that are involved in controlling the function of other proteins through the phosphorylation of hydroxyl groups of serine and threonine amino acid residues on these proteins.

Rb, retinoblastoma protein or gene name **RB**, is a tumor suppressor protein that is dysfunctional in several major cancers.

Reactive oxygen species (ROS): Chemically reactive molecules containing oxygen.

Sarcoma: Cancer of the connective tissue *e.g.,* rhabdomyosarcoma (cancer of the skeletal muscle), fibrosarcoma (cancer of the fibroblast), osteosarcoma (cancer of the bone cell).

Second Messenger: Small molecule (*e.g.,* Ca^{2+}, InsP_3) that is formed in the cytosol in response to an extracellular signal and relays the signal to the interior of the cell.

Selectins: Cell surface adhesion molecules.

Sialolithiasis: Salivary gland stone.

Signal transduction: Relaying of signal by conversion from a physical or a chemical form to another, usually refers to the translation of an extracellular signal into a response in cell biology.

Stem cells: The undifferentiated biological cells that can differentiate into specialized cells and can divide through mitosis to produce more stem cells.

Synaptojanin: A protein that involves in vesicle uncoating and it is an important regulatory lipid phosphatase.

Syndecan-4: A heparan sulfate proteoglycan that localizes in cellular membranes and regulates cell-matrix interactions.

T cell: Lymphocyte responsible for cell-mediated immunity.

Telomerase/telomere terminal transferase: A ribonucleoprotein that is an enzyme for adding DNA sequence repeats ("TTAGGG" in all vertebrates) to the 3' end of DNA strands in the telomere regions at the ends of eukaryotic chromosomes.

Tumor necrosis factors (TNF): A group of cytokines that lead to cell death/apoptosis.

Urolithiasis: The condition where urinary stones are formed and located anywhere in the urinary system.

Vesicle trafficking: A process in eukaryotic cells that involves movement of important biochemical signal molecules from synthesis-and-packaging locations in Golgi body to specific 'release' locations on the inside of the plasma membrane of the secretory cell, in the form of Golgi membrane-bound micro-sized vesicles, termed *membrane vesicles* (MVs).

Zinc finger: A loop of polypeptide chain held in a hairpin bend and bound to a zinc atom; it is a DNA-binding structural motif that is present in many gene-regulatory proteins.

Index

www.ingramcontent.com/pod-product-compliance
Lightning Source LLC
Chambersburg PA
CBHW050812220326
41598CB00006B/188